南昌大学“双一流”博士点建设专项经费支持

系统动力学上机实验指导

SYSTEM DYNAMICS EXPERIMENTAL INSTRUCTION

刘静华 ◎ 著

图书在版编目（CIP）数据

系统动力学上机实验指导/刘静华著.—北京：经济管理出版社，2023.11
ISBN 978-7-5096-9461-9

Ⅰ.①系… Ⅱ.①刘… Ⅲ.①系统动态学—实验—高等学校—教学参考资料 Ⅳ.①N941.3-33

中国国家版本馆 CIP 数据核字(2023)第 221943 号

组稿编辑：杜　菲
责任编辑：杜　菲
责任印制：许　艳
责任校对：王淑卿

出版发行：经济管理出版社
（北京市海淀区北蜂窝 8 号中雅大厦 A 座 11 层　100038）
网　　址：www. E-mp. com. cn
电　　话：（010）51915602
印　　刷：北京晨旭印刷厂
经　　销：新华书店
开　　本：720mm×1000mm/16
印　　张：15.5
字　　数：261 千字
版　　次：2024 年 8 月第 1 版　　2024 年 8 月第 1 次印刷
书　　号：ISBN 978-7-5096-9461-9
定　　价：88.00 元

联系地址：北京市海淀区北蜂窝 8 号中雅大厦 11 层
电话：（010）68022974　　邮编：100038

前　言

系统动力学（System Dynamics，SD）出现于1956年，创始人为美国麻省理工学院（MIT）的福瑞斯特（J. W. Forrester）教授，为分析生产管理及库存管理等企业问题而提出的系统仿真方法，最初称为工业动态学。它是一门分析与研究信息反馈系统的学科，也是一门认识系统问题和解决系统问题的交叉综合学科。从系统方法论来说，系统动力学是结构的方法、功能的方法和历史的方法的统一。它基于系统论，吸收了控制论、信息论的精髓，是一门综合自然科学和社会科学的横向学科。

系统动力学运用“凡系统必有结构，系统结构决定系统功能”的系统科学思想，根据系统内部组成要素互为因果的反馈特点，从系统的内部结构来寻找问题发生的根源，而不是用外部的干扰或随机事件来说明系统的行为性质。

本教材是在刘静华教授近20年来从事系统动力学教学和科研以及实践教学的基础上凝结而成的，是多年教学与科研实践的积累，学生反响良好。其中的实验内容部分借鉴了其他老师的资料，上机实验解答是教师和学生教与学的结晶。本教材主要是用实例的形式，按照教师上课的学时来设计，以短小精悍的习题和系统动力学理论同步配套的形式呈现，同时辅以教学视频，极大地方便教师备课和授课指导，特别适用于学生将理论应用于实际，提高解决实际问题的能力，尤其是4个上机作业更强化了学生综合运用管理知识的能力。本教材的第一部分8个题目，与系统动力学理论教学同步配套使用，侧重于系统动力学Vensim软件的每次课后作业的求解和应用；第二部分 16 学时，进一步强化了 Vensim 综合建模应用，尤其适用于管理统计优化中系统动力学上机的复习与提升。

本教材得到了南昌大学“双一流”博士点建设专项经费的支持。特别感谢南昌大学公共政策与管理学院徐兵院长和喻登科副院长在本书出版的过程中给予的大力支持。由于时间仓促，书中难免存在一些不当之处，敬请读者批评指正。

目 录

第一部分
上机实验题目

上机实验题目 1

1. 请给出系统动力学建模的 4 个原件。

2. 请绘制系统动力学建模的 SD 4 类图形（SFD—流位流率图、CLD—因果环图、结构图、时域图）。

3. 请绘制系统动力学建模的两种类型因果关系图。

4. 请画出系统动力学建模的 5 种基本结构。

5. 请画出系统动力学建模的 6 种典型行为。

6. 请列出系统动力学建模的 8 种描述。

7. 请给出系统动力学建模的指数增长的结构图形、仿真方程、仿真图形、6 个案例。

上机实验题目 2

请写出 4 类仿真函数的名称、仿真图形、仿真和仿真曲线。

上机实验题目 3

1. 请画出指数增长的 SD 通用模型。

2. 请画出 GDP 增长的 SD 模型，假设 GDP（1950）= 1，Rate = 0.04，仿真出 1950~2100 年的流位 GDP 图形，并给出 2000 年、2050 年、2100 年 GDP 的值。

3. 请画出 GDP 增长的 SFD、CLD、结构图、时域图。

4. 请写出 GDP 的差分式、积分式、递推式。

5. 请计算当 T=？时，GDP=2。

6. 请用微分式推导出增长率 g 为变量时，流位 GDP 翻倍的公式是否符合 70/100g，并用第 5 问验证。

7. 请在国家统计局网站查出 1950~2021 年的 GDP 数据，用 Excel 做出趋势图，选用平均法或其他统计法算出 g，并用 SD 通用模型建模，比较仿真数据和实际数据的拟合程度，选择合适的增长率 g。

上机实验题目 4

1. 建立世界模型Ⅱ的因果关系图。

2. 建立世界模型Ⅱ的流位流率系（用表格的形式，注意变量用斜体）。

3. 建立世界模型Ⅱ流率基本入树 T_1 的建流位控制污染流率二部分图；结构模型：入树 T_1（t）仿真方程；污染量 POL、年污染排放量 POLG、年污染治理量 POLA 曲线；仿真检验结果分析。

4. 总结建立表函数的基本步骤及提升规律。

5. 建立世界污染量入树 T_1（t）资本入树 T_2（t）组合模型二部分图；入树 T_1（t）T_2（t）组合结构模型；入树 T_2（t）仿真方程；资本 CI、年资本投入 CIG、年资本折旧 CID 曲线；T_1（t）T_2（t）组合仿真检验结果分析。

6. 建立世界污染量入树 T_1（t）资本 CI 入树 T_2（t）资源 NR 入树 T_3（t）组合模型：资源 NR 入树 T_3（t）二部分图；入树 T_3（t）仿真方程；设关联数入树 T_1（t）T_2（t）T_3（t）组合仿真检验；资源 NR 及年资源消耗量 NRUR 曲线；T_1（t）T_2（t）T_3（t）组合仿真检验结果分析。

7. 建立世界污染量入树 T_1（t）资本 CI 入树 T_2（t）资源 NR 入树 T_3（t）农业资本比重 CIAF 入树 T_4（t）组合模型：农业资本比重 CIAF 入树 T_4（t）二部分图；入树 T_1（t）T_2（t）T_3（t）T_4（t）组合结构模型；入树 T_4（t）仿真方程；T_1（t）T_2（t）T_3（t）T_4（t）组合仿真检验结果分析。

8. 建立世界污染量入树 T_1（t）资本 CI 入树 T_2（t）资源 NR 入树 T_3（t）农业资本比重 CIAF 入树 T_4（t）入树 T_5（t）组合模型：入树 T_1（t）T_2（t）

T_3（t）T_4（t）T_5（t）组合结构模型：入树 T_5（t）仿真方程；人口 P 及年出生人口 BR、年死亡人口 DR 仿真曲线；T_1（t）T_2（t）T_3（t）T_4（t）T_5（t）组合仿真检验结果分析。

上机实验题目 5

逐层增枝检验建模的题目见期刊《系统工程理论与实践》2017 年 9 月刘静华文章。

1. 建立 T_1 入树模型及仿真方程，写出仿真图形 T_1（t）和 R_1（t）。
2. 建立 T_1-T_2 入树模型及仿真方程，写出仿真图形 T_2（t）和 R_2（t）。
3. 建立 T_1-T_3 入树模型及仿真方程，写出仿真图形 T_3（t）和 R_3（t）。
4. 建立 T_1-T_7 入树模型及仿真方程，写出仿真图形 T_7（t）和 R_7（t）。
5. 建立 T_1-T_9 入树模型及仿真方程，写出仿真图形 T_8（t）和 R_8（t）、T_9（t）和 R_9（t）。

上机实验题目 6

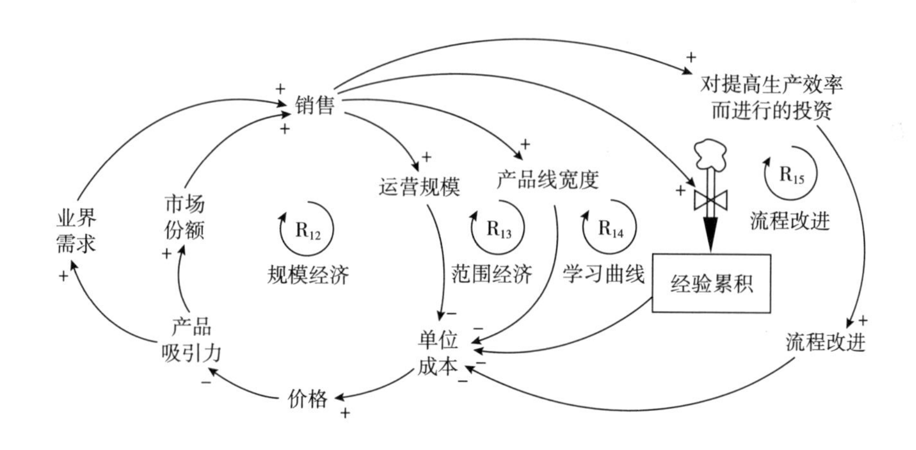

规模经济、范围经济、学习曲线和流程改进

注：图中涉及的每一个效果都包括两个正反馈过程：一是随着市场份额的增加而导致的销售额的增加；二是随着市场总规模的扩大而导致的销售额的增加。

根据上面的图，查找相关的养猪文献，以一个养猪场为例，试着用系统动力学建模，给出仿真方程及仿真图形。

1. 规模经济。分析小型、中型、大型养猪场的规模数量？

根据猪周期和历史年每月的平均价格计算利润。

2. 产品线宽度。哪些品种好饲养？

3. 经验积累。需要哪些经验，如何获得这些经验，需要多长时间（精确到月）？

4. 对提高效益而进行的投资。随着规模的扩大，面临哪些问题？如何筹措？

5. 风险因素。养猪面临的疫病及关键管理技术有哪些？请着重分析非洲猪瘟的影响及关键管理制度。

6. 用系统动力学建模，给出仿真方程及仿真图形。

上机实验题目 7

1. 请给出疾病传染模型 PPT，包括假设、适用、模型、仿真方程、符号的解释、反馈过程描述、传染病的基本特征，并以此顺序写出疾病传染的简单模型：SI 模型。

2. 请写出 S 增长的 Logistic 模型三种表达的公式、Gompertz 模型公式、Richards 模型（1959）的净出生率公式、Weilbull 模型公式、Rayleigh 分布公式。

3. 请写出创新传播中新观念和新产品的建模，包括模型、公式、符号的解释。

4. 请介绍创新传播的 Logistic 模型，包括公司、客户、用途、图形、实际曲线、模型得出过程、符号表示、不足。

5. 请给出 Bass 模型，包括优点、适用、建模目的、模型、两个反馈过程、符号解释、仿真方程、三种条件下的相变图、模型的行为特征、评论 Bass 模型、拓展。

6. 请给出重置消费模型，包括产品淘汰和重置模型、仿真曲线、重置购买行为建模、重置消费模型的行为特征。

上机实验题目 8

试比较如下 3 种模型的建模方程和思路，并查阅相关文献和资料，建立中国人口模型。

1. 第 1 种人口模型。

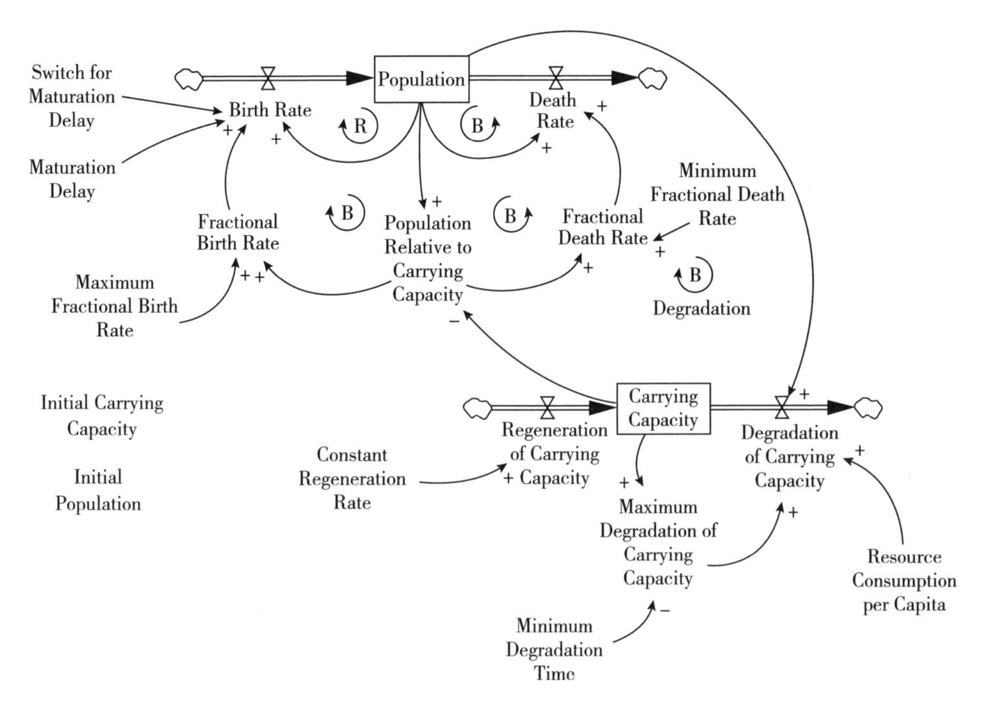

(1) Population = Initial Population.

(2) Birth Rate = Switch for Maturation Delay * DELAY3(Fractional Birth Rate * Population, Maturation Delay) + (1 − Switch for Maturation Delay) * Fractional Birth Rate * Population.

Units: People/Year

Comment: Births are proportional to the population. When the Switch for Maturation Delay = 0, those born immediately add to the population and can reproduce, die, and consumer the carrying capacity. When the switch = 1, there is a third-order maturation delay with an average delay time of the Maturation Delay before births enter the population stock.

(3) Death Rate = Fractional Death Rate * Population.

Units: People/Year

Comment: Deaths are proportional to the population.

(4) Switch for Maturation Delay = 0.

Units: Dimensionless

Comment: 1 = Maturation Delay between births and entering the population. 0 = no maturation delay.

(5) Maturation Delay = 20 constant.

Units: Year

Comment: The average maturation delay.

(6) Fractional Birth Rate = Maximum Fractional Birth Rate * (1−(1/(1+exp(−7 * (Population Relative to Carrying Capacity−1))))).

Units: 1/Year

Comment: The fractional birth rate is a declining function of the population relative to the carrying capacity. A logistic function is used.

(7) Maximum Fractional Birth Rate = 0. 04.

Units: 1/Year

Comment: The maximum fractional net birth rate.

(8) Population Relative to Carrying Capacity = Population/Carrying Capacity.

Units: Dimensionless

Comment: The ratio of population to carrying capacity determines the fractional birth and death rates.

(9) Fractional Death Rate = Minimum Fractional Death Rate * (1 + Population Relative to Carrying Capacity^2).

Units: 1/Year

Comment: The fractional death rate is an increasing function of the ratio of population to carrying capacity. A power function is assumed.

(10) Minimum Fractional Death Rate = 0. 01.

Units: 1/Year

Comment: The minimum fractional death rate.

(11) Carrying Capacity = Initial Carrying Capacity.

Units: People

Comment: The carrying capacity defines the equilibrium or maximum sustainable population. It is consumed and degraded by the population and can also regenerate.

(12) Regeneration of Carrying Capacity = Constant Regeneration Rate.

Units: People/Year

Comment: Regeneration of the carrying capacity. Equal to a constant rate set by the user.

(13) Degradation of Carrying Capacity = MIN(Maximum Degradation of Carrying Capacity, Population * Resource Consumption per Capita).

Units: People/Year

Comment: The carrying capacity of the environment is consumed or degraded in proportion to the population. The minimum function ensures that degradation falls to zero as the carrying capacity falls to zero (carrying capacity can never be negative).

(14) Constant Regeneration Rate = 0.

Units: People/Year

Comment: Exogenous constant regeneration rate, set by the user.

(15) Maximum Degradation of Carrying Capacity = Carrying Capacity/Minimum Degradation Time.

Units: People/Year

Comment: The maximum degradation rate is determined by the carrying capacity and the minimum degradation time. This formulation captures the fact that the carrying capacity must remain nonnegative and that damage to the environment falls as there is less undamaged environment remaining.

(16) Minimum Degradation Time = 1.

Units: Year

Comment: The minimum time constant for the degradation of the environment.

(17) Resource Consumption per Capita = 0.

Units: People/Person/Year

Comment: Resource consumption per capita, expressed in people - equivalent units of carrying capacity consumed per person per year.

（18） Initial Carrying Capacity = 1e+006.

（19） Initial Population = 1000.

2. 第 2 种人口模型。

（1） S 形增长。

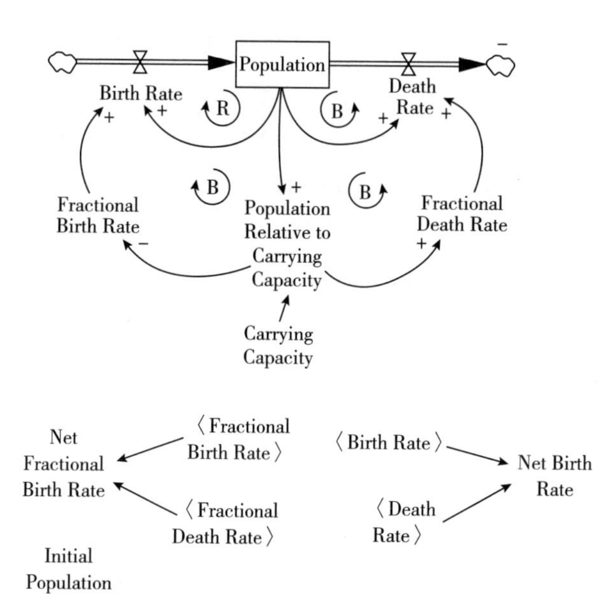

（2） 建模。

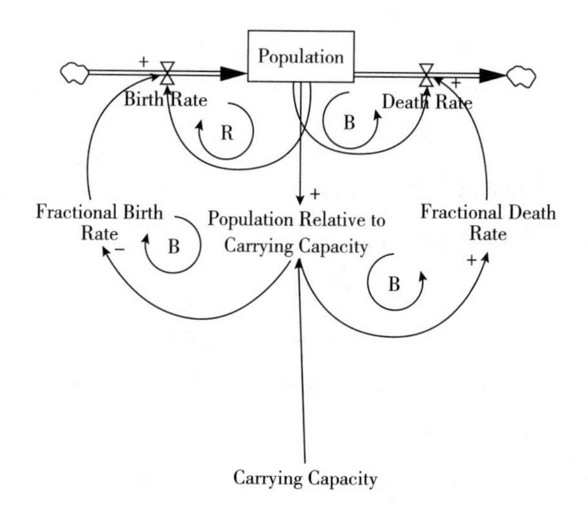

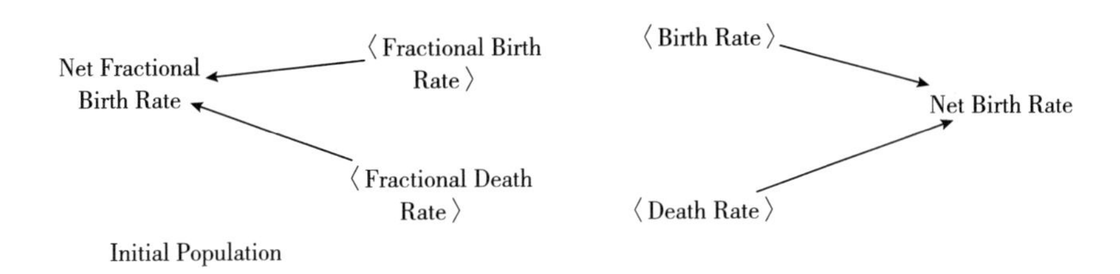

（3）仿真方程。

Time=（0，100）Time Step=0. 125 Unit for time=Year

①Population=Initial Population.

Units：People

Comment：The population is increased by births and decreased by deaths.

②Birth Rate=Fractional Birth Rate * Population.

Units：People/Year

Comment：Births are proportional to the population.

③Fractional Birth Rate=1-(1/(1+exp(-7 * (Population Relative to Carrying Capacity-1)))).

Units：1/Year

Comment：The fractional birth rate is a declining function of the population relative to the carrying capacity. A logistic function is used.

④Death Rate=Fractional Death Rate * Population.

Units：People/Year

Comment：Deaths are proportional to the population.

⑤Fractional Death Rate=0. 25+0. 25 * Population Relative to Carrying Capacity^4.

Units：1/Year

Comment：The fractional death rate is an increasing function of the ratio of population to carrying capacity. A power function is assumed.

⑥Net Fractional Birth Rate=Fractional Birth Rate-Fractional Death Rate.

Units：1/Year

Comment：The net fractional birth rate is fractional births less fractional deaths.

⑦Net Birth Rate=Birth Rate-Death Rate.

Units：People/Year

Comment：The net birth rate is births less deaths.

⑧Initial Population = 2.

Units：People

Comment：The initial population.

⑨Population Relative to Carrying Capacity = Population/Carrying Capacity.

Units：Dimensionless

Comment：The ratio of population to carrying capacity determines the fractional birth and death rates.

⑩Carrying Capacity = 1000.

Units：People

Comment：The carrying capacity defines the equilibrium or maximum sustainable population.

3. 第3种人口模型。

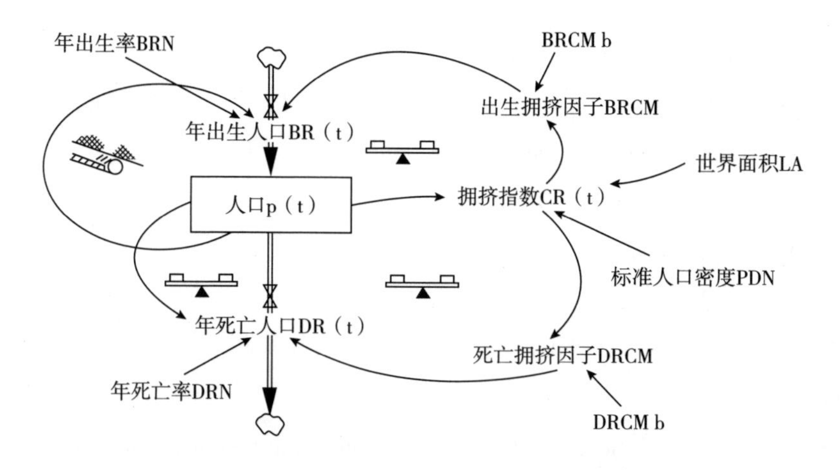

年出生人口 BR(t)原每年按 0.04 出生率增加，现为 BR(t) = p(t)×0.04×出生拥挤因子 BRCM(拥挤指数 CR(t))。

其中，BRCM(CR(t))为表函数，自变量为拥挤指数 CR(t)。

年死亡人口 DR(t)原每年按 0.028 死亡率减少，现为 DR(t) = p(t)×0.028×出生拥挤因子 DRCM(拥挤指数 CR(t))。

其中，DRCM(CR(t))为表函数，自变量为拥挤指数 CR(t)。

拥挤指数 CR(t)=(人口 P(t)/135×10^6 平方公里)/(26.5 人/平方公里)

世界面积 LA=135×10^6 平方公里(1.35 亿平方公里)

1958 年人口密度 PDN=26.5 人/平方公里

拥挤指数 CR(1958)=1，

当 t<1958 时，CR(t)<1；

当 t>1958 时，CR(t)>1。

由拥挤指数 CR(t)至 BR(t)为负因果关系，70 多年统计数规律，由 BR(t)=p(t)×0.04×出生拥挤因子 BRCM(拥挤指数 CR(t))。

建立出生拥挤因子 BRCM(拥挤指数 CR(t))表函数。

出生拥挤因子 BRCM(t)=BRCMb（拥挤指数 CR(t))

拥挤指数 CR（t）	0	1（1958 年）	2	3	4	5
出生拥挤因子 BRCM	1.05	1	0.9	0.7	0.6	0.55

由拥挤指数 CR(t)至 DR(t)为正因果关系，70 多年统计数规律，由 DR(t)=p(t)×0.04×出生拥挤因子 DRCM(拥挤指数 CR(t))建立出生拥挤因子 DRCM(拥挤指数 CR(t))表函数。

死亡拥挤因子 DRCM=DRCM(拥挤指数 CR(t))

拥挤指数 CR（t）	0	1（1958 年）	2	3	4	5
死亡拥挤因子 DRCM	0.9	1	1.2	1.5	1.9	3

仿真区间：1900（年）至 2100（年）

仿真步长：DT=1（年）

p（1900）=1.65×10^9（人）

4. 三种建模方法的区别。

上机实验题目 9

使用通用存量管理结构，完成下题。

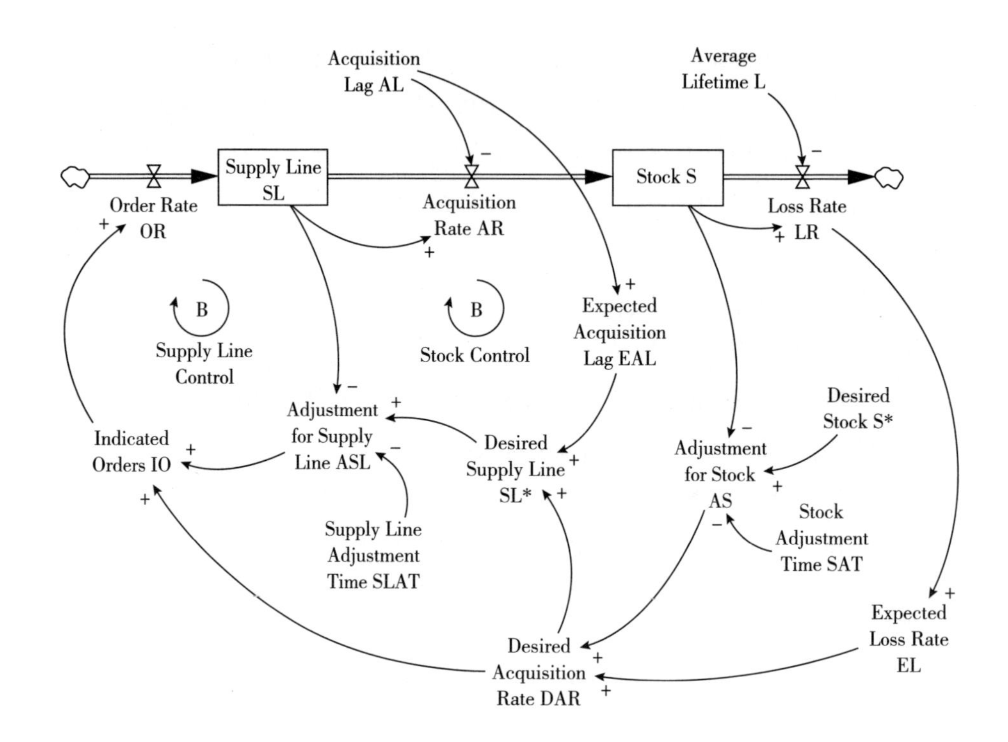

查知网文献或外文文献，建一个振荡和崩溃的模型，要求有实际背景、模型，方程、仿真图形及结果分析，实际背景见下表的 14 个系统（要求本科生 3 人使用一个系统，研究生 1~2 人使用一个系统，不得重复）。

系统	存量	供给线	损失速率	获得速率	订货速率	典型行为
库存管理	库存	订购的物品	向顾客交货	供应商发货到达	订购商品	商业周期
资本投资	资本厂房	建造中的厂房	折旧	建造完成	新合同	建筑周期
设备	设备	订购的设备	折旧	设备交付	新设备订购	商业周期
人力资源	雇员	空缺和实习生	解雇和离职	雇用速率	空职产生	商业周期
现金管理	现金余额	未决贷款申请	支出	借款速率	贷款申请速率	现金流周期
市场营销	顾客基数	潜在顾客	叛离投奔竞争者的顾客	新顾客补充	联系新顾客	顾客基数的起伏循环
肉猪饲养	肉猪存量	幼猪和怀孕种猪	屠宰速率	成熟速率	饲养速率	肉猪周期
农产品	库存	田野中的农作物	消费速率	收割速率	栽培速率	商品周期

续表

系统	存量	供给线	损失速率	获得速率	订货速率	典型行为
商业房地产	建筑物存量	开发中的建筑物	折旧	完工速率	开发速率	房地产的兴衰循环
用电热器烹调	壶的温度	电热线圈中的热量	扩散到空气中	从电热线圈扩散到壶中	炉温设定	烹调过火
驾驶	与下一辆车的距离	汽车动量	摩擦损耗	速度	油门和刹车	走走停停的交通状况
淋浴	水温	水管中的水温	排水速率	从莲蓬头中流出的水	水龙头设置	过热或者过冷
人体能量级别	血液中的葡萄糖	消化道中的淀粉和糖类	新陈代谢	消化	食物消化	能级循环
饮酒	血液中的酒精	胃中的酒精	酒精新陈代谢	酒精从胃向血液中扩散	酒精消费速率	醉酒

上机实验题目 10

1. 查询近 5 年的生猪出栏价格和猪肉价格（精确到月或天）。

2. 参照 Sterman3 模型和相关的参考文献建养猪周期的 Sd 模型。

3. 将非洲猪瘟作为外生变量，分析下面的专业判断是否正确：猪市场有周期性，一般为 2~3 年或 3~5 年。

上机实验题目 11

You can choose handwritten or typed answers. Graphs can be drawn by hand, just make sure that you are sufficiently accurate. Computer output should be printed.

1st Basic structure：Instantaneous cause and effect relationship

1.1 Illustrate an instantaneous cause and effect relationship in a CLD (causal loop diagram) and in a SFD (Stock and Flow Diagram) and show how such a relationship can be described in a graph with cause on the horizontal axis and effect on the vertical axis.

Hint: Recall that an instantaneous cause and effect relationship does not involve accumulation (the cause may be a stock, but the effect is never a stock).

1.2 Think about the following cause and effect relationships and discuss if they can be properly represented by an instantaneous relationship. If, and only if, you think an instantaneous relationship is appropriate, use a graph to illustrate what the relationship may look like (same type of graph as in the last part of question 1.1). Comment on whether the graphs you draw should be linear or nonlinear. Also try to formulate the instantaneous cause and effect relationships mathematically.

-Effect of interest rate on interest payments

-Effect of per unit production costs on profits

-Effect of production on size of inventory

-Effect of amount of water in a funnel on the outflow from the funnel

-Effect of births on population size

-Effect of population size on births

-Effect of hours studied per day on learning per day

-Effect of velocity on distance travelled

-Effect of force on acceleration

-Effect of acceleration on velocity

2nd Basic structure: One stock and two flows

2.1 Which of the variables below are stocks and which are flows? Give your answer in terms of SFDs where you name the stocks and the flows.

Hint: Use the snapshot method, that is, imagine that you take a picture of the situation. What can be "seen" in the picture is normally a stocks and what cannot be "seen" is normally a flow (in a picture you can see the water level in the lake, but you cannot see how fast the river flows into the lake as it crosses a thin border between the two.

Sometimes a better hint is to consider if something can accumulate over time (or can be stored), if so, it is a stock.

(a) A lake, a river flowing into the lake and a river flowing out.

(b) Oil production, underground oil reservoir, oil storage.

(c) Students in the lab, students entering, students leaving.

(d) Ordering of equipment, scrapping of equipment, equipment in place.

(e) Withdrawals, bank account, deposits.

(f) Mature fish, maturation, natural death, recruitment, young fish.

(g) Customers using a product and customers no longer using a product.

(h) Knowledge of words, learning words, forgetting words.

(i) Production of bottles, production of lemonade, amount of bottles with lemonade.

(j) Perceived price, perceived price last week, change in perceived price.

2.2

(a) Suggest units for all the variables in question 2.1 Many answers could be correct, for example for the content of a bathtub (stock): *cubic meters*, *litres*, *gallons*, and for the water flowing in or out (flows): *litres per minute*, *litres per hour*, *cubic meters per day*.

Make sure that stocks and connecting flows measure the same thing!

(b) Explain why it does not matter for the analysis whether you for instance measure all flows in "per day" or "per month". What should determine what time unit to use in a model?

(c) Why do you have to measure all in-and outflows to *all* stocks in a model with the same time unit (for example per second or per minute). Hint. Recall the difference equation used in the simulation software (Euler) with its one and only Timestep (*dt*) that is used to calculate the development of all stocks from one point in time to the next.

2.3 Graphical integration. In the figure below you see a variable inflow to and a constant outflow from a stock. Produce a diagram showing how the stock develops (remember the three questions to ask: direction, size, and shape). Assume that the stock is zero to begin with (time=0). Make sure that you number the axis and that you are accurate.

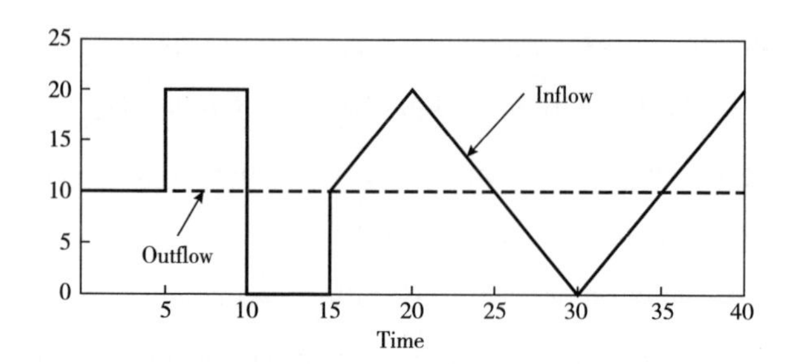

2.4 Explain on the basis of stock and flow development in question 2.3 why:

(a) Stocks show different behaviour patterns than flows.

(b) Stocks create delays.

(c) Stocks serve as a "memory" or summary of past net flows.

(d) Use your own words to explain what accumulation is, and how accumulation differs from instantaneous cause and effect. Does it make sense to talk about two types of cause and effect? Explain.

2.5 Think through the following situation. A country has built up a large debt. However, after years with deficits on the balance of trade, the country has managed to reduce the trade deficit to zero. Has the government solved the debt problem?

Hint: Use a stock and flow representation, and assume that it is only the trade deficit that matters for debt accumulation.

2.6 Look at the following diagram, which shows arrivals and departures from the hospital in Schwarzarch, at the two first weeks of the skiing season, from Dec. 31 to Jan. 13.

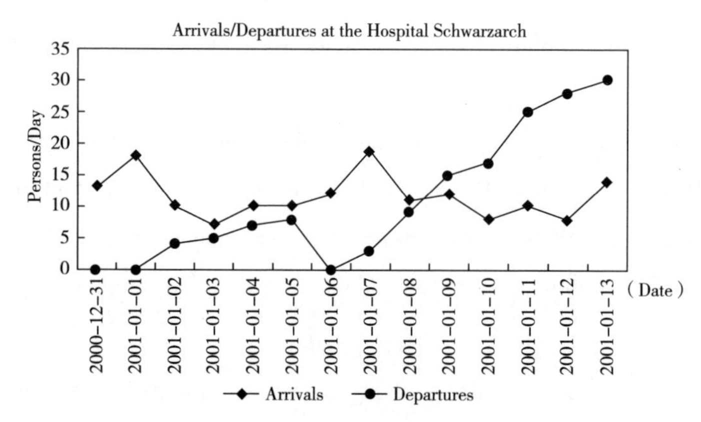

Draw a SFD and use this to explain your answers to the following questions:

(a) What day did most patients arrive?

(b) What day did fewest patients arrive?

(c) What day did most patients depart?

(d) What day did fewest patients depart?

(e) What day were there most patients in the hospital?

(f) What day were there fewest patients in the hospital considering the entire period from the morning of 2000-12-31 to the afternoon of 2001-01-13?

2.7 Think of two persons N and S. N thinks of the world as consisting of only instantaneous cause and effect relationships, S thinks of the world as consisting of only stock and flow relationships. Think of a system where Y is caused by X.

(a) What would N assume about X when she observes a steady increase in Y?

(b) What would S assume about X when she observes a steady increase in Y?

(c) If in reality Y is a stock, explain the frustration of N.

(d) If in in reality Y is not a stock, explain the frustration of S.

(c) Explain what is mcant by an "cvent based world view".

(f) Explain what is meant by the phrase "endogenously generated behaviour".

2.8 The purpose of this task is to develop your ability to conduct and to report from a project, for instance your thesis work. You will see a similar question in all later assignments. The question should give you practise in going through the five essential steps P'HAPI that are frequently repeated in the lecture notes: Problem definition, Hypothesis, Analysis, Policy, and Implementation, see the first lecture notes in File Storage. You should recognize the first three letters to represent the scientific method (P'HA), the fourth represents analysis (operations research) to identify good policies (P), and the last represents good management in order to implement policies (I). A manufacturing business has problems with its inventory of finished goods. Over the last year the inventory has declined steadily and at the end of the year it is nearly empty. Over the entire last year production has been 80% of sales. Write a couple of sentences only (!) about each of the following points. Do not show any diagrams, however draw a stock and flow diagram for yourself to help you reason about the problem. Hint: When you have your first draft ready, read the text again and see if you are able to identify each of the elements in P'HAPI in your own writing. If not, rewrite and read again until you are comfortable with your text.

P'-what the problem is and why it is important.

H-describe a hypothesis for why the problem has occurred (model structure).

A-analyse your choice of structure (reasons for your hypothesis).

Analyse model behaviour (graphical integration) and discuss it in light of observed Behaviour.

P-what policy recommendations follow from your tested hypothesis?

I-what challenges must be overcome to implement the policy?

上机实验题目 12

You can choose handwritten or typed answers. Graphs can be drawn by hand, just make sure that you are sufficiently accurate. Computer output should be printed.

1st Basic structure: One stock and two flows (continued from Assignment 1)

A benefit of using simple examples from physics is that physical laws are well established and you all have experience with the example systems. These two conditions are not so easily met in examples from the social sciences, examples that will be dealt with later.

2.9 Many high school students in physics classes have difficulties understanding the relationships between acceleration, speed and distance. Some of you may remember a formula for distance saying that:

Distance = 1/2 * *Acceleration* * *Time*2.

See if you can explain this formula by using stock and flow diagrams plus graphical integration (Do NOT use Vensim at this stage!). Follow the steps below.

(a) First consider acceleration and speed. The word "accelerate" means "to increase speed". Draw a stock and flow diagram. Assume that the speed is zero at time zero. Acceleration is constant and equal to 10 meters/second 2 (approximately the acceleration of an object that falls towards the earth with no air resistance) . Use graphical integration (ask the 3 questions) and draw a graph showing the development of speed over a 4-second period.

(b) Next consider speed and distance. Note that speed is a measure of how long distance an object travels over a given period of time. Draw a stock and flow diagram. Assume that the distance travelled is zero at time zero. Assume that the speed develops as you found in point a) . Use graphical integration and draw a graph showing the development of distance over the same 4-second period.

(c) Compare the result of your graphical integration with the above formula. Do you get the same numbers for distance after four seconds with both methods? Can you explain why distance depends on time squared and can you explain the term 1/2 in the formula?

(d) Did the stock and flow diagram help you get a better intuitive understanding of the mathematical formula? Give reasons for your answer whatever it is. If you understand speed and distance, you understand the basics of rocket science.

2. 10 Using Vensim to simulate.

(a) Build in Vensim the model of speed and distance from question 2. 9. Print your Vensim SFD and equations for the model. Make sure that all variables have correct units. Note that in Vensim you need two different names for Speed, you may use the terms Speed for the stock and "Change in distance" for the flow that influences distance.

(This is one quite rare case where a stock is also the exact flow for another stock) .

(b) Simulate the model with the same initial values and the same constant acceleration as in question 2. 9.

-Create a time graph with scale from 0 (ZERO) to 80 for all three variables.

-Why is it usually preferable to use a scale starting at zero?

-Create an output table and choose report *Interval* equal to 1 (second) .

-Choose Run Specs. . . from the Model menu and set unit of time to "Other" and write seconds. Set Length of simulation from 0 to 4 (seconds) . Choose Euler's Method and DT=0. 001 (second) . In Stella the Timestep is denoted DT. Simulate and show acceleration, speed and distance in the time graph and in the output table. Compare to results from question 2. 9 and consider the accuracy of the simulated time behaviour.

(c) Set DT = 1 second and simulate. Recall the equation for Euler integration: Stock (t) = Stock (t−1) +Timestep * (Inflow (t−1) −Outflow (t−1))

Use this equation to explain why the simulation shows correct results for Speed and too low values for Distance.

(d) Go to Model>Run Specs... and select integration method Runge−Kutta 4 (RK4) rather than Euler's method. Keep DT = 1 (second) . Comment on accuracy, you do not have to explain.

(e) Keep the Runge−Kutta 4 method and let acceleration increase linearly from zero to 10 meters/second 2 over the 4−second interval. Model acceleration as a linear and instantaneous function of the variable called *time* (*time* does not have to be modelled; it is a variable that exists in the Vensim program itself) . Simulate for different DTs and report results for the highest DT that does not lead to changes in the second decimal for distance after 4 seconds. You may change the scale on the vertical axis.

2. 11 Use the P'HAPI framework to write a short abstract for a project studying the increase of CO_2 in the atmosphere. Assume that the expressed goal is to stabilize the level of CO_2 by 2030 and that the project should give advice on how emissions should develop.

Focus on the hypothesis for accumulating CO_2 in the atmosphere. Write about two lines or less for each of the five steps in P'HAPI. Follow the same pattern that was suggested in a similar task in Assignment 1; see lecture notes and the handout on P'HAPI.

In this assignment, you should team up with one other student. Copy the other student's abstract into your own answer sheet and comment on how well the other student has described each of the P'HAPI points. Give reasons for your comments, do not simply say good, excellent, bad, wrong, incorrect etc. Make sure that you comment on the first and only version of the abstract that you receive from the other student. (This is a useful exercise both when you revise your own writing and when reading material produced by others.) After commenting on the other student's abstract, you may revise your own abstract (the one you gave to the other student) before you hand in both abstracts.

2nd Basic structure: First order linear feedback

Reinforcing loops

3.1 Identify reinforcing loops responsible for growth in the following examples (could be more than one loop). Give your answers in terms of causal loop diagrams. Remember to denote link polarity with+and−signs, and include the R−letter to denote that the loop is reinforcing. (Do not use verbs in your variable names, links and polarities indicate in what direction the variables will change, see textbook Chapter 5.1 to 5.3 and lecture notes for hints on drawing causal loop diagrams).

Drawing by hand is fine. You can also draw CLDs in Vensim: Place one or several Modules (rectangle with smooth corners) in the diagram sheet, double click in the white area outside the module, then select the modules, and then choosc "Name only" in the area popping up on the right−hand side. Now you can copy and paste modules with name only in your CLD. You can place and draw instantaneous cause and effect arrows between the module names. Right click on the arrow's circle to choose Polarity. To denote reinforcing (R) or balancing (B) loop, click the T button and choose the Loop label. To get a nice diagram choose the Map View (rather than x_2).

−Population Growth.

−A quarrel (include level of anger and provoking comments).

−A war between two countries (provocation and retaliation).

−Economic growth (include machines, labour, technology, production).

−Spread of an epidemic (consider those infected and the infection rate).

−Market growth for a new product (assume that production costs and product prices vary (decrease) with accumulated production.

3. 2 In a bacteria culture, there is a net fractional growth rate of 3. 5% per minute (instantaneous growth rate). Assume that the initial number of bacteria is 1000.

(a) Write an equation for net growth in the number of bacteria.

(b) Use a structure graph from stock to net flow *to reason about* how the stock of bacteria will develop over time. Make sure you number the axis and that you are accurate.

Hints: Remember that the structure graph is not a behaviour-over-time graph! For any value of the stock, you can see how large the net flow is, and consequently you can reason about how the stock will develop from the stock value you consider. The stock needs time to change from one point in time to the next.

(c) Use the 70-rule to find out how many minutes it takes for the bacteria culture to double in numbers.

(d) Model the bacteria system in Vensim. Simulate the model where time runs in minutes, and find out how many bacteria there will be after 1 hour (60 minutes), 6 hours and 24 hours. Use Runge-Kutta 4 (RK4) and set DT=4. Which is the largest DT that gives the correct answer measured by the second decimal for the number of bacteria after 1 hour.

(e) For a time-horizon of 60 minutes, find the exact number of bacteria by using the formula for development over time for a linear reinforcing system, see the lecture notes.

Try out different time steps DT when using the Euler integration method. Reason about the need for accuracy given that there is uncertainty in the estimate of the instantaneous growth rate.

(f) Would you say development after a 24 hours simulation is realistic? Perform a structural test and explain what may be wrong with the model?

(g) Assume from now on that a predator is present, which eats bacteria. Include eating by this predator in your Stella model. This predator eats exactly 17. 5 bacteria per minute-as long as there are bacteria available. Thus, the eating by the predator does not vary with the stock of bacteria. Use a revised structure graph for net growth to explain what happens if there are 1000 bacteria initially. Test your answer by simulating in Vensim.

(h) What happens if there are only 400 bacteria initially? Use the structure graph to explain your answer. Test your answer by simulating in Vensim.

(i) Can you identify an unstable equilibrium point (or repellor) for this system with a predator present?

Balancing loops

3.3

(a) Draw a stock and flow diagram for a system that consists of a stock and a net inflow.

The net inflow is determined by the deviation between a constant desired stock and the actual stock, the difference is divided by an adjustment time. The adjustment time is constant and equal to 5 time units. Write an equation for the net flow.

(b) For what values of the desired stock and the actual stock is the net inflow zero? Use the equation for Net flow to explain in what direction the stock will develop if the desired stock is higher than the actual stock.

(c) Assume that the desired stock is constant and equal to 10 units. Construct and use a structure graph for the net flow to explain in what direction the stock will develop for different values of the stock.

(d) In the diagram below sketch the development for the stock when the desired stock develops as shown in the figure. Assume that the initial stock value is zero. Use the method with tangents and the 63% rule. Show the stock development and *show the tangents* that you use! Be as accurate as you can, but do not use a calculator or Vensim.

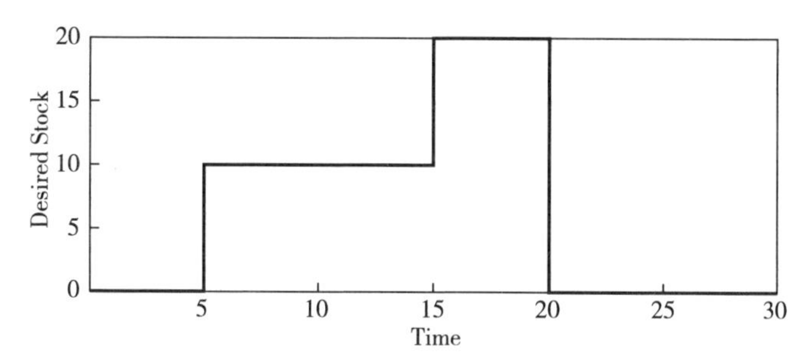

(e) Look at the development of Stock in part d) and make a rough sketch of how the net flow develops (hint: Use the formula for the net flow to reason).

(f) Do not start working on this question before you have finished the preceding ones. Model the above system in Vensim with adjustment time = 5 time units and simulate the development over time. To model the desired stock use Step functions that add to or subtract from the current value of the desired stock at the appropriate points in time. Make sure that you make a proper choice of DT and/or integration method.

Note: If you get clearly different results in f) and e), that is a hint to you as a modeller that something is wrong. Either you do not understand behaviour well in e), or you have made a mistake when simulating in f). Hence you see that it is important to develop intuition for behaviour in order to detect errors in simulation models, and to understand and explain simulation results.

(g) Simulate the model with the desired stock = 1.0 * time. Then simulate the model with Desired Stock = 10+SINWAVE (Amplitude, Period), where Amplitude = 5 and Period = 4.

Comment on the relationship between the desired stock and the actural stock.

(SINWAVE is a function of time. If you do not know about sine waves from before, it is sufficient that you look the time graph for the desired stock to see how it works.)

3.4

(a) Draw a CLD that captures how a person fills a bucket with water towards some desired level. Draw a SFD for the same system.

(b) Draw a CLD that captures how a farmer determines how many cows to slaughter in a process where she wants to double the number of cows. Make assumptions about births of new calves. Draw a SFD for the same system.

上机实验题目 13

You can choose handwritten or typed answers. Graphs can be drawn by hand, just make sure that you are sufficiently accurate. Computer output should be printed.

1st Basic structure: First order linear feedback (continued from Assignment 2)

3.5 In economics, a typical assumption is that prices are set such that demand and supply are in equilibrium. A dynamic model is needed to understand how prices change over time in cases of imbalances between supply and demand. Models may differ between markets: Department stores typically operate with posted prices that change only slowly while prices set in auctions may change from minute to minute. Consider the Stella model described by the equations below (you may look up Sterman's textbook page 816 for further details). A more complete model would also include effects of price on supply and demand. That is left out in most of this assignment.

STOCKS:

Anchor(t) = Anchor(t-dt) + (Updating_of_anchor) * dt

INIT Anchor = 10

INFLOWS:

Updating_of_anchor = Gap/Time_to_update_anchor

AUXILIARY VARIABLES AND CONSTANTS

Time_to_update_anchor = 1

Gap = Price-Anchor

Price = Anchor * Adjustment_for_demand_supply_balance

Adjustment_for_demand_supply_balance = 1.05

The time unit is months. Notice that this is yet another example of a feedback process of the anchoring and adjustment type; the anchor is an average of easily observable recent prices and adjustments are made for the current supply and demand balance.

(a) Use Stella to create a SFD from these equations.

(b) Is the model linear or nonlinear? Are there counteracting and/or reinforcing loops?

(c) First assume that demand exceeds supply such that there is an upward pressure on price: Adjustment_for_demand_supply_balance = 1.05.

Before drawing a structure graph, which loop will dominate with this parameter value? What behaviour will follow?

(d) Draw a structure graph of the net flow to analyse behaviour for the case in point c), with Adjustment_for_ demand_supply_balance = 1.05.

Hint: Show structure graphs for each of the two loops separately before you draw a graph for the net flow. Use the structure graph for the net flow to reason about the behaviour that this system will produce.

(e) Next assume that Adjustment_for_demand_supply_balance = 0.95. Construct a new structure graph for the net flow and use it to reason about behaviour.

(f) Simulate the model for the two values for Adjustment_for_demand_supply_balance (0.95 and 1.05) and show the behaviour of the price anchor in one and the same diagram (double click the graph and tick off Comparative to show both runs in one diagram. To reset the comparative graph, choose Model>Restore>Graphs and tables).

Are the results consistent with your analysis? If not, reconsider your analysis and/or your simulation model.

(g) How would you summarize what you have learned here? Can a reinforcing loop create exponential decay? Have you learnt something about loop dominance?

2nd Basic structure: Nonlinearities

4.1 Consider a model describing the filling of an empty glass with water. Assume that the water comes from a faucet (infinite source).

(a) Model this system. Assume that the initial water level is 0 litre and that the desired water level is 1 litre. First assume that the inflow is given by the following decision rule (could also be called strategy or policy):

Inflow = (Desired water level−Water level) /Adjustment time

where the adjustment time is 2 seconds. First, characterize this system and say what behaviour this type of system produces.

(b) In your opinion, is the equation for the inflow a good description of how a person would fill a glass from a normal kitchen faucet?

Hint: perform an "extreme condition test", where you assume that the glass is very big and that the desired water level is 100 litres. Is the decision rule realistic for the kitchen faucet?

(c) Assume that the kitchen faucet has a maximum flow of 0.1 litre/second. Use the following function to model the inflow.

Inflow = MIN (0.1 litre/second, Gap/Adjustment time).

This MIN-function always selects the lower of the two possible values separated by the comma. Draw a structure graph for the inflow where the MIN-function limits the inflow. Use the structure graph to reason about how the water level in the glass develops, draw a time graph.

Hint: Remember the lessons from graphical integration (open loop) and the tangent method for a closed loop first order linear system.

(d) Simulate the model with the linear policy with and without the MIN-function. Show both results in onc graph (click Comparative).

(e) Use Stella's graphical-function to model the nonlinear decision rule for the case where the inflow is described by the MIN-function.

4.2 Build a Stella model of the reindeer simulator. Central variables in the model are *total amount of lichen* and *number of reindeer*. Use the following information and look carefully at the units; they are in tonnes and not in millimetres. *Units should match in each and every equation*:

Initial lichen density = 1000 tonnes/km^2

CC = 1200 tonnes/km^2

(CC is carrying capacity; the highest lichen density that gives positive growth)

Area = 5 km^2

Eating per reindeer per year = 0.4 tonne/reindeer/year

Growth in lichen density = MSY * 4 * (Lichen density/CC) * [(CC-Lichen density)/CC]

MSY = 100 tonnes/km^2/year

(MSY is maximum sustainable yield or maximum growth in lichen density)

Use the Euler method for simulation and DT = 1 year to replicate the settings of the reindeer simulator.

(a) Show a SFD of the model and make sure that you represent all variables mentioned in the text.

(b) Use the equations to construct a structure graph for growth of lichen (total growth for the entire area, NOT growth in lichen density) as a function of lichen density (usually we make the structure graph as a direct function of a stock; in this case it is more practical to use the stock of lichen relative to carrying capacity). For what lichen densities is growth zero and for what lichen density is growth at its maximum? You may check your answer by simulating the model with an increasing number of reindeer over time. Hint: Tick off for Scatter to graph lichen growth versus lichen thickness; sometimes referred to as a phase plot.

(c) With your own words, discuss if this graph for growth makes sense biologically (do not spend time looking up biological literature, simple arguments suffice).

(d) When lichen growth is at its maximum (MSY), how many reindeer can this growth rate feed every year?

(e) Assume that there is no reindeer and thus no grazing. Use the structure graph for lichen growth to reason about lichen behaviour when lichen density is less than 300 tonnes, when it is around 600 tonnes, and when it is above 900 tonnes. Use what you have learnt from graphical integration of open-loop systems and from structure graphs of linear reinforcing and balancing minor feedback loops.

(f) Simulate the model, still with no reindeer and thus with no grazing. Show lichen density. First start with an initial lichen density of 110 tonnes/km^2, and next with 600 tonnes/km^2 (use comparative graphs). Characterise and explain the difference you see in behaviour.

(g) Now allow for reindeer and grazing. Assume that the number of reindeer is constant and equal to 1250 and consider three situations with initial lichen density equal to respectively 300, 600 and 900 tonnes/km^2. Construct a structure graph for the net growth of lichen and use this graph to discuss behaviour for the three starting points for lichen.

(h) Set the number of reindeer constant and equal to 1250 and simulate the model starting with initial lichen density equal to respectively 300, 600 and 900 tonnes/km^2. (use a comparative graph with all three). Is behaviour consistent with your analysis in g) and can you explain behaviour by comparing in-and outflows?

(i) Using the structure graph once more, formulate a *hypothesis* for a policy for grazing as a function of lichen density.

Desired grazing = f (Lichen density)

The goal for your policy should be to stabilise lichen density at 600 tonnes/km^2. In other words, the policy should be such that no matter what the starting point is for lichen density, lichen density should end up at the desired point.

Hint: Draw a policy for desired grazing as a function of lichen density in the structure graph. Do not think about the needed number of reindeer for now.

(j) Now model your hypothesized policy and simulate to test it. Formulate the grazing policy by using a graphical function. Start the simulator with initial lichen density of 110 tonnes/km^2 and with 900 tonnes/km^2 (use comparative graphs). Also simulate the model with these initial lichen densities and a constant number of reindeer equal to 1250. Compare the results.

(k) Change your policy (graphical function) so that it either becomes more or less aggressive. Do this by changing the slope (steepness) of the function for Grazing. Still the end result should be that grazing ends up equal to maximum growth. Compare to point j). What are the implications of aggressiveness for implementation of policies?

Simulate with an initial lichen level of 110 tonnes/km^2 only. Show the development of the number of reindeer for both cases.

4.3 Write an abstract about the problem of managing a lichen pasture that has been overgrazed by reindeer. Maximum 18 lines in total with a font size of 12.

Follow the structure of P'HAPI, present the problem, a hypothesis (the dynamic model)

for why lichen is declining (needs to be succinct) and analyse your hypothesis (structure and behaviour). Use insights from the policy questions in 4.2 when discussing a hypothesis for a policy and when analysing this hypothesis. Regarding implementation, discuss needs for cognitive change among decision makers by referring to likely misperceptions and by referring to the question of policy aggressiveness.

In this assignment, you should team up with one other student. Copy the other student's abstract into your own answer sheet and comment on how well the other student has described each of the P'HAPI points. Give reasons for your comments, do not simply say good, excellent, bad, wrong, incorrect etc. Make sure that you comment on the first and only version of the abstract that you receive from the other student. After commenting on the other student's abstract, you may revise your own abstract (the one you gave to the other student) before you hand in both abstracts. (This is a useful exercise both when you revise your own writing and when reading material produced by others.)

上机实验题目 14

You can choose handwritten or typed answers. Graphs can be drawn by hand, just make sure that you are sufficiently accurate. Computer output should be printed.

1st Basic structure: Major feedback loops (with delays)

Reinforcing loops

4.4 Consider the following model. A population is split in two stocks: Young Population and child Producing Population. Both populations are equally split on males and females. Assume that the child producing population gives birth to one child for every couple every year (Births Per Person = 0.5 person per year, i. e. it takes a couple (two persons) to produce one child). The parameter Years Giving Birth denotes the length of the period in which couples have children and thus it determines the total number of children per couple. Assume that Years Giving Birth = 4 years such that each person gives birth to 2 children and each couple gives birth to 4 children. Ignore what happens after the child producing period is over (the Ending rate flows into the cloud).

Births = Producing Population * Births Per Person

Starting = Young Population/Age When First Child

Ending = Producing Population/Years Giving Birth

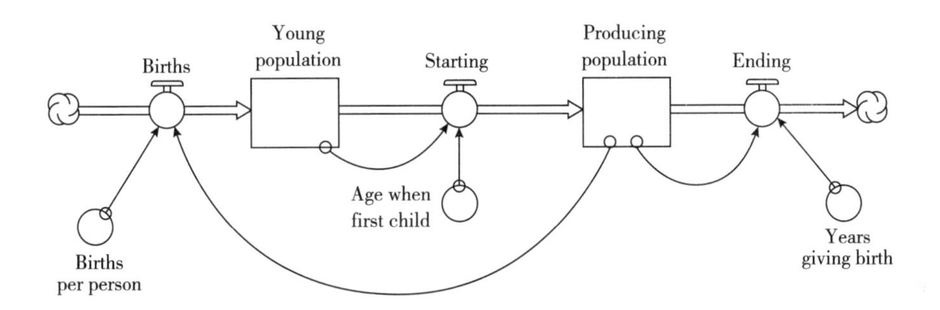

(a) Draw a causal loop diagram for the model shown in the above SFD (make use of the two rules for causal loop layout!). Identify and characterize all feedback loops in this model.

(b) Model the system. Set the Young Population = 100 million persons and Producing Population = 22 million persons (unit is *million* persons). Set Age When First Child = 15 years and simulate behaviour over a 30-year period. Show the variable Young Population. Is the system linear or non-linear? Explain behaviour over time in terms of model structure and parameters. Run sensitivity test for each parameter to check your explanations for how the different parameters influence behaviour. The usual advice is to explain first and then test your explanation by simulating. If you run first and then explain, you may end up with incorrect explanations that only feel right!

(c) Next, set Young population = 0. What behaviour do you expect for Young population compared to what you found in question b? Comment both on the early (before 7.5 years) and the late development (after 22.5 years) for both Young and Producing populations. You may also see what happens after 30 years and again compare to question b). Hint: To help you distinguish transient and steady-state behaviour, model and print the proportion of young population to the total population. Use comparative graphs and show the proportions for both point b) and point c) .

(d) Reset Young population to 100 million. Set Age When First Child = 25 years and simulate again and show the result together with the result from b), use comparative graph. Explain with words why the two developments differ.

(e) Repeat the above comparison (Age When First Child = 15 and 25 years) for a value of the parameter Years Giving Birth of 2. 2 rather than 4. 0. How many children will a couple on average get? Compare the results to those in b) and d) and explain the difference.

Delays

4. 5 Assume that Perceived Information is an information delay of the variable Information.

(a) Show how this information delay can be modelled by the SMTHN function (Click on SMTHN in the list of Built-ins and then OK and you get the get to see what parameters are needed, or consult the help menu on SMTH) . What are the delay times for the *individual* first order delays that make up the higher order delays when the order (number of stocks) is 1, 3 and 6? Model and show SFT for a third order information delay where you show all the individual stocks (i. e. do not use a SMTH-function) . Is the 6th order delay a linear or a non-linear system?

(b) Can the choice of Order for a delay have implications for the choice of time-step DT in a model? Discuss for the case of N = 3.

(c) Use the third order model that you made in a) . Assume that Information steps up from zero to 10 at week 2 and a total delay time of 12 weeks. Simulate and show Information and the values of all three stocks in one graph. Give a rough explanation of the early and late development for each of the stocks. Reason according to what you have learnt in graphical integration and about transients.

(d) Compare the behaviour of a first and a third order information delay for the same step in Information as in c) . Both delays should have the same 12-week delay time. Given that the total delay times are the same, is it logical that the third order delay gives the higher Perceived Information in the long run?

(e) Next change the variable Information to become a pulse stretching over one week. Do this by subtracting a negative step with height 10 units in week 3 from the step already in place (use two STEP-functions). Simulate with N = 1, 3, and 6. Comment on how information about the pulse is perceived for orders N = 1 and N = 6. Specifically, comment on how the two different delays represent information gathering and processing for decision-making.

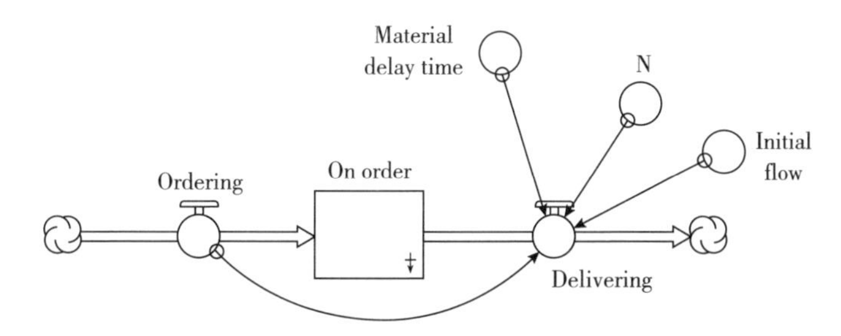

(f) Next, build the above model in Stella, where Delivering is a N'th order material delay of Ordering. Use the function DELAYN. Choose order N = 6 and a Material Delay Time of 12 weeks. In the DELAYN function the entry "initial" denotes the initial flow through all N individual stocks (implying that initially what is on order is evenly spread on all the N individual stocks). The stock on order contains the sum of all N individual stocks and is equal to the initial flow times the material delay time. (Note that the stock on order is only there to keep track of what is on order, the stock does not directly influence delivering. You cannot read out the sum of the N stocks in the DELAYN function.) Let ordering be a pulse of 100 units from time 2 to 3 weeks. Try to explain behaviour before you simulate and write down your explanation after you have tested it by simulating.

(g) Set ordering = 10 units per week and make sure that the initial On Order corresponds to this ordering rate, N = 6. Let Material Delay Time increase in a step from 12 to 14.5 weeks in week 10. What could the reason be for such a change? Explain behaviour of on order.

Balancing loops

4.6 Consider the Stella model below. This is a model of a fishery with open access (all fishers can take part in the fishery, there is no quota system to limit harvest). Fishing effort depends on fishing capacity and capacity utilization. Utilization goes up if income per unit effort exceeds unit operating costs. The catch depends on the effort and the CPUE (catch per unit effort) . The CPUE reflects the density of fish and says how easy it is to find and catch the fish. Unit profits reflect income per unit effort minus operating costs minus per unit capacity costs. Unit profits influence the desired capacity (ignore the help variables) and investments in fishing capacity.

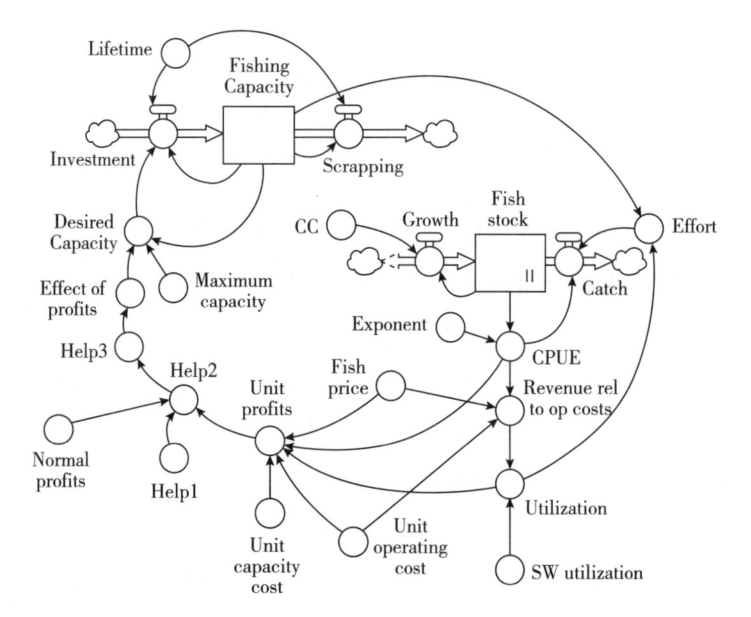

(a) Capture the essential feedback loops in a CLD, make use of the two rules for CLDs.

(b) Without using Stella, reason about how this fishery may develop if initially there is only a little fishing capacity and there is a lot of fish (close to carrying capacity) . Draw curves for how you think the fishing capacity and the fish stock will develop. Explain why development will be as you show in the drawing. Since you do not know parameter values, you should think in terms of typical developments. Assume that the fish price is higher than the initial unit costs.

4.7 P'HAPI

Write an abstract for a project dealing with the problem of "becoming drunker than intended". Keep P'HAPI in mind. Argue for the importance of the problem. Present your hypothesis of both physiology and decision rules of inexperienced drinkers. Analyse both structure and behaviour. Then discuss policy and implementation challenges.

In this assignment, you should team up with one other student. Copy the other student's abstract into your own answer sheet and comment on how well the other student has described each of the P'HAPI points. Give reasons for your comments, do not simply say good, excellent, bad, wrong, incorrect etc. Make sure that you comment on the first and only version of the abstract that you receive from the other student. After commenting on the other student's abstract, you may revise your own abstract (the one you gave to the other student) before you hand in both abstracts. (This is a useful exercise both when you revise your own writing and when reading material produced by others.)

第二部分

上机实验解答

上机实验解答 1

1. 系统动力学建模的 4 个原件。

（1）瞬时因果关系。

（2）影响因子、常数或辅助变量（Auxiliary）。

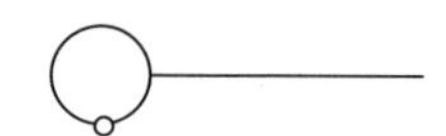

（3）流位。

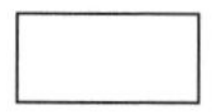

（4）流率。

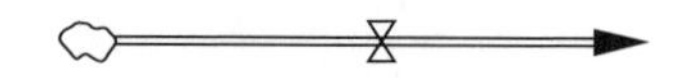

2. 系统动力学建模的 SD 4 类图形（SFD—流位流率图、CLD—因果环图、结构图、时域图）。

（1）SFD——流位流率图。

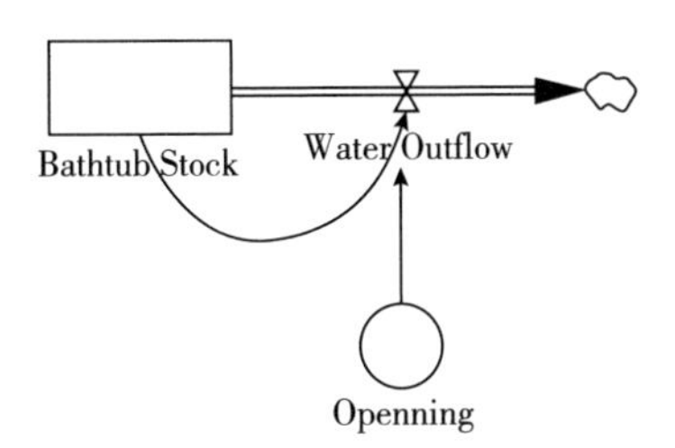

（2）CLD——因果环图。

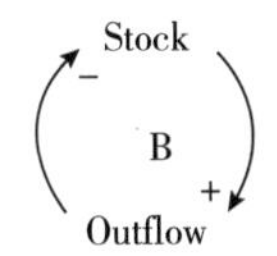

（3）结构图。

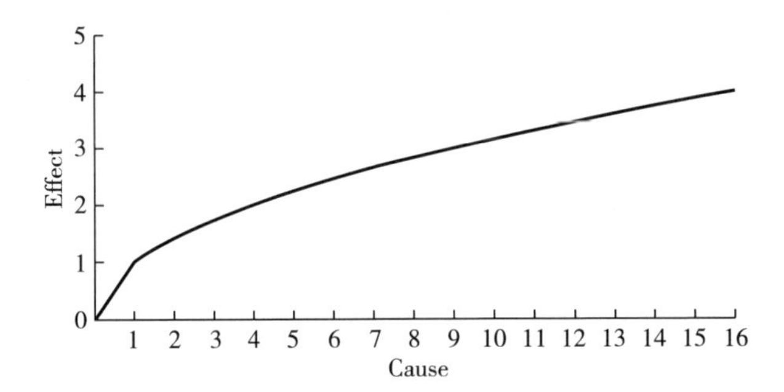

（4）时域图。

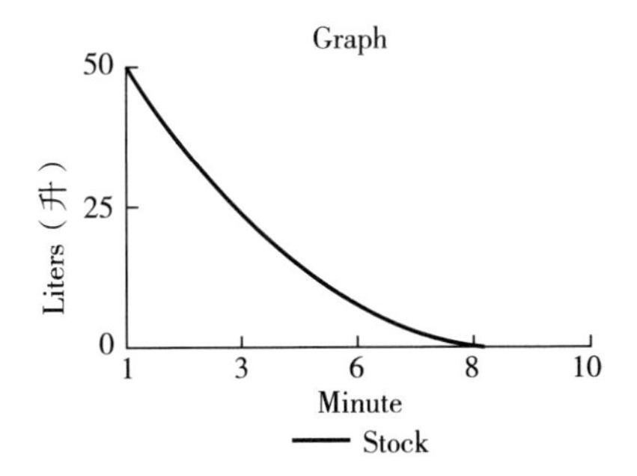

3. 系统动力学建模的两种类型因果关系图。

（1）瞬时因果关系。

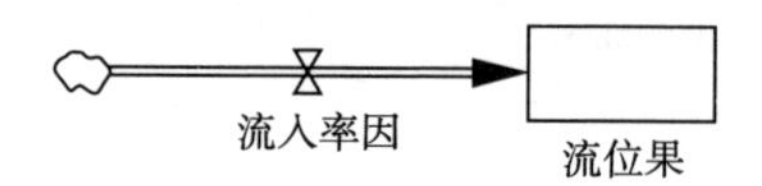

（2）累积因果关系。

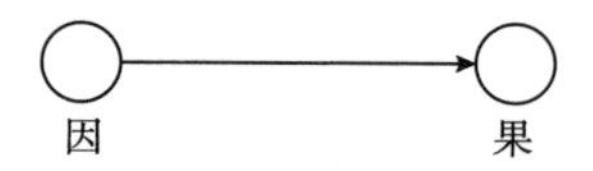

4. 系统动力学建模的 5 种基本结构。

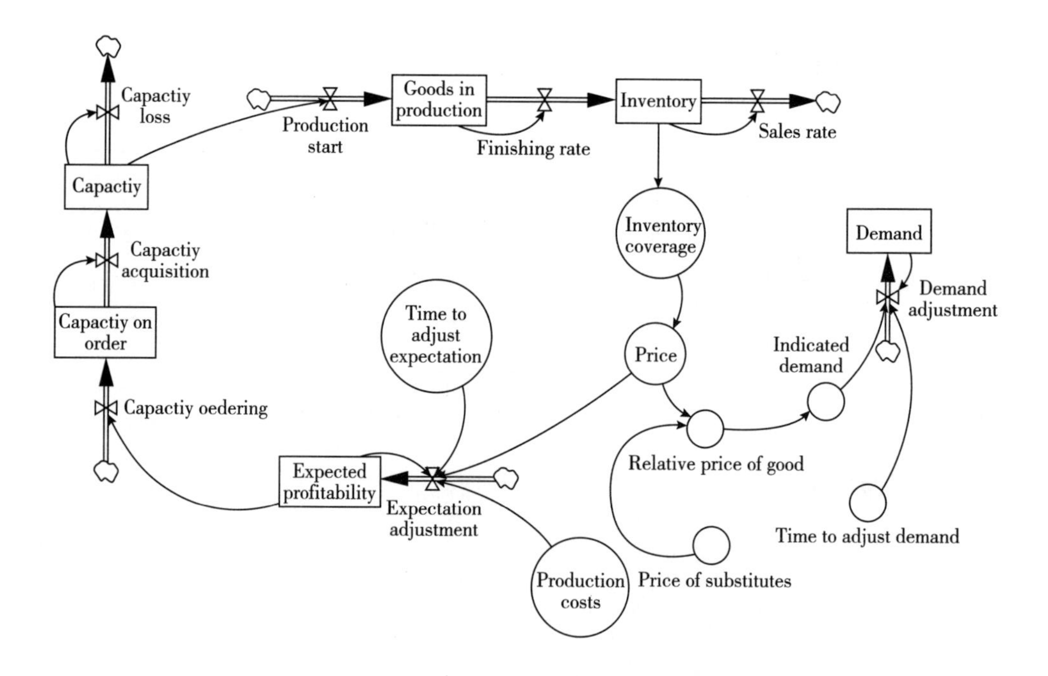

5. 系统动力学建模的 6 种典型行为。

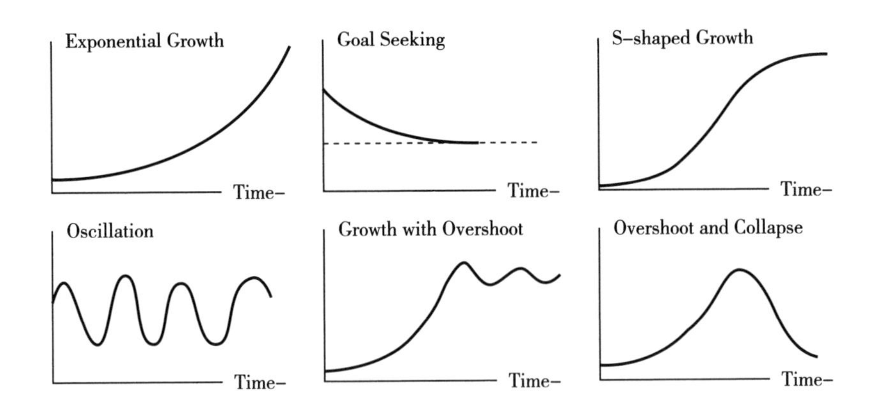

或

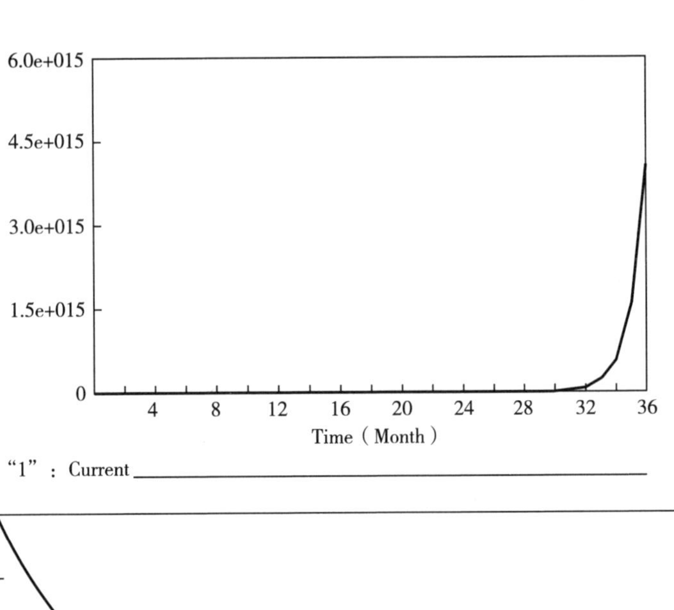

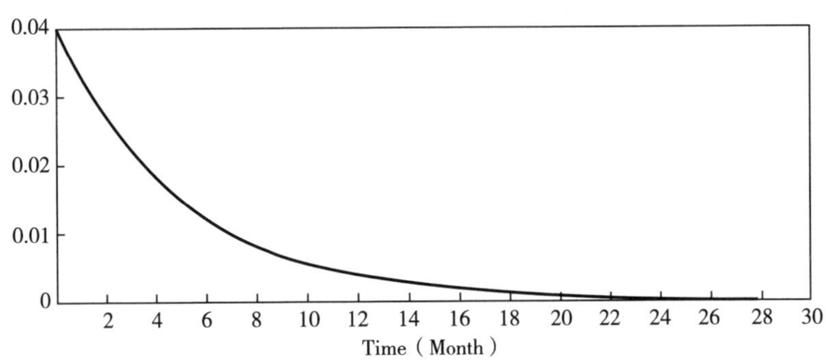

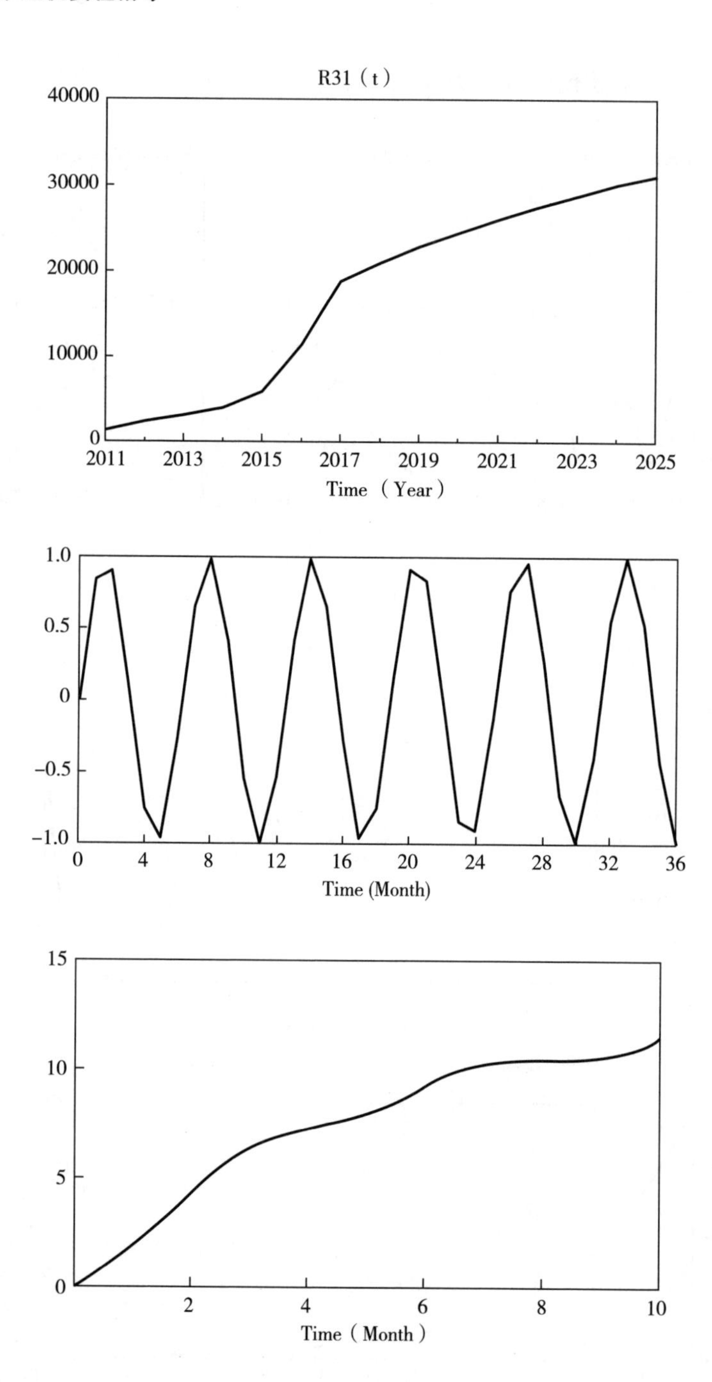
R31（t）
40000
30000
20000
10000
0
2011
2013
2015
2017
2019
2021
2023
2025
Time （Year）
1.0
0.5
0
-0.5
-1.0
0
4
8
12
16
20
24
28
32
36
Time (Month)
15
10
5
0
2
4
6
8
10
Time（Month）

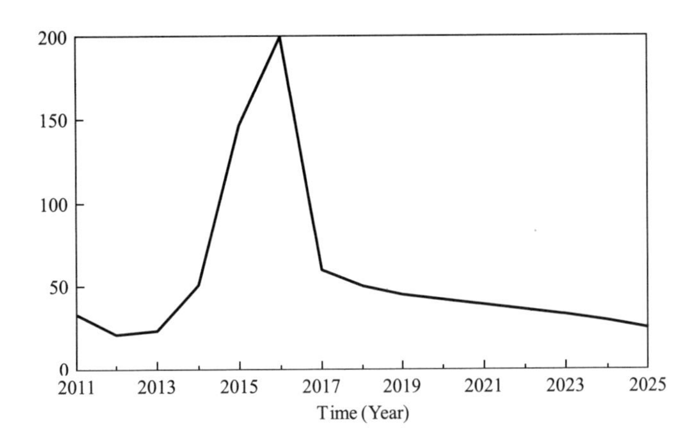

6. 系统动力学建模的 8 种描述。

（1）Verbal：Water flows fast into the container and the stock increases.

（2）Hydraulic metaphor/analogy.

（3）Causal loop diagram.

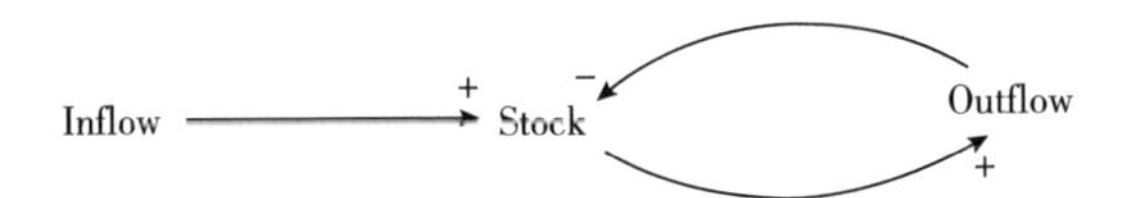

（4）Stock and flow diagram.

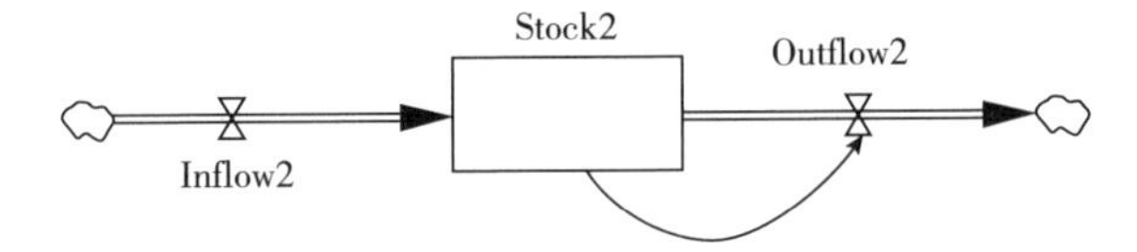

（5）Structure graph.

（6）Integral equation：$Stock=\int_{t_0}^{t}(Inflow-Outflow)\,\mathrm{d}t+Stock_0$.

（7）Differential equation：$\mathrm{d}(Stock)/\mathrm{d}t=Inflow-Outflow$.

（8）Differens equation：$Stock_t=Stock_{t-1}+Timestep*(Inflow_{t-1}-Outflow_{t-1})$.

7. 系统动力学建模的指数增长的结构图形、仿真方程、仿真图形、6 个案例。

（1）指数增长的结构图形。

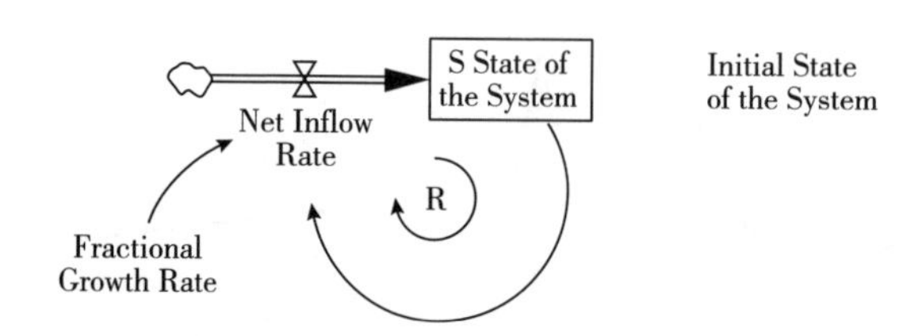

（2）仿真方程。

S State of the System = Initial State of the System

Time = （0，1000） Time step = 1

Net Inflow Rate = a Fractional Growth Rate * S State of the System

a Fractional Growth Rate = 0.007

Initial State of the System = 1

（3）仿真图形。

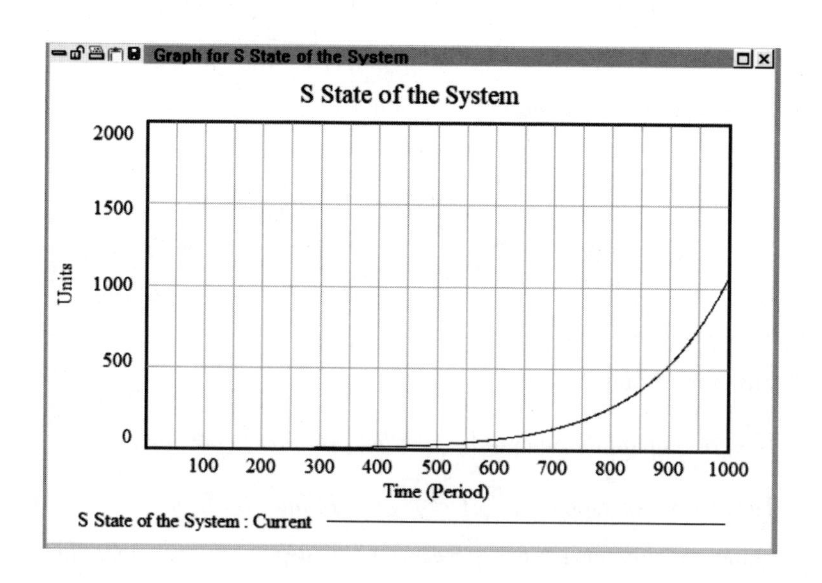

（4）6个案例。

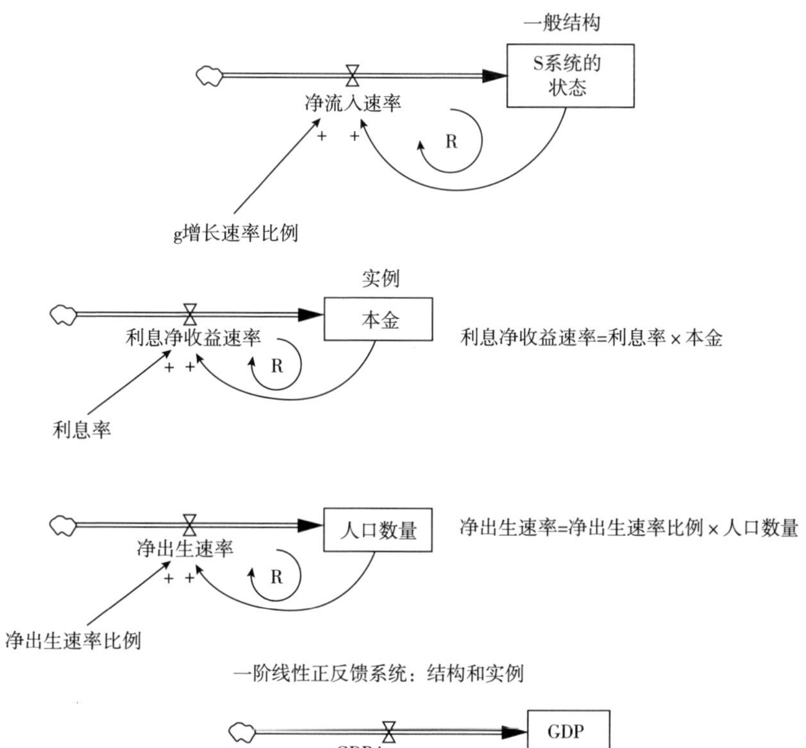

一阶线性正反馈系统：结构和实例

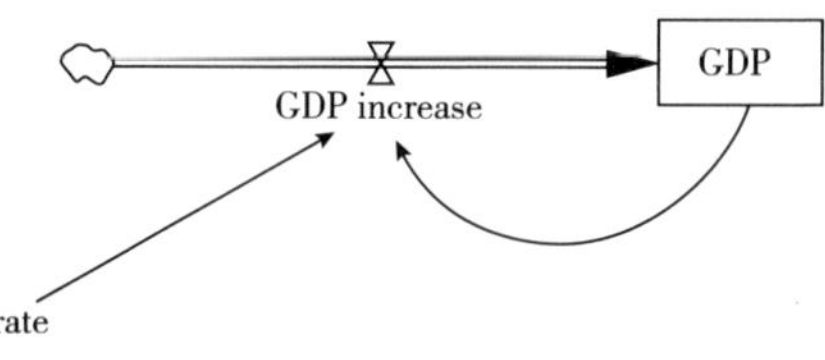

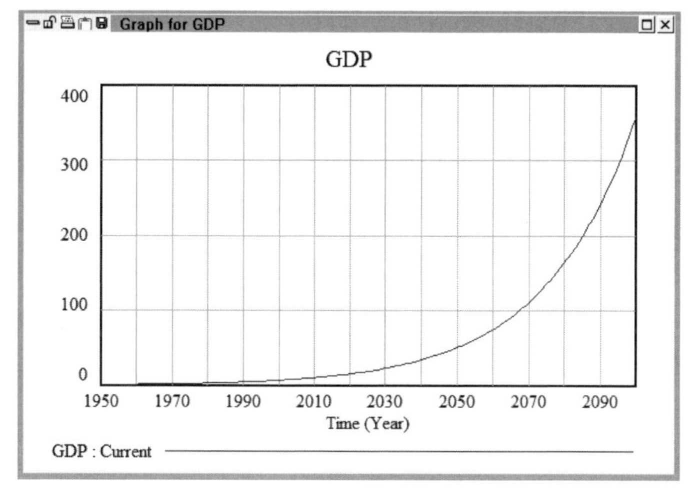

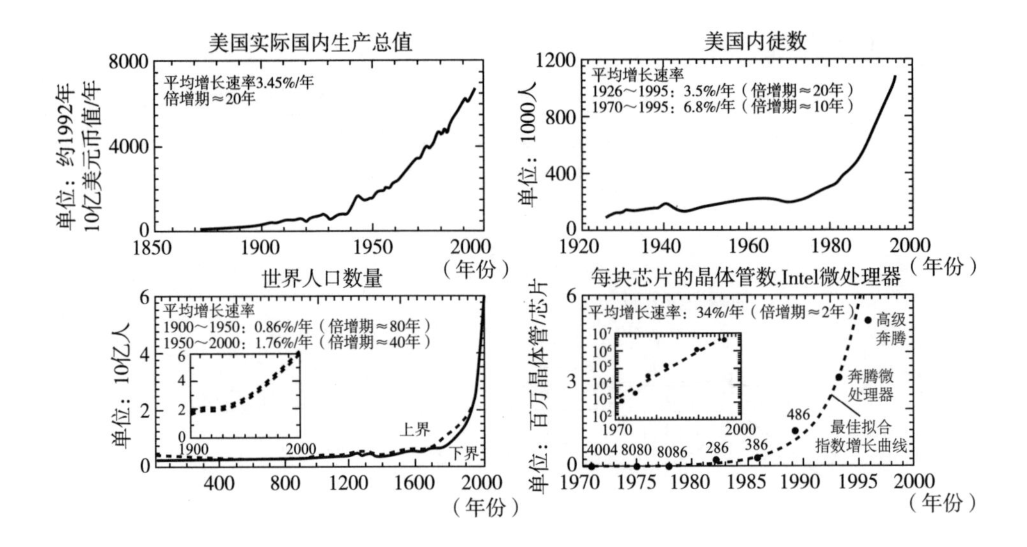

上机实验解答 2

4 类仿真函数的名称、仿真图形、仿真和仿真曲线。

（1）数学函数。

1）1Y1＝SIN（Time）（正弦函数）。

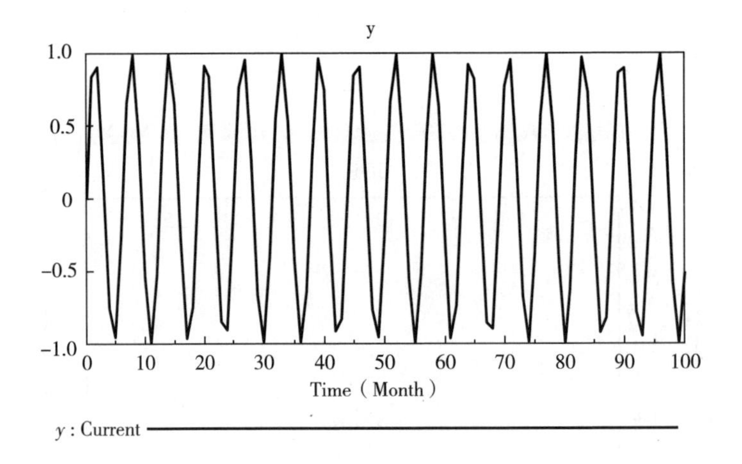

“1Y1” ＝SIN（Time）Time＝（0，100）

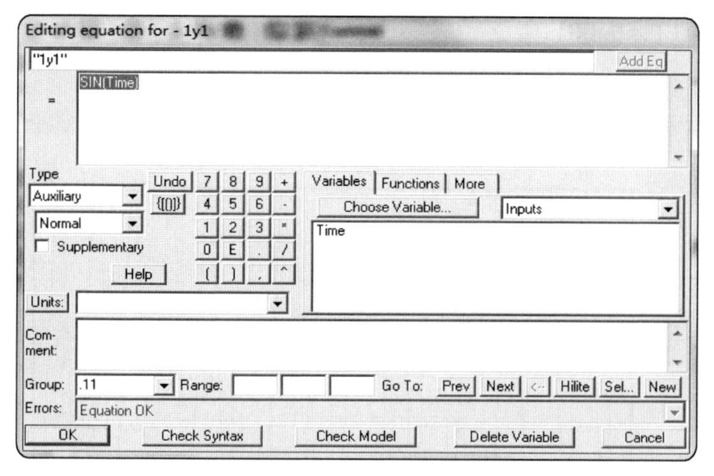
Editing equation for - 1y1
"1y1"
Add Eq
SIN(Time)
Type
Auxiliary
Normal
Supplementary
Undo
Help
Variables
Functions
More
Choose Variable...
Inputs
Time
Units:
Com-
ment:
Group:
.11
Range:
Go To:
Prev
Next
Hilite
Sel...
New
Errors:
Equation OK
OK
Check Syntax
Check Model
Delete Variable
Cancel

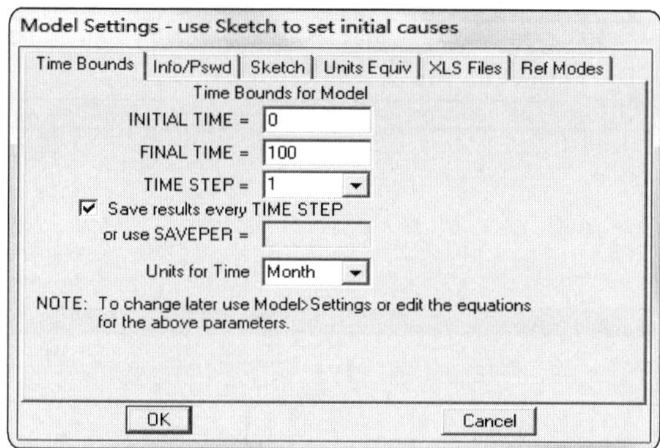
Model Settings - use Sketch to set initial causes
Time Bounds
Info/Pswd
Sketch
Units Equiv
XLS Files
Ref Modes
Time Bounds for Model
INITIAL TIME = 0
FINAL TIME = 100
TIME STEP = 1
Save results every TIME STEP
or use SAVEPER =
Units for Time Month
NOTE: To change later use Model>Settings or edit the equations
for the above parameters.
OK
Cancel

TIME STEP = 1

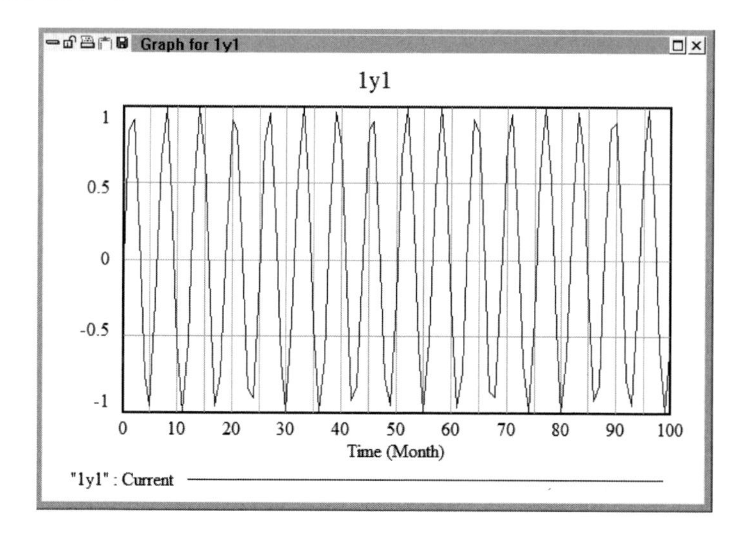
Graph for 1y1
1y1
1
0.5
0
-0.5
-1
0
10
20
30
40
50
60
70
80
90
100
Time (Month)
"1y1" : Current

TIME STEP＝0. 001

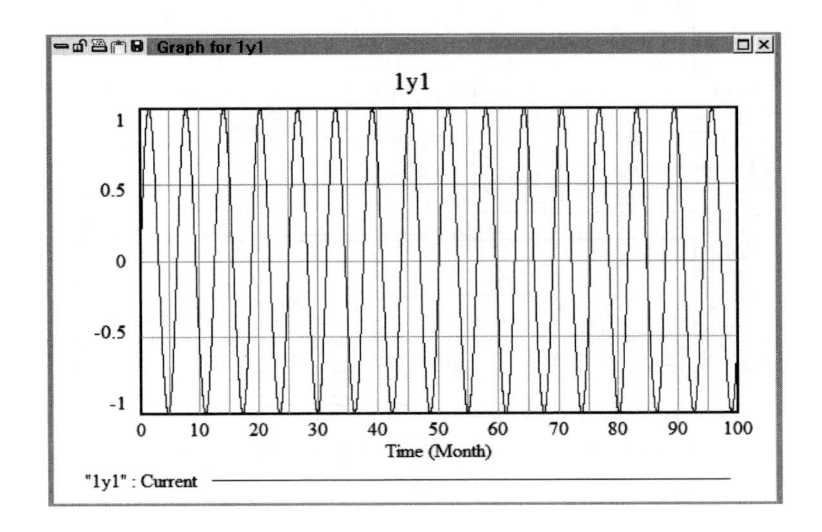

TIME STEP＝2

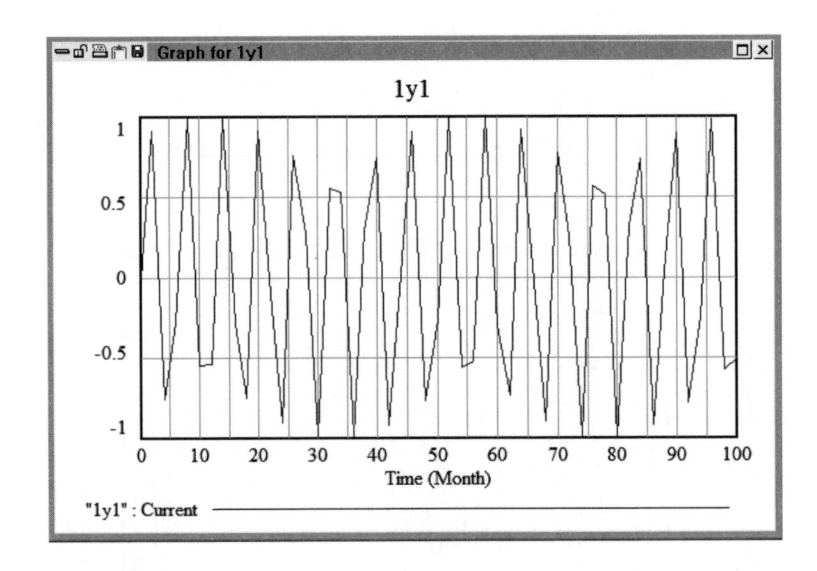

设置中可调节步长，0. 1 是光滑曲线，1 是不光滑曲线。

2）1Y2＝SIN（Time+1. 5708）（余弦函数）Time＝（0，100）。

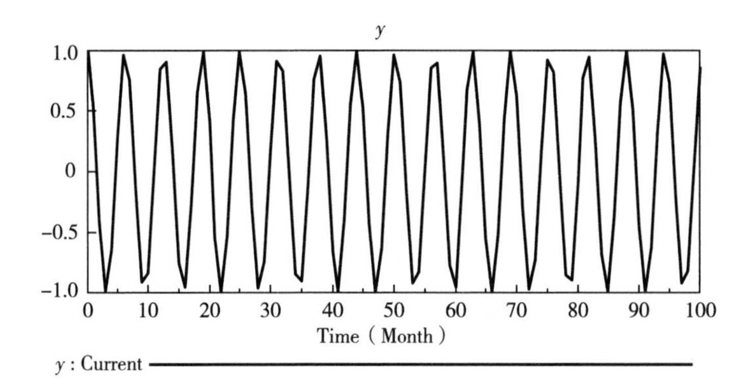

3）EXP（X）定义 2。

EXP（X）= eX，e 是自然对数的底，e=2.7182…，X 的值必须小于 36，人们常用指数函数去描述系统，有了数字函数将会带来很大方便。

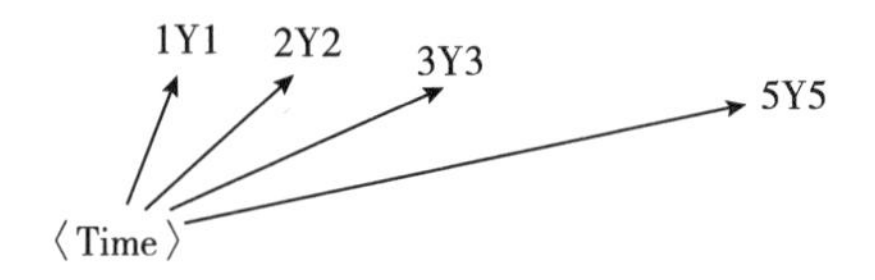

“1Y3” =EXP（Time）

Time=（0，12）

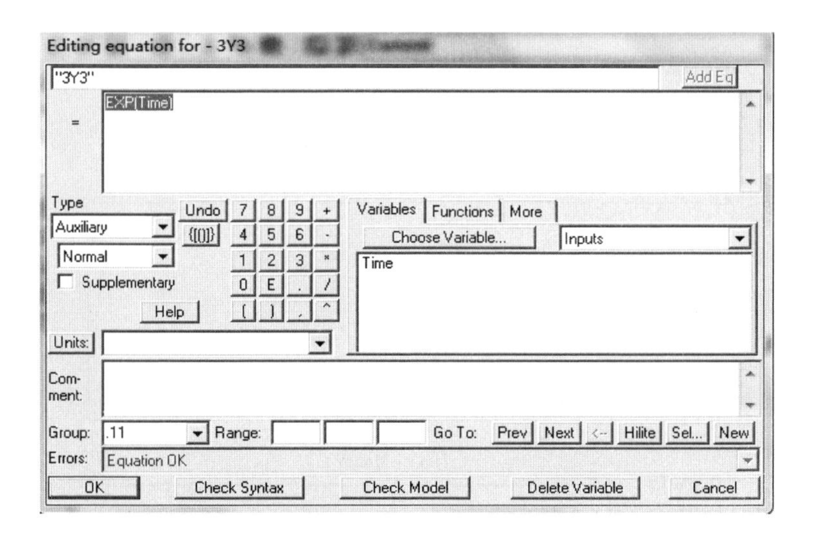

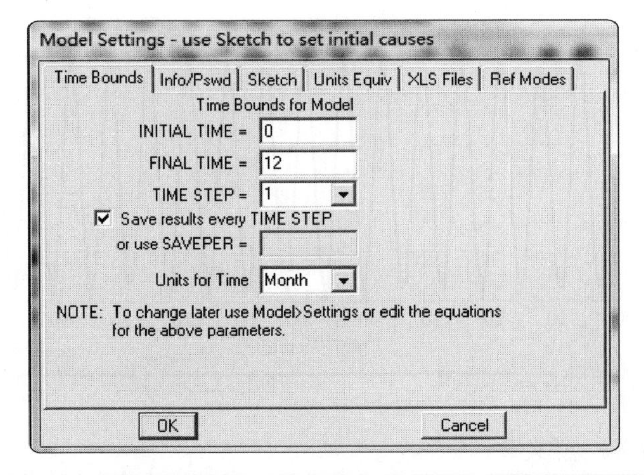

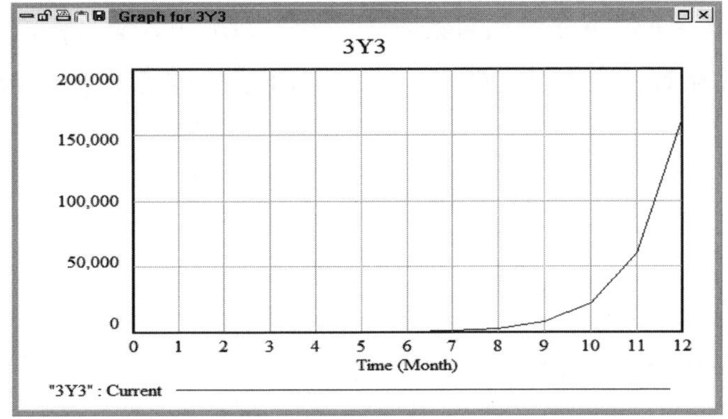

4）LN（X），变量 X 大于零。

即以 e 为底的对数函数，它与 EXP（X）互为反函数，这样可以用 EXP（X）和 LN（X）来计算非以 e 为底的幂函数和对数函数。

1Y4=LN（Time），自然数底对数函数。

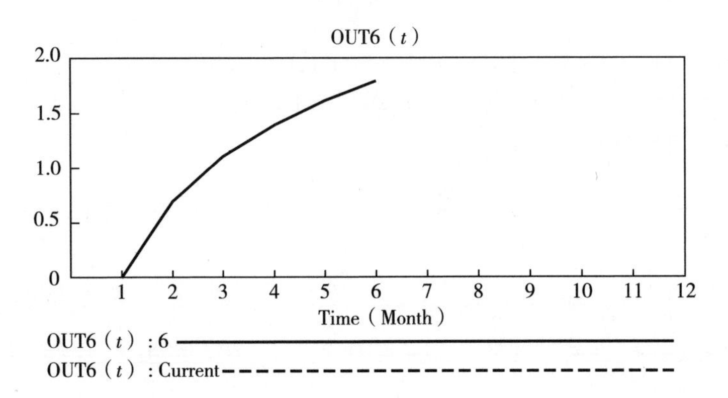

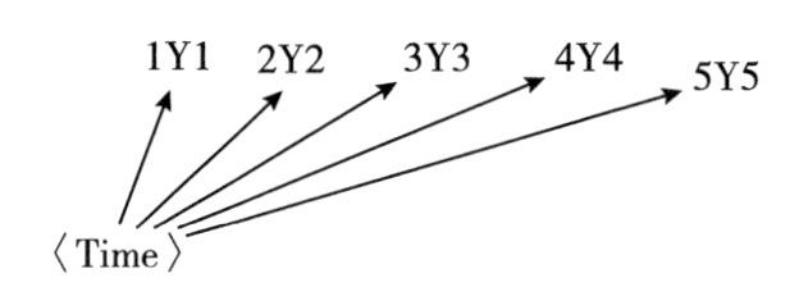

“1Y4” =LN（Time）

TIME=（1，12）

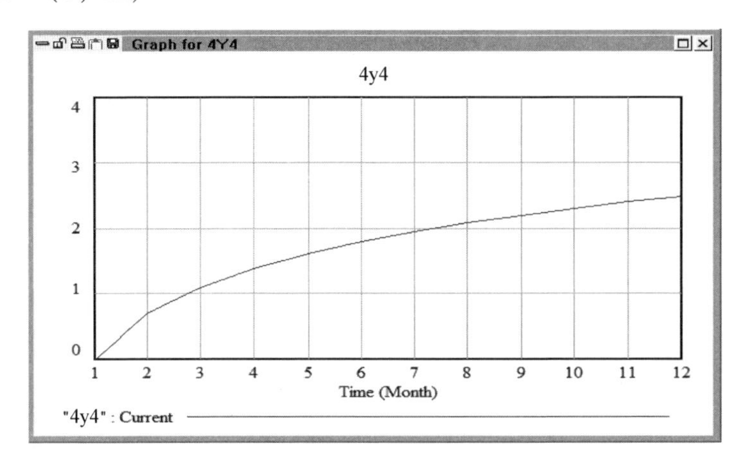

5）开方 SQRT（X）= $\sqrt{X}$，X 必须是非负数。

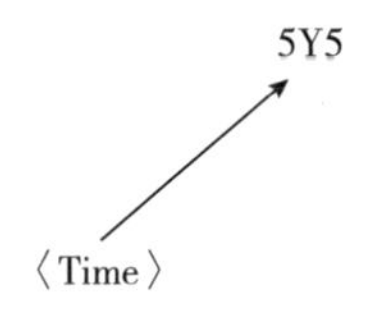

“1Y5” =SQRT（Time）Time=（0，100）

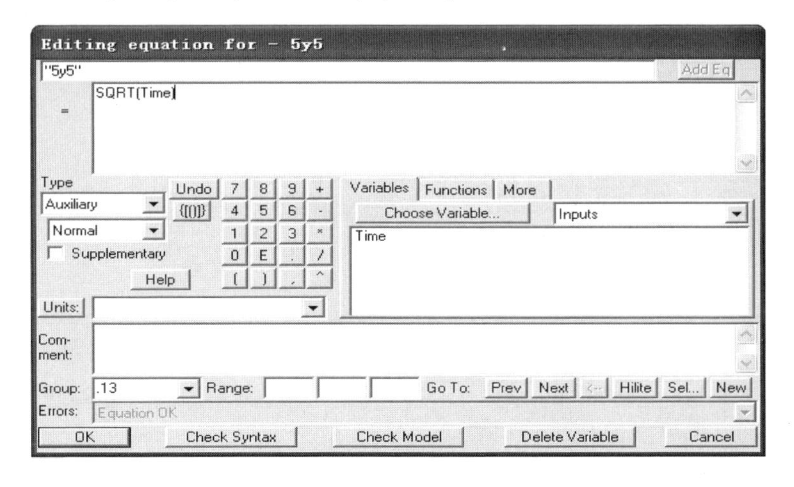

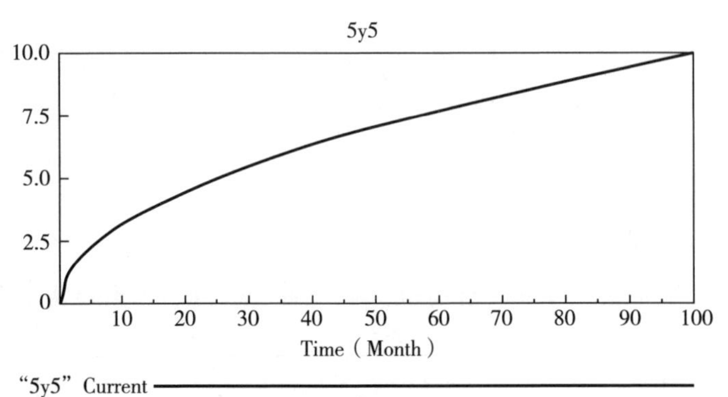

6）ABS（X）=｜X｜，对 X 取绝对值。

1Y6=｜X（t）｜=ABS（-Time）绝对值函数

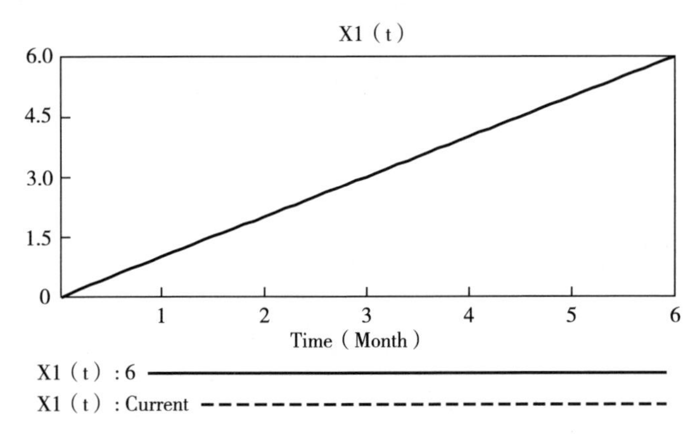

"1Y6" =ABS（-Time）Time=（0，6）

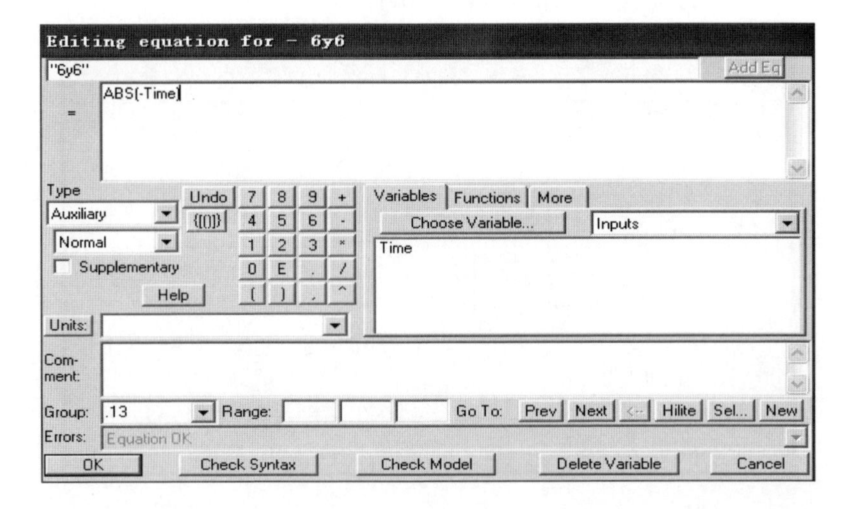

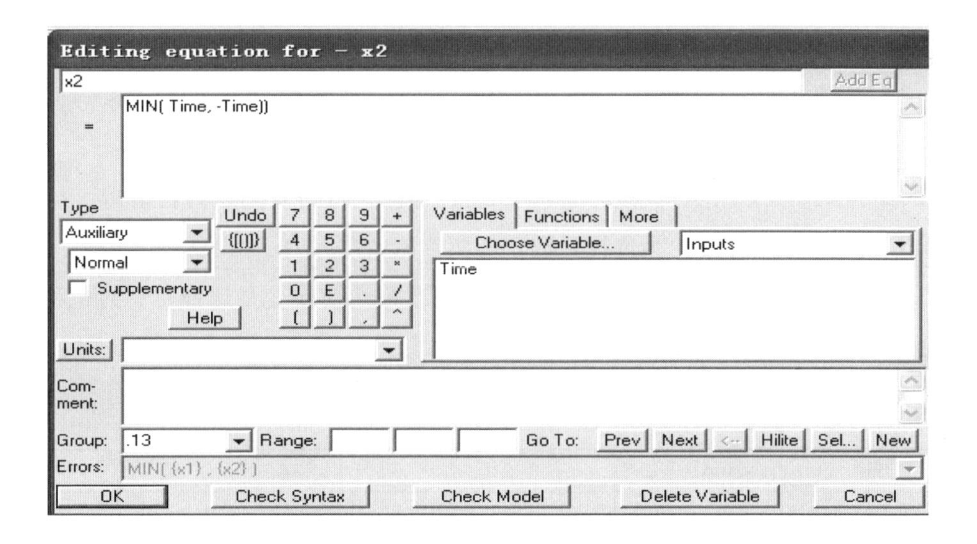

（2）逻辑函数。

逻辑函数的作用类似于其他计算机语言中的条件语句，Vensim PLE 的逻辑函数有最大函数、最小函数和选择函数三种。

1）最大函数。

定义：若 $\text{MAX}(P,Q)=\begin{cases}P，若 P\geqslant Q\\Q，若 P\leqslant Q\end{cases}$，其中 P，Q 是变量或常量，则 MAX（P，Q）为最大函数。

X1 = MAX（Time，-Time）

Time	1	2	3	4	5	6	7
OUT6（t）	1	2	3	4	5	6	7

2）最小函数。

定义：若 $MIN(P, Q)=\begin{cases}Q，若 P \geqslant Q \\ P，若 P \leqslant Q\end{cases}$，则 MIN（P，Q）为最小函数。

MIN 同 MAX 一样，可以从 MIN（P，Q）基本功能中派生出各种用法。

X2=MIN（Time，-Time）

Time	1	2	3	4	5	6	7
OUT6（t）	-1	-2	-3	-4	-5	-6	-7

3）选择函数。

定义：若 $IF\ THEN\ ELSE(C, T, F)=\begin{cases}T，C 条件为真时 \\ F，否则 C 为逻辑表达式\end{cases}$，则 IF THEN ELSE（C，T，F）为选择函数。

IF THEN ELSE 函数常用于仿真过程中作政策切换或变量选择。有时也称条件函数。

X3=IF THEN ELSE（Time>3，6，2）选择函数

Time	0	1	2	3	4	5	6
OUT4（t）	2	2	2	2	6	6	6

（3）测试函数。

设计这一函数的目的主要是用于测试系统动力学模型性能，所以称为测试函数。

在给出测试函数以前，必须重申一个概念，系统动力学的变量皆是时间 TIME 的函数，所以当仿真时间 TIME 发生变化时，各变量值都随之发生变化。不过，各变量与 TIME 的依赖关系存在差别，有的是以 TIME 为直接自变量，有的则是间接变量。测试函数以 TIME 为直接自变量，但在函数符号中常缺省。

1）阶跃函数。

定义：STEP（P，Q）$=\begin{cases}0，若\ TIME\leqslant Q\\P，若\ TIME>Q\end{cases}$，其中，P 表示阶跃幅度；Q 表示 STEP 从零值阶跃变化到 P 值的时间，则 STEP（P，Q）为阶跃函数。

S1

STEP（100，10）

Z1＝STEP（100，10）（阶跃函数）

S1＝STEP（100，10），TIME＝（0，100），TIME STEP＝1

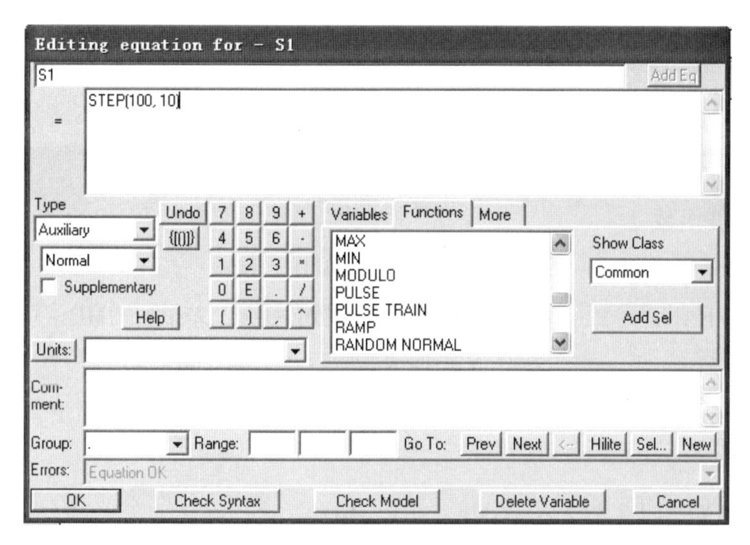

S1＝STEP（100，10），TIME＝（0，100），TIME STEP＝0.001

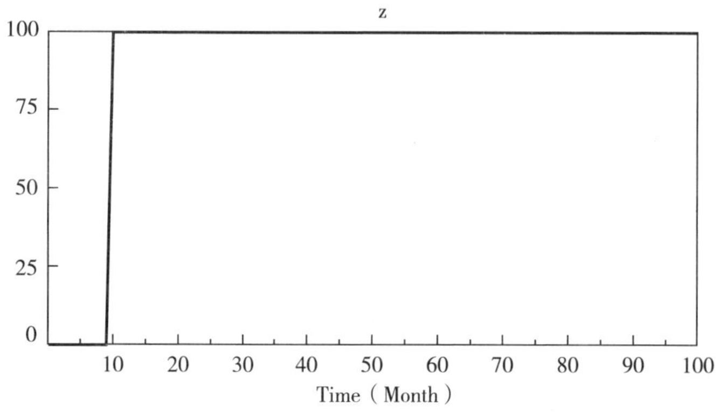

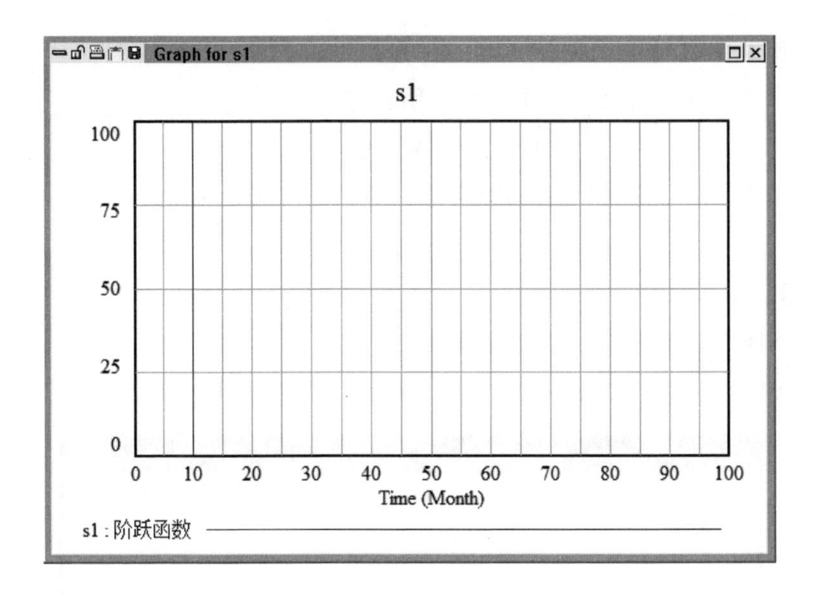

2）斜坡函数。

定义：$RAMP(P, Q, R) = \begin{cases} 0, & 若\ TIME \leqslant Q \\ P*(TIME-Q), & 若\ R \geqslant TIME > Q \\ P*(R-Q), & 若\ TIME \geqslant R \end{cases}$，其中，P为斜坡斜率，Q为斜坡起始时间，R为斜坡结束时间，则RAMP（P，Q，R）为斜坡函数。

Z2=RAMP（1，3，8），TIME=（1，10）

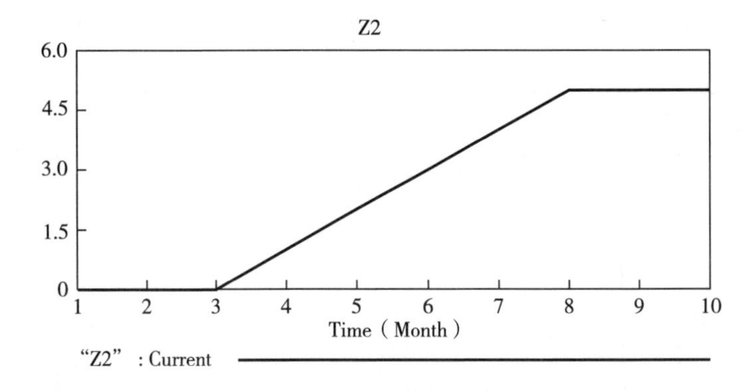

3）脉冲函数。

①宽脉冲函数，若PULSE（Q，R）随TIME变化则产生脉冲。

其中，Q 表示一个脉冲出现的时间；R 表示脉冲宽度。则 PULSE（Q，R）为脉冲函数。

Z3＝PULSE（3，6），TIME＝（0，100），TIME STEP＝1

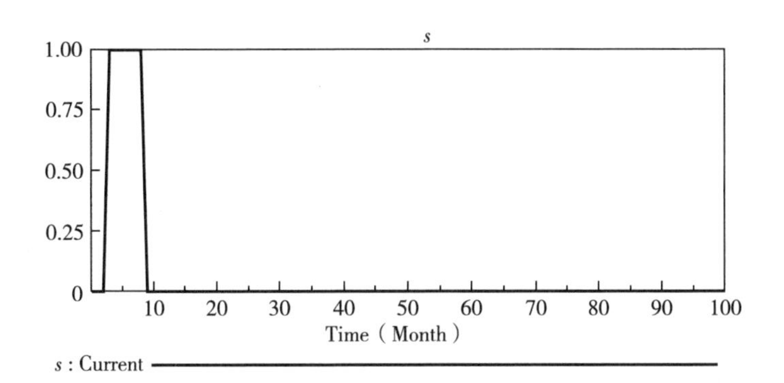

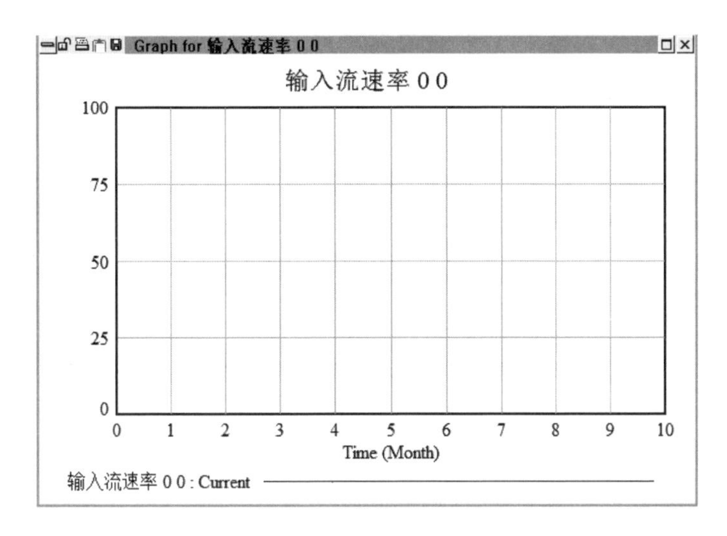

②尖脉冲函数。

PULSE TRAIN（｛start｝，｛duration｝，｛repeattime｝，｛end｝）

Duration----长短（高）

Repeattime---尖脉冲间隔

PULSE TRAIN（2，1，3，8），TIME STEP＝（1，10）

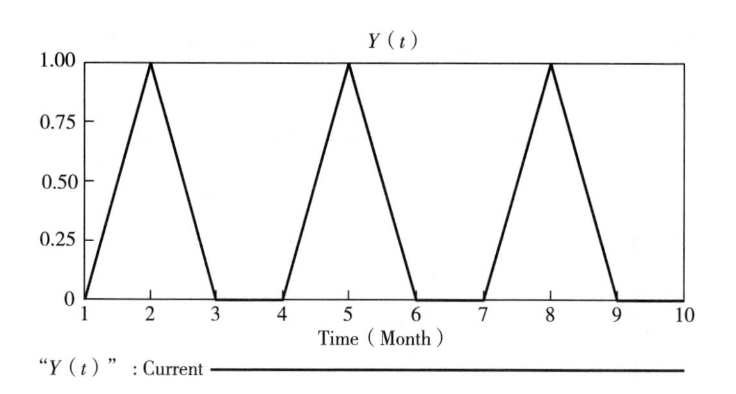

4）随机函数。

定义：RANDOM UNIFORM（A，B，S）产生在区间（A，B）内的均匀分布随机数，其中，S 给定随机数序列就确定，S 取不同的值产生随机数序列也不同，RANDOM UNIFORM（A，B，S）为均匀分布随机函数。

Z4＝RANDOM UNIFORM（2，4，8），TIME STEP＝（0，100）

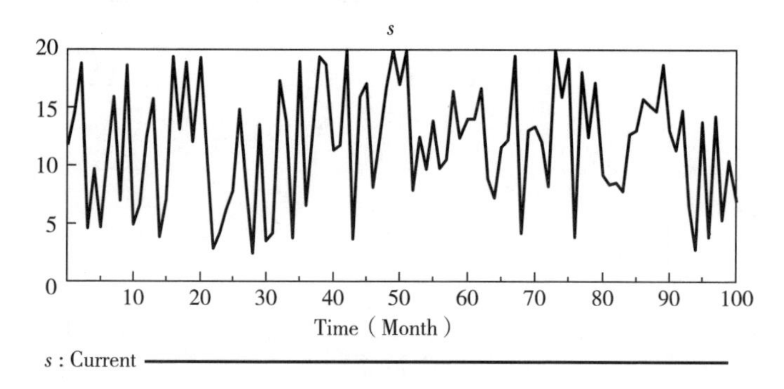

Z5＝RANDOM NORMAL（1，20，30，6，10），TIME STEP＝（0，100）

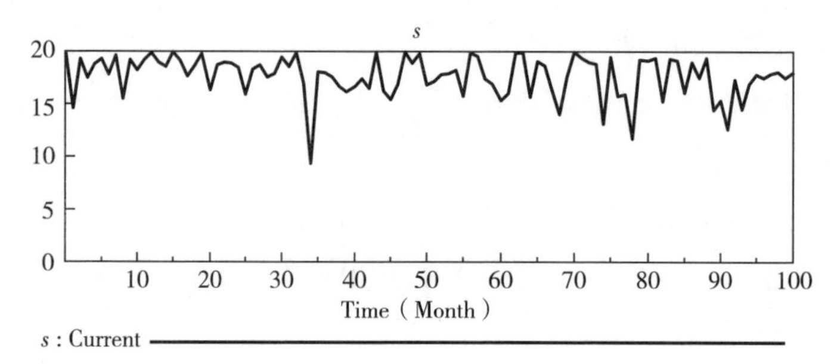

（4）延迟函数。

定义：量变化需要经过一段时间的滞后才能得到响应，这种现象称为延迟。刻画延迟现象的函数则称为延迟函数。

延迟是系统动力学中一个重要的概念，因为在系统中存在大量延迟现象，如培训后学员要经过一段时间才能发挥作用；投资要经过一段时间才能成为新的增产能力；人得病，有潜伏期；污染物排放到江河中，要经过扩散才能产生污染等。另外，延迟函数的构造丰富了系统动力学理论。

延迟函数的分类：发生的物流线上的延迟称为物流延迟；发生在信息流线上的延迟称为信息延迟。

例1：一阶物流延迟 Vensim 函数

OUT(t)′=DELAY1I(“IN(t)”，1，0)

一阶物流通用延迟函数（见 Vensim 函数库）

OUT（t）=DELAY1I（{in}，{dtime}，{init}）

{in} =流入率 IN（t）

{dtime} =延迟时间 DEL

{init} =延迟流位初始值为 LEV（to）

Time	1	2	3	4	5	6	7
IN（t）	0	6	8	7	0	0	0

Y=DELAY1I（“IN（t）”，1，0）

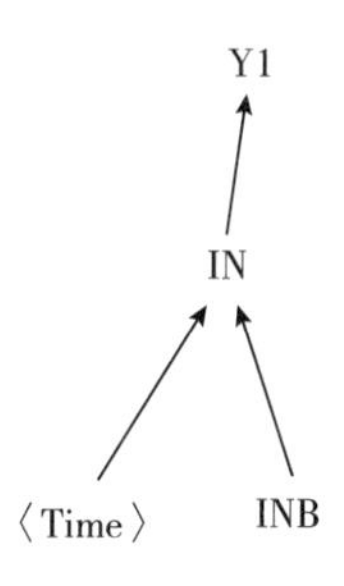

Y1=DELAY1I（IN，1，0）

IN=INB（Time）

INB=［（0，0）-（10，10）］，（1，0），（2，6），（3，8），（4，7），（5，0），（6，0），（7，0）lookup

TIME=（1，6）

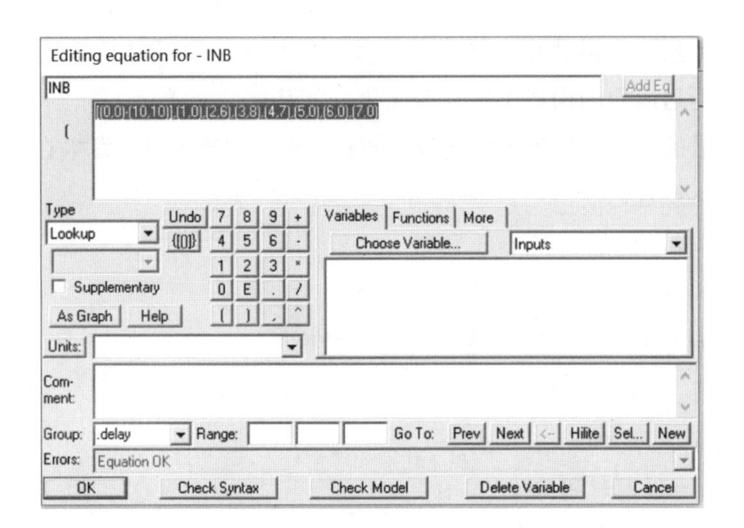

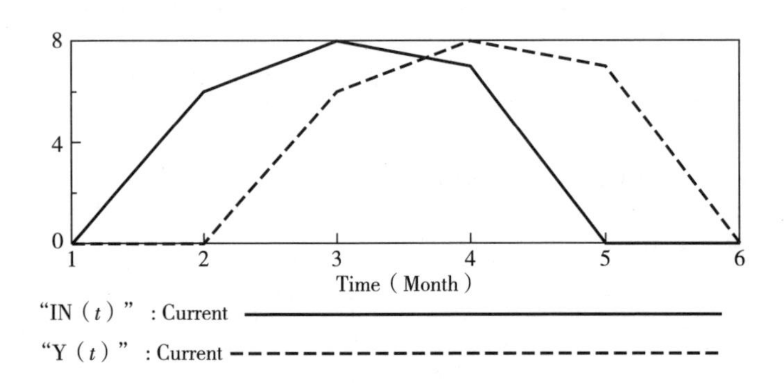

三阶物流通用延迟函数（见 Vensim 函数库）

OUT（t）=DELAY3I（｛in｝，｛dtime｝，｛init｝）

｛in｝=流入率 IN（t）

｛dtime｝=延迟时间 DEL

｛init｝=延迟流位初始值为 LEV（to）

Time	1	2	3	4	5	6	7
IN（t）	0	6	8	7	0	0	0

DELAY3I（“IN（t）”，3，0）

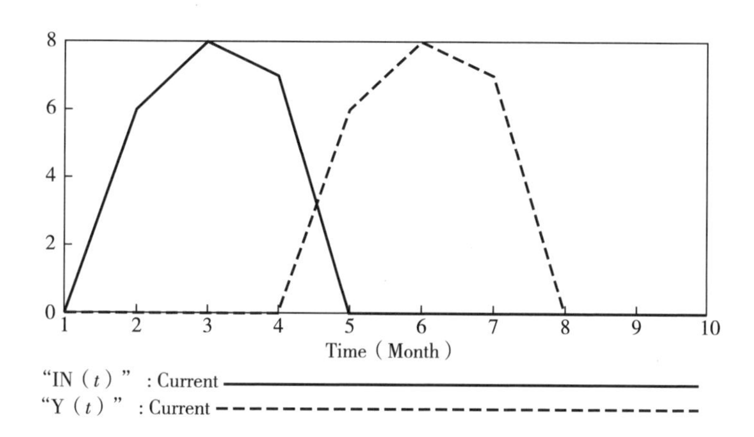

物流延迟——一阶物流延迟原理流图

物流延迟——一阶物流延迟方程

L　LEV（t+DT）=LEV（t）+DT＊（IN（t）-OUT（t））

N　LEV（to）=IN（to）＊DEL

R　IN（t）

R　OUT（t）=LEV（t）/DEL

C　DEL——延迟时间

C　DT，initial=0，final=12，saveper=1

例 2：延迟时间

DEL=DT

DT=1，initial=0，final=12，saveper=1

Time	1	2	3	4	5	6	7
IN（t）	0	6	7	8	0	0	0

例 1 仿真揭示：OUT（t）=IN（t-1）

Time	1	2	3	4	5	6	7
IN（t）	0	6	7	8	0	0	0
OUT（t）	0	0	6	7	8	0	0

仿真揭示：OUT（t）=IN（t-1）曲线

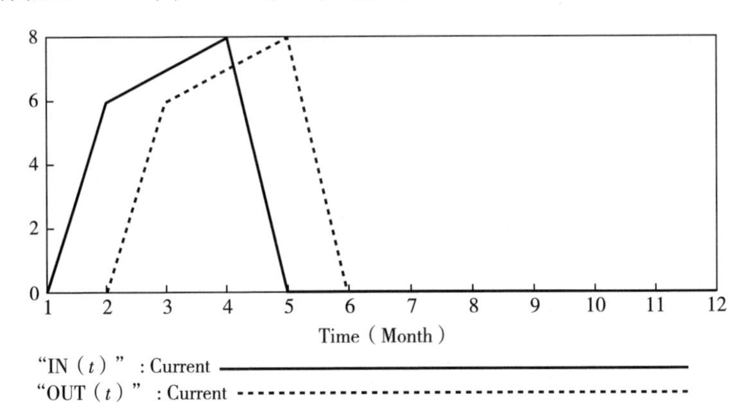

一阶延迟原理（证明 OUT（t+1）=IN（t））

已知

LEV1（t+DT）=LEV1（t）+DT＊（IN（t）-OUT（t））

OUT（t）=LEV1（t）/DEL

延迟时间 DEL=仿真步长 DT=1

则，由

LEV1（t+1）=LEV1（t）+1＊（IN（t）-OUT（t））

OUT（t）=LEV1（t）/1

得

LEV1（t+1）=IN（t）

而 OUT（t+1）=LEV1（t+1）

所以

OUT（t+1）=IN（t）

例 3：延迟时间

DEL=DT

DT=1，initial=0，final=12，saveper=1

IN（t）=STEP（5，1）

仿真延迟效应

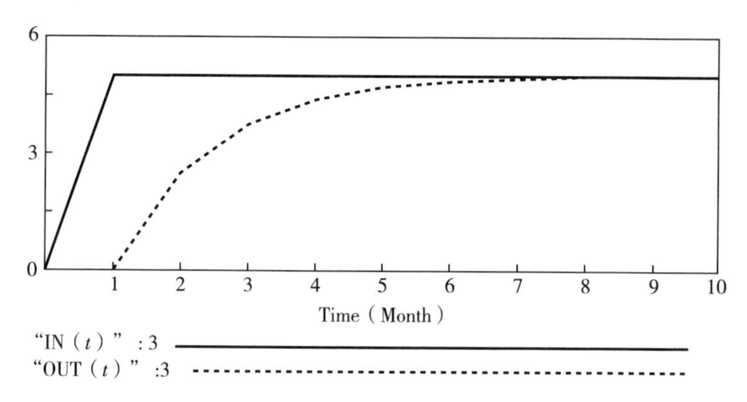

三阶物流延迟原理

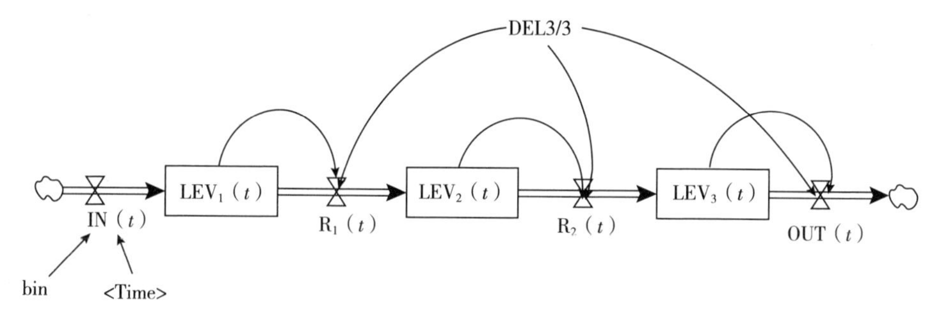

物流延迟——三价物流延迟一般式

L　LEV1(t+DT)=LEV1(t)+DT∗(IN(t)-R1(t))

N　LEV(to)=IN(to)∗(DEL/3)

R　IN(t)

R　R1(t)=LEV1(t)/DEL/3

C　DEL/3

L　LEV2(t+DT)=LEV2(t)+DT∗(R1(t)-R2(t))

N　LEV2(to)=LEV1(to)

R　R2(t)=LEV2(t)/(DEL/3)

L　LEV3(t+/DT)=LEV3(t)+DT∗(R2(t)-OUT(t))

N　LEV3(to)=LEV2(to)

R　OUT(t)=LEV2(t)/(DEL/3)

C　DT=0. 125，initial=0，final=12，saveper=0. 125

例 4：延迟时间

DEL=DT

DT=1，initial=0，final=12，saveper=1

Time	1	2	3	4	5	6	7
IN（t）	0	6	7	8	0	0	0

例 2 仿真揭示：

R1（t）=IN（t−1），R2（t）=IN（t−2）

OUT（t）=IN（t−3）

Time	1	2	3	4	5	6	7	8
IN（t）	0	6	7	8	0	0	0	0
R1（t）	0	0	6	7	8	0	0	0
R2（t）	0	0	0	6	7	8	0	0
OUT（t）	0	0	0	0	6	7	8	0

例 4 仿真揭示：

R1(t)=IN(t−1)，R2(t)=IN(t−2)

OUT(t)=IN(t−3)

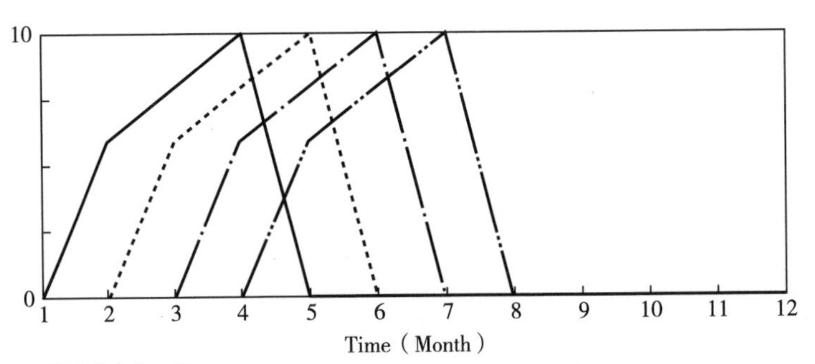

例 5：DEL/3=2

IN(t)=RAMP(1，2，10)

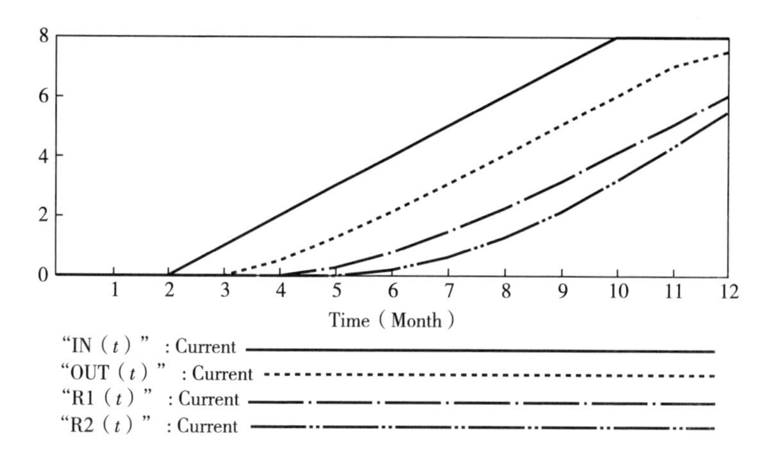

习题

1. 设延迟时间 DLE=2DT 等，进行一阶物流通用延迟函数仿真结果分析。
2. 设延迟时间 DLE=2DT 等，进行三阶物流通用延迟函数仿真结果分析。

一阶信息延迟

F11=SMOOTHI（IN，1，0）Auxiliary

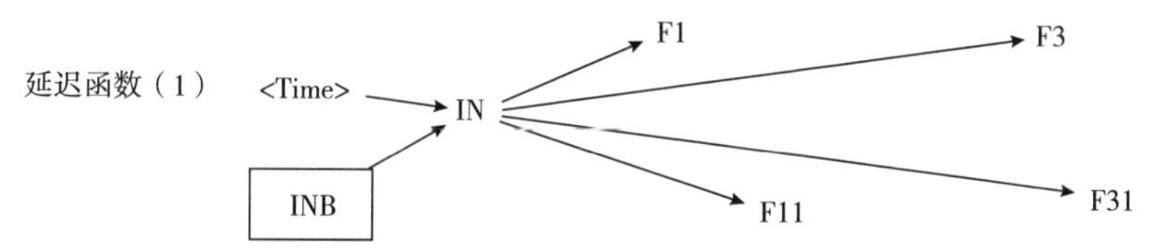

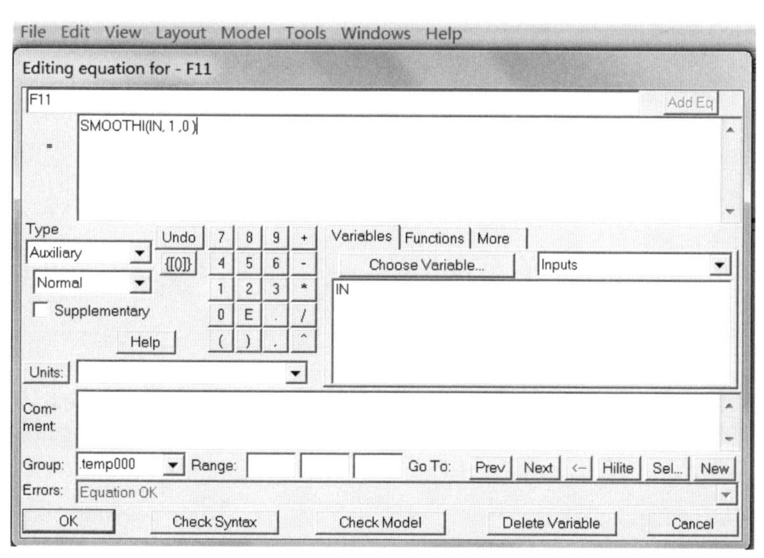

3. 三阶信息延迟函数。

F31 = SMOOTH3I（IN，3，0）Auxiliary

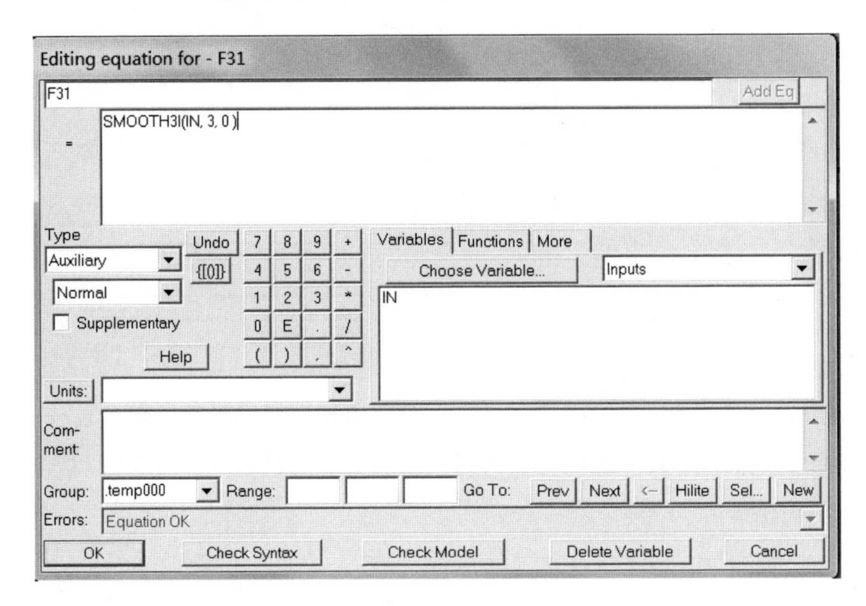

4. 三阶信息延迟原理。

（1）模型第一种画法。

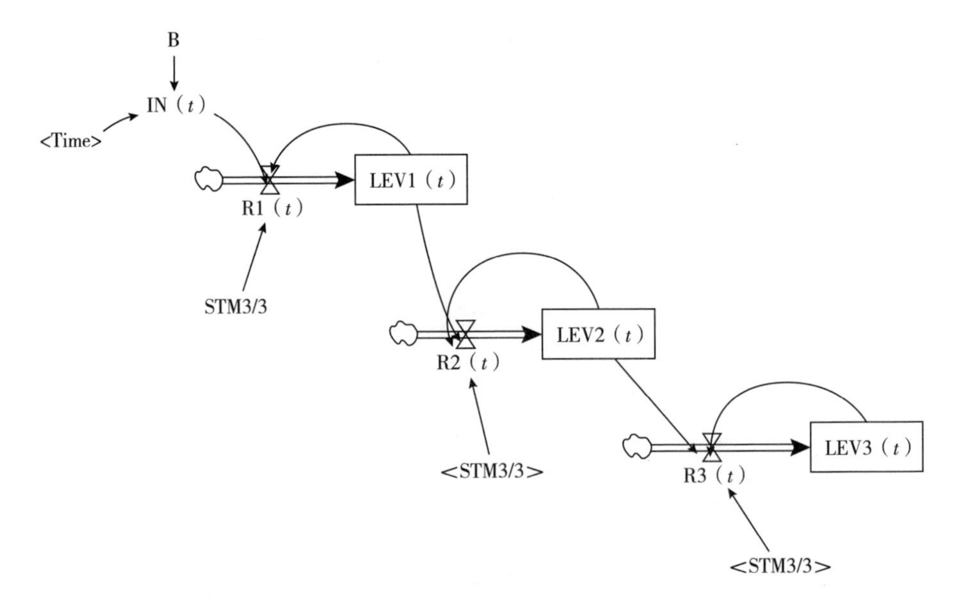

（2）模型第二种画法。

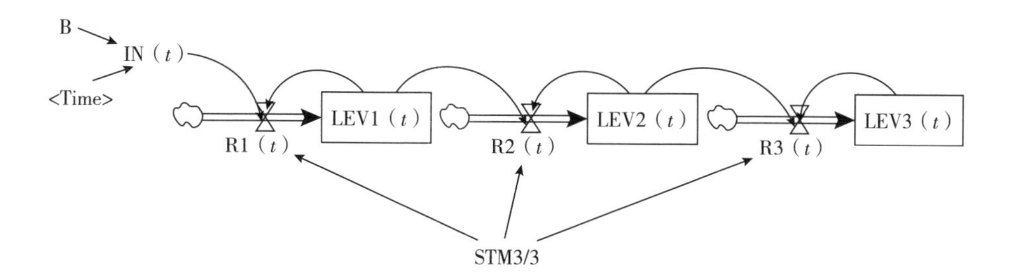

方程：

L　L1(t+1)= L1(t)+1 * R1(t)

N　L1(to)= IN(to) 0

R　R1(t)= (IN(t)-L1(t))/STM

A　IN(t)

A　OUT1(t)= L1(t)

L　L2(t+1)= L2(t)+1 * R2(t)

N　L2(to)= L1(to) 0

R　R2(t)= (L1(t)-L2(t))/STM

A　OUT2(t)= L2(t)

L　L3(t+1)= L3(t)+1 * R3(t)

N　L3(to)= L2(to)0

R　R3(t)= (L2(t)-L3(t))/STM

A　OUT3(t)= L3(t)

C　STM = 1

上机实验解答 3

1. 指数增长的 SD 通用模型。

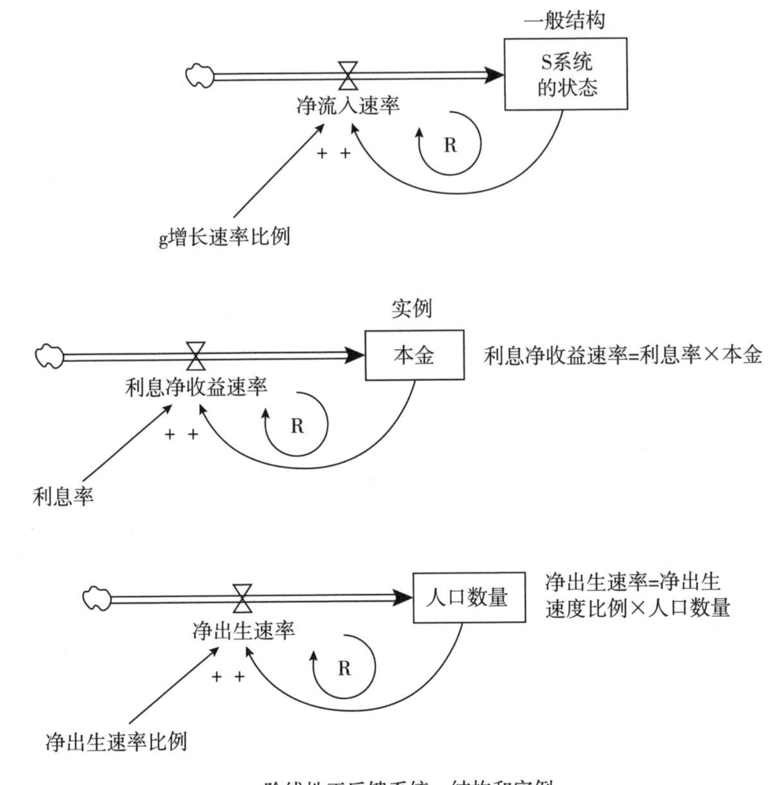

一阶线性正反馈系统：结构和实例

2. GDP 增长的 SD 模型，假设 GDP（1950）= 1，Rate = 0.04。

仿真出 1950~2100 年的流位 GDP 图形，并给出 2000 年、2050 年、2100 年 GDP 的值。

（1）GDP 增长的 SD 模型。

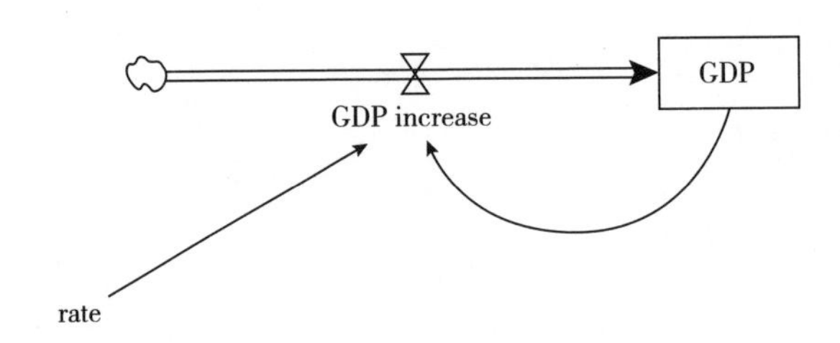

（2）1950~2100 年流位 GDP 图形。

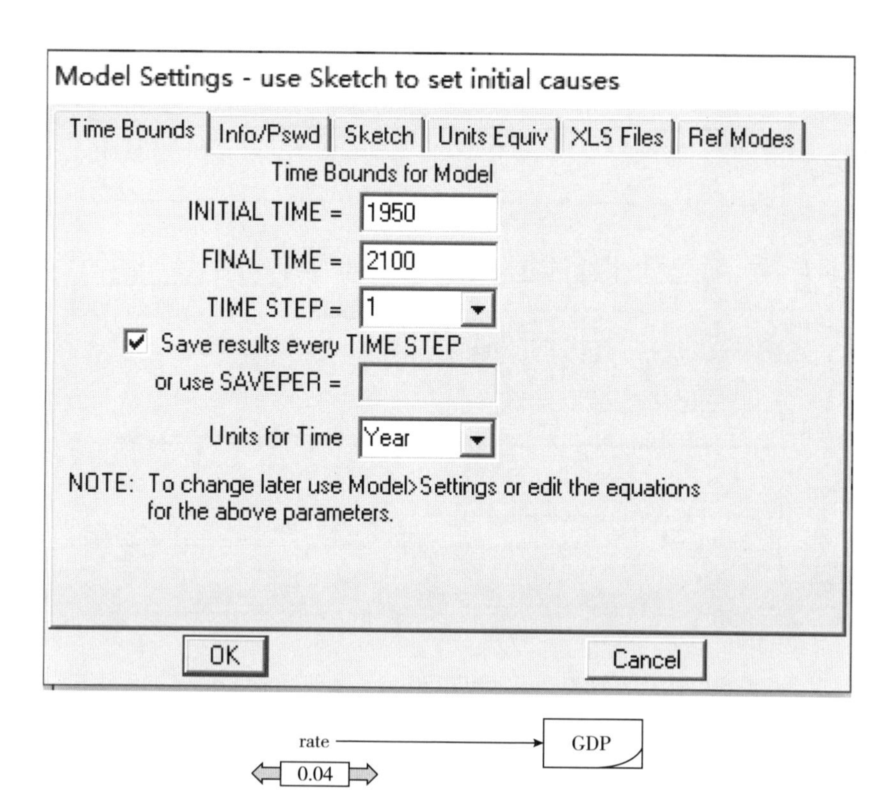

Model Settings - use Sketch to set initial causes
Time Bounds
Info/Pswd
Sketch
Units Equiv
XLS Files
Ref Modes
Time Bounds for Model
INITIAL TIME = 1950
FINAL TIME = 2100
TIME STEP = 1
Save results every TIME STEP
or use SAVEPER =
Units for Time Year
NOTE: To change later use Model>Settings or edit the equations for the above parameters.
OK
Cancel
rate
0.04
GDP

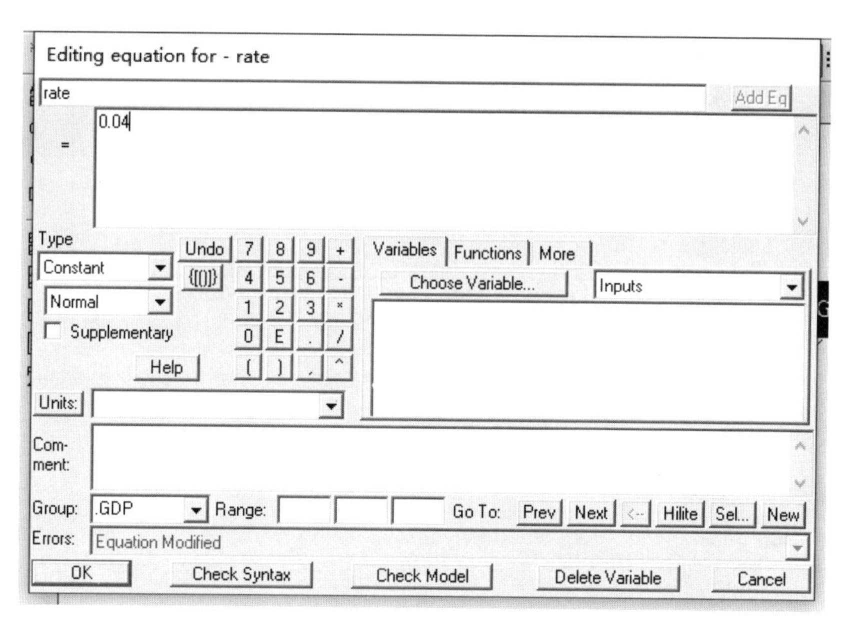

Editing equation for - rate
rate
Add Eq
=
0.04
Type
Constant
Normal
Supplementary
Undo
Help
Variables
Functions
More
Choose Variable...
Inputs
Units:
Com-ment:
Group: .GDP
Range:
Go To:
Prev
Next
Hilite
Sel...
New
Errors: Equation Modified
OK
Check Syntax
Check Model
Delete Variable
Cancel

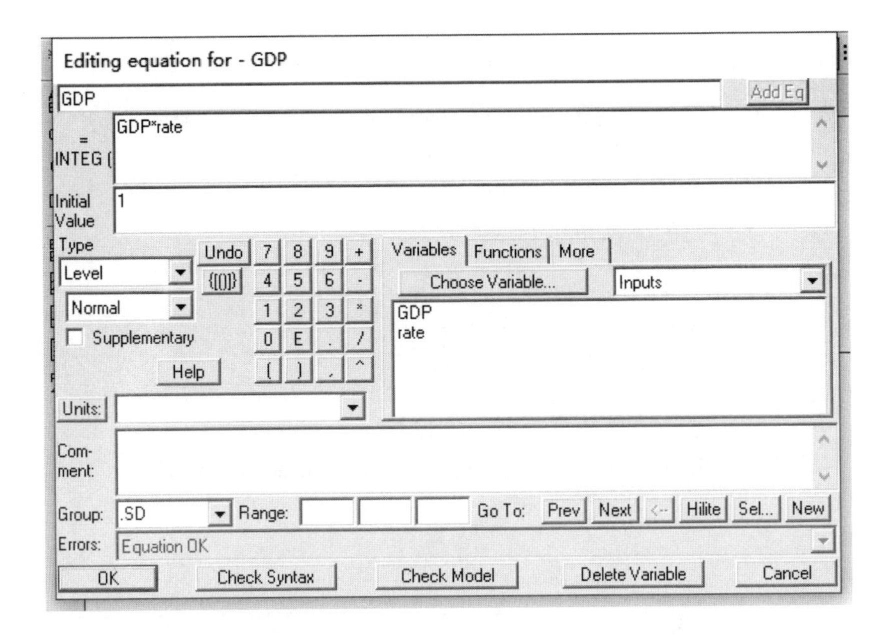

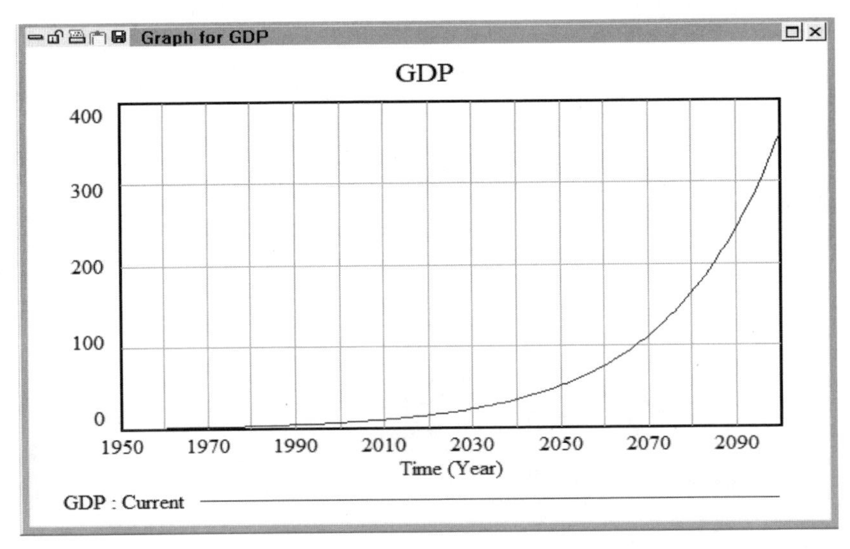

（3）2000 年、2050 年、2100 年 GDP 值。

2000 年 7. 106

Table

Time (Year)	1999	2000	2001	2002	2003	2004	2005	2006	2007	20
"GDP" Runs:	Current									
GDP	6.833	7.106	7.390	7.686	7.994	8.313	8.646	8.992	9.351	9.

2050 年 50. 50

Time (Year)	2049	2050	2051	2052	2053	2054	2055	2056	2057	20
"GDP" Runs:	Current									
GDP	48.56	50.50	52.52	54.62	56.81	59.08	61.44	63.90	66.46	69

2100 年 358. 92

Time (Year)	2092	2093	2094	2095	2096	2097	2098	2099	2100
"GDP" Runs:	Current								
GDP	262.26	272.75	283.66	295.00	306.80	319.08	331.84	345.11	358.92

3. GDP 增长的 SFD、CLD、结构图、时域图。

(1) GDP 增长的 SFD。

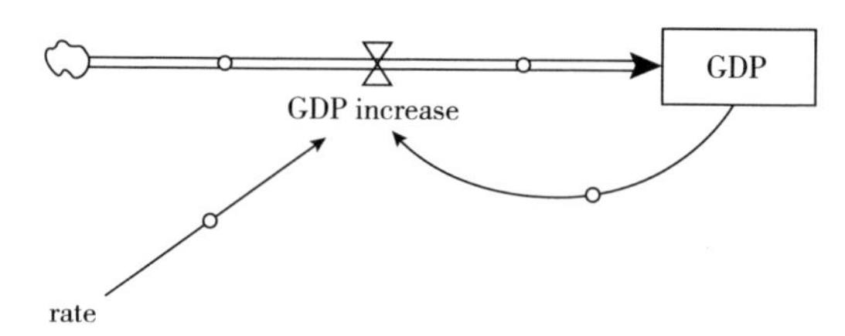

(2) GDP 增长的 CLD。

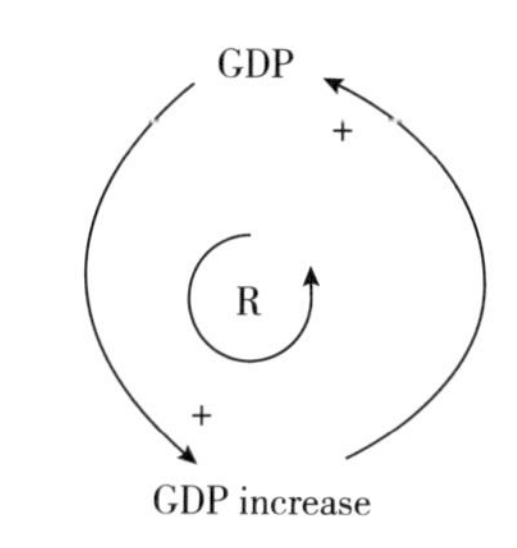

(3) GDP 增长的结构图。

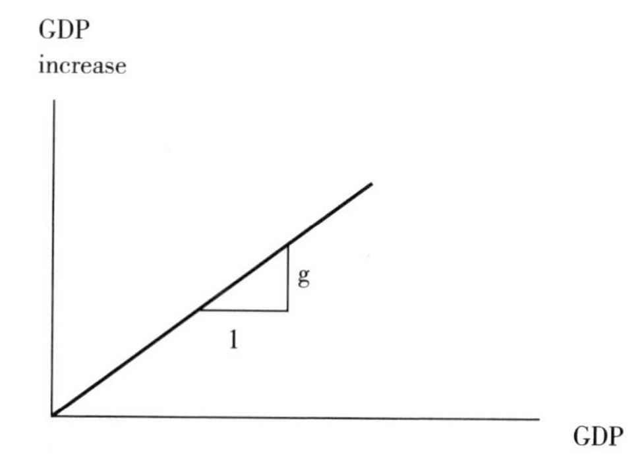

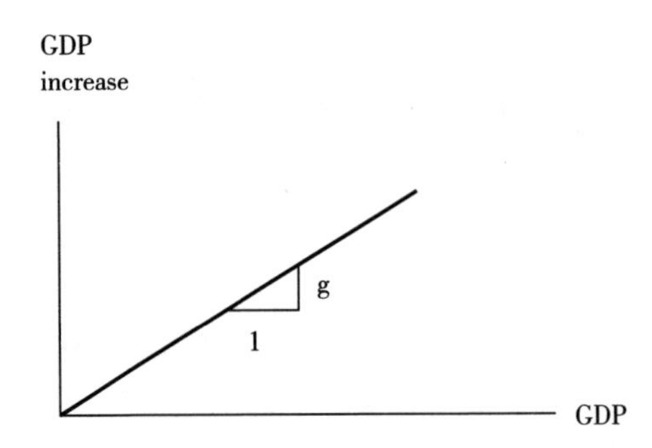

（4）GDP 增长的时域图。

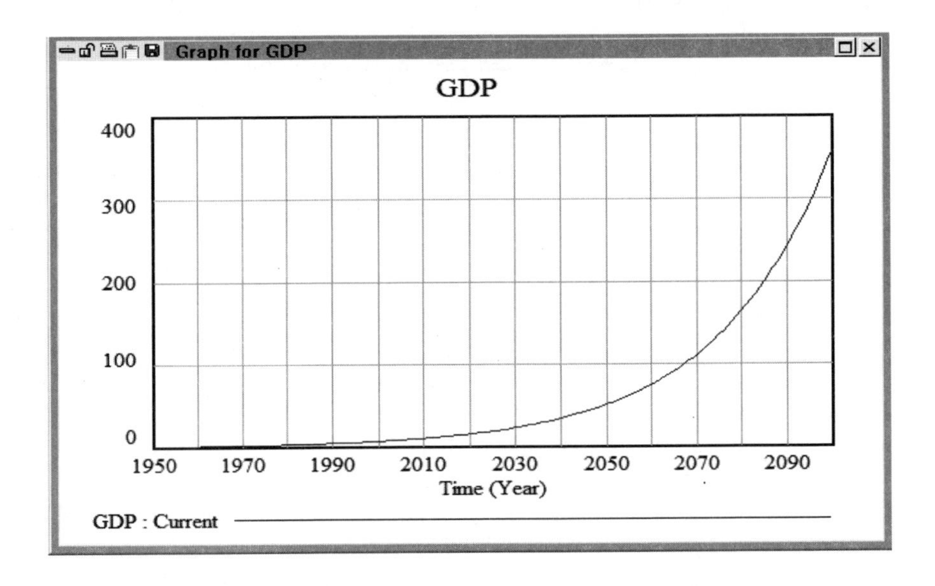

4. GDP 的差分式、积分式、递推式。

（1）差分式。

$$\frac{\mathrm{d}GDP}{\mathrm{d}t} = GDP\ increase = GDP * rate$$

（2）积分式。

$$GDP = \int_{1950}^{2100} GDP\ increase dt$$

（3）递推式。

$$GDP_t = GDP_{t-1} + Timestep * GDP\ increase_{t-1}$$

5. T=？时，GDP=2。

Time (Year)	1965	1966	1967	1968	1969	1970	1971	1972	1973	19
"GDP" Runs:	Current									
GDP	1.800	1.872	1.947	2.025	2.106	2.191	2.278	2.369	2.464	2.

T=1968。

6. 请用微分式推导出增长率 g 为变量时，流位 GDP 翻倍的公式是否符合 70/100g，并用第 5 问验证。

（1）推导过程。

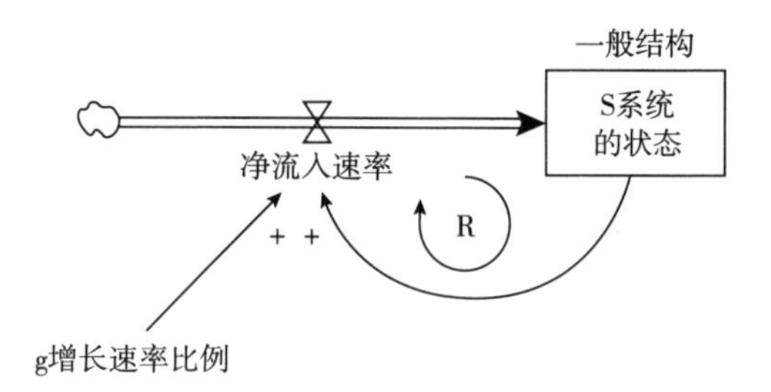

$S(t)=S(0)e^{(gt)}$

$\frac{ds}{dt}=gs\Rightarrow\frac{ds}{s}=gdt\Rightarrow\int\frac{ds}{s}\int gdt\Rightarrow \ln s=gt+c$

$=>s=c^{*}\cdot e^{(gt)}$

当 $t=0$ 时，$e^{(gt)}=1$，则 $c^{*}=S(0)$

$=>S(t)=S(0)e^{(gt)}$

当 $S(t)=2S(0)$ 时，求 $t_d=?$

即 $2S(0)=S(0)e^{(g_d^t)}=>t_d=\ln 2/g$，由于 $\ln 2=0.6931=>t_d=0.7/g=70/100g$，即为 70 倍增法则。

（2）验证过程。

GDP=GDP ∗ g∧（t-1950）t=1968

而 g=rate=0.04

70/100g=17.5

则 T=1950+17.5=1967.5，近似等于 1968，所以符合。

7. 1950~2021 年的 GDP 数据，用 Excel 做出趋势图，选用平均法或其他统计法算出 g，并用 SD 通用模型建模，比较仿真数据和实际数据的拟合程度，选择合适的增长率 g。

g=0. 133

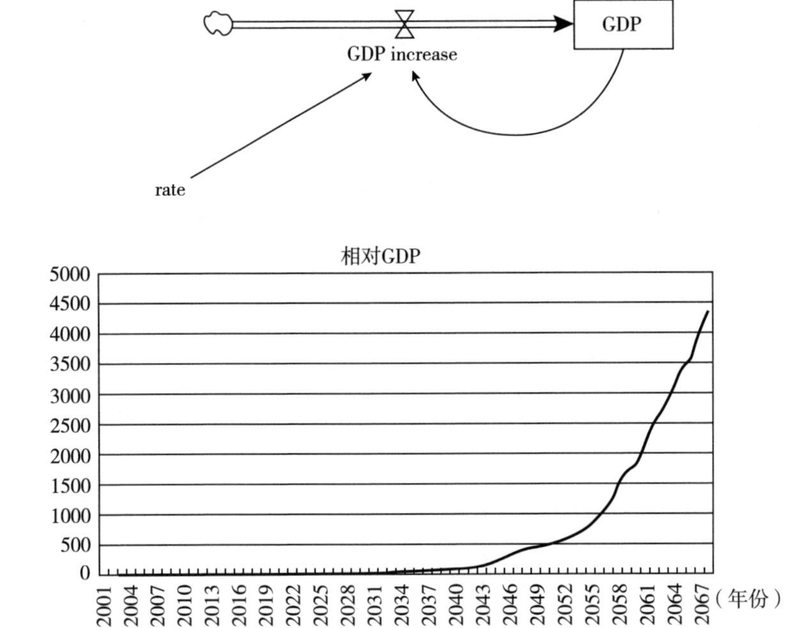

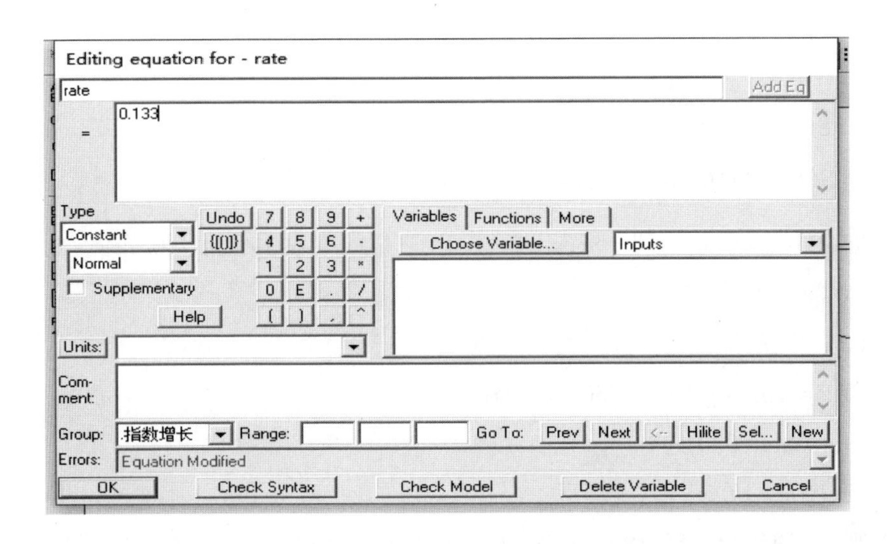

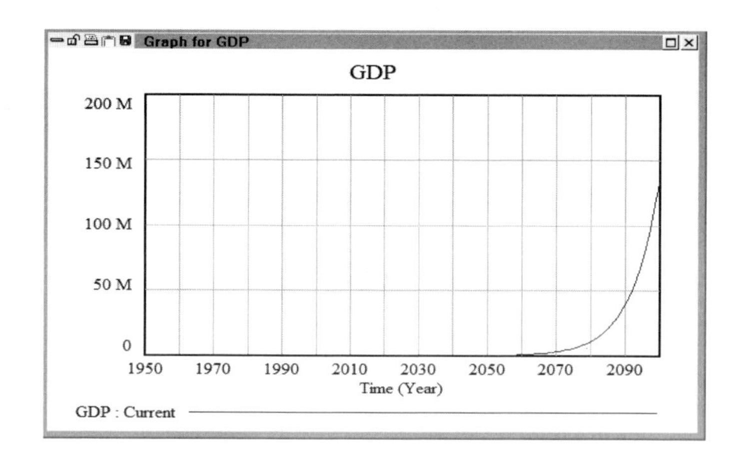

Time (Year)	1993	1994	1995	1996	1997	1998	1999	2000	2001	2(
"GDP" Runs:	Current									
GDP	214.72	243.28	275.64	312.30	353.83	400.89	454.21	514.63	583.07	6(

Table

Time (Year)	2011	2012	2013	2014	2015	2016	2017	2018	2019	2(
"GDP" Runs:	Current									
GDP	2,032	2,302	2,609	2,956	3,349	3,794	4,299	4,871	5,519	6,

1952~2018 年我国 GDP 增长趋势。

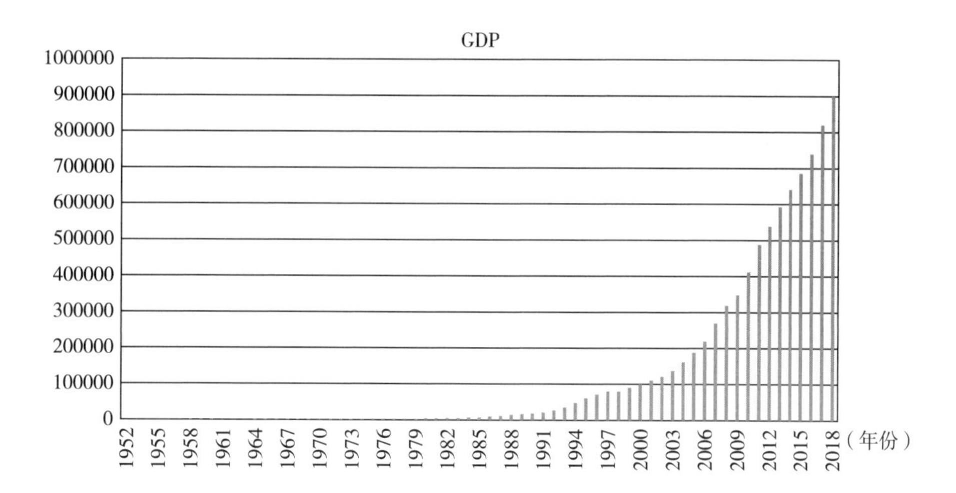

经计算，g 约为 0.11。

（1）设置。

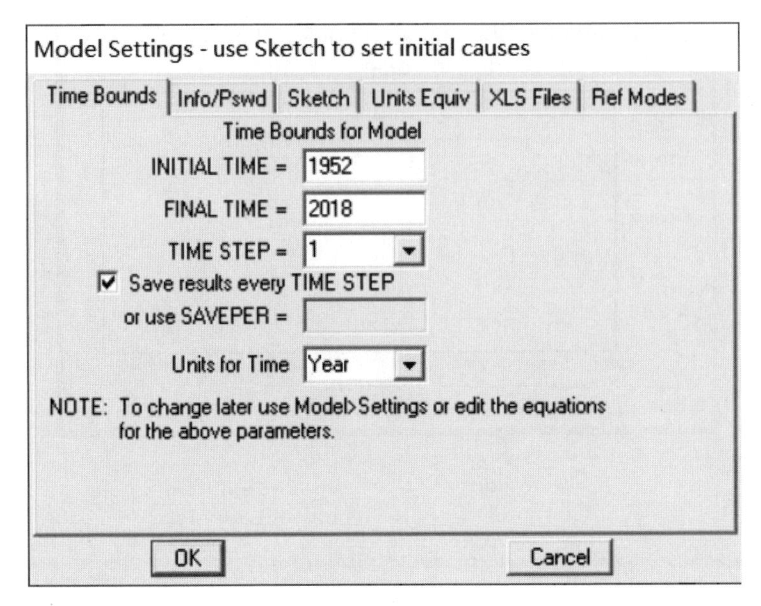
Model Settings - use Sketch to set initial causes
Time Bounds
Info/Pswd
Sketch
Units Equiv
XLS Files
Ref Modes
Time Bounds for Model
INITIAL TIME = 1952
FINAL TIME = 2018
TIME STEP = 1
Save results every TIME STEP
or use SAVEPER =
Units for Time Year
NOTE: To change later use Model>Settings or edit the equations for the above parameters.
OK
Cancel

（2）建立模型。

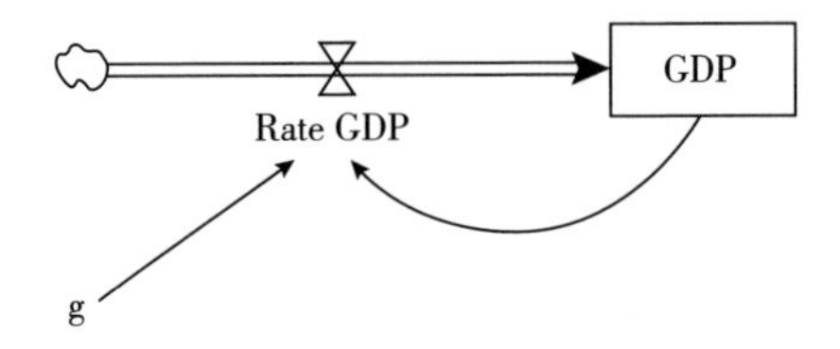
GDP
Rate GDP
g

（3）方程入树。

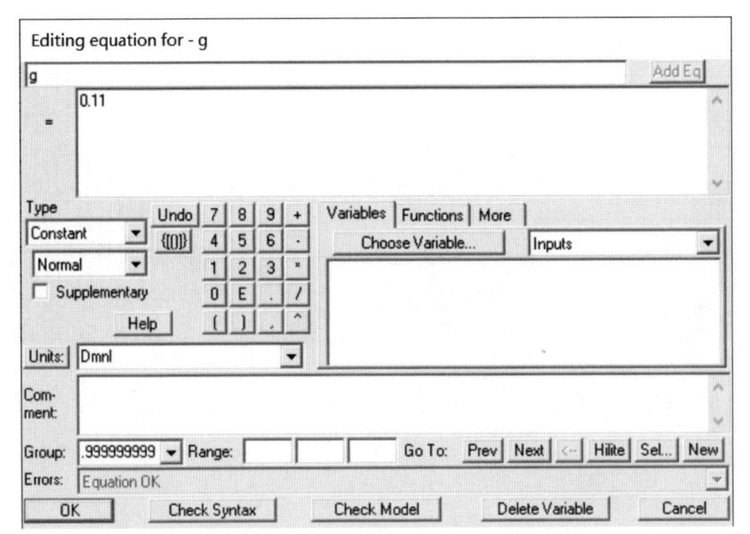
Editing equation for - g
g
Add Eq
0.11
Type
Constant
Normal
Supplementary
Undo
Help
Variables
Functions
More
Choose Variable...
Inputs
Units:
Dmnl
Comment:
Group:
Range:
Go To:
Prev
Next
Hilite
Sel...
New
Errors:
Equation OK
OK
Check Syntax
Check Model
Delete Variable
Cancel

其余部分与第 2 题中方程类似。

（4）仿真图形及结果。

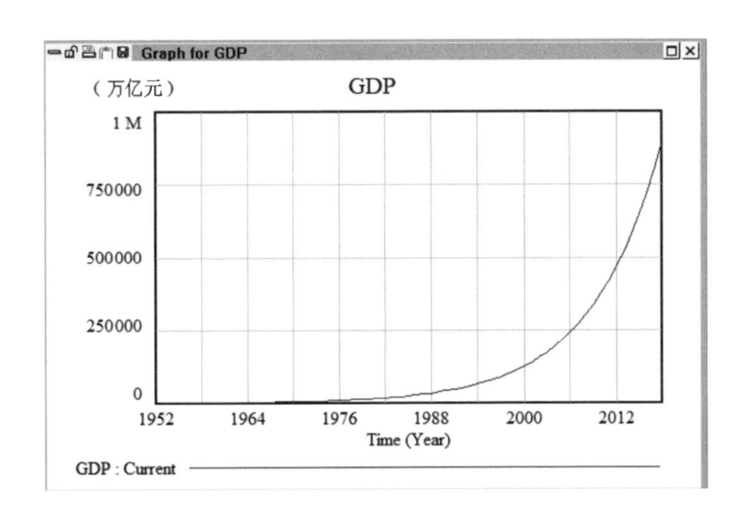

Table

Time (Year)	1952
"GDP" Runs:	Current
GDP	679.09

Table

Time (Year)	2013	2014	2015	2016	2017	2018
"GDP" Runs:	Current					
GDP	519,637	579,396	646,026	720,319	803,156	895,519

（5）分析。仿真结果中的 2018 年 GDP 值与比实际值 900309. 5 略少，而且，实际的 GDP 值在某一时间段内会下降，而仿真结果中的值只会增加。

实际的 g 应当是 0. 115 左右。

1）实际数据的拟合程度，选择合适的增长率 g。

E10 fx

	A	B	C	D	E	F
1	年份	GDP（亿元）	增长率			
2	1978	3678.7	0.11466007			
3	1979	4100.5	0.118790391			
4	1980	4587.6	0.075442497			
5	1981	4933.7	0.090560837			
6	1982	5380.5	0.123278506			
7	1983	6043.8	0.210198881			
8	1984	7314.2	0.247381805			
9	1985	9123.6	0.137204612			
10	1986	10375.4	0.172639127			
11	1987	12166.6	0.247217793			
12	1988	15174.4	0.132723534			
13	1989	17188.4	0.100934351			
14	1990	18923.3	0.165246019			
15	1991	22050.3	0.233915185			
16	1992	27208.2	0.308399674			
17	1993	35599.2	0.363744129			
18	1994	48548.2	0.243230439			
19	1995	60356.6	0.172690311			
20	1996	70779.6	0.113356108			
21	1997	78802.9	0.063635983			
22	1998	83817.6	0.066202086			
23	1999	89366.5	0.108537315			
24	2000	99066.1	0.10306452			
25	2001	109276.3	0.102530009			
26	2002	120480.4	0.133597664			
27	2003	136576.3	0.181869768			
28	2004	161415.4	0.152299595			
29	2005	185998.9	0.177579545			
30	2006	219028.5	0.23656967			
31	2007	270844.0	0.187032018			
32	2008	321500.5	0.083974986			
33	2009	348498.5	0.180106084			
34	2010	411265.2	0.178687621			
35	2011	484753.2	0.112146346			
36	2012	539116.5	0.095166629			
37	2013	590422.4	0.092084413			
38	2014	644791.1	0.064607747			
39	2015	686449.6	0.078882849			
40	2016	740598.7	0.138036564			
41	2017	842828.4				
42						
43		平均增长率	0.151492966			
44						

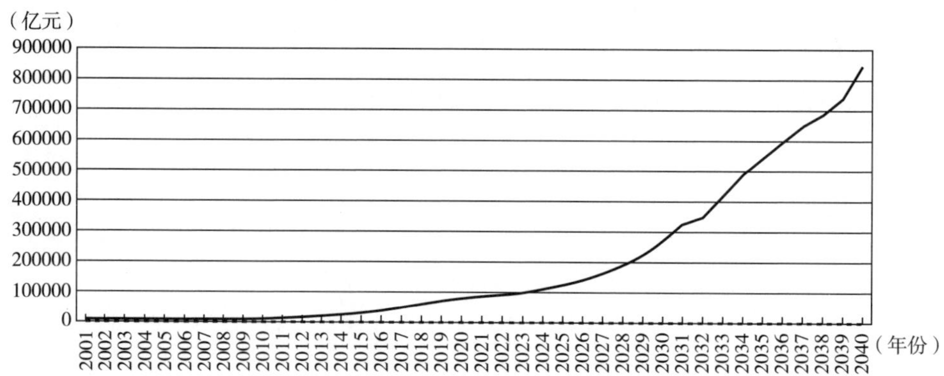

2）平均法。

平均法 g=0. 151492966

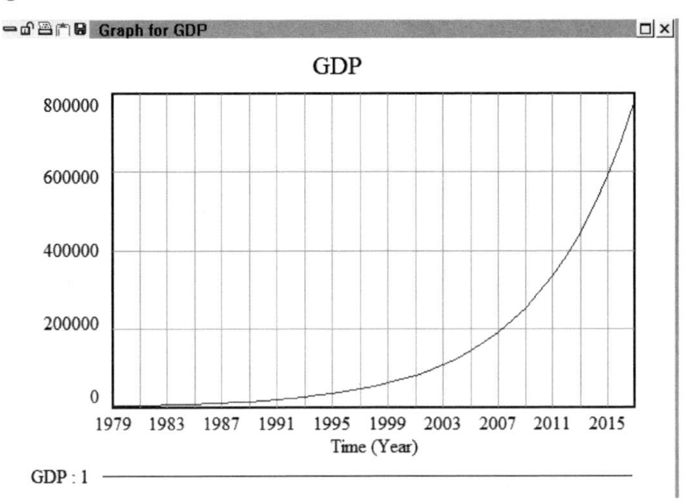

建模法 GDP 为 782750. 5272

实际 GDP 为 842828. 4

可得 g=0. 153735994

3）移动平均法。

1952~2020 年 GDP 实际数据在 GDP. xls 文件中，实际趋势图如下：

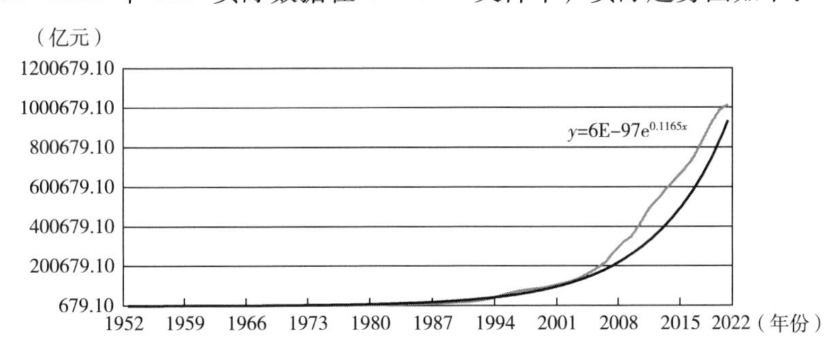

注：1950 年、1951 年两年数据缺省。

平均增长率 $g_1=0.116829347$，其中，$g_2=\frac{1}{69}\sum_{n=1}^{69}\left(\sqrt[n]{\frac{GDP(n)}{GDP(0)}}-1\right)=0.088957$，

$g_3=\sqrt[69]{\frac{GDP(69)}{GDP(0)}}-1=0.111767$，在 Vensim 中求出仿真数据并与实际曲线拟合

如下（仿真数据见 GDP. xls 文件）：

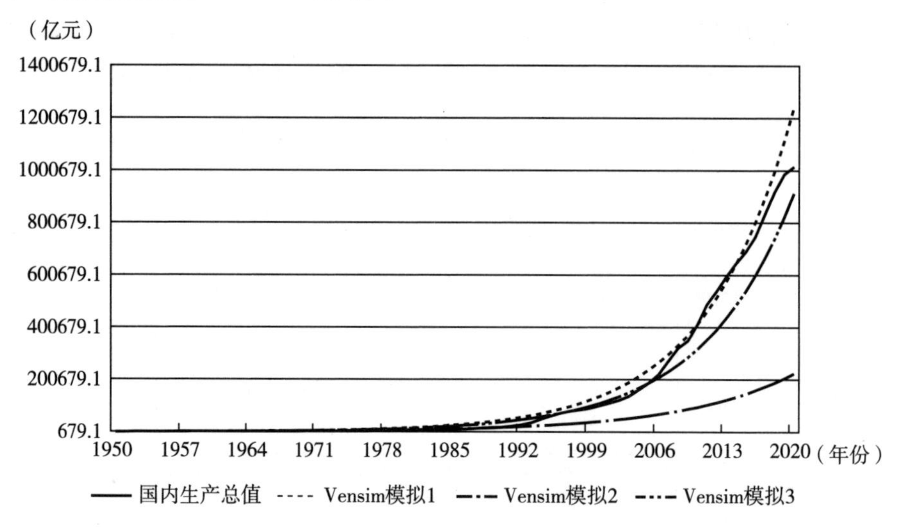

g1 = 0. 116829 得到的曲线在前中期拟合程度最好，后期实际数据偏低。

上机实验解答 4

1. 世界模型Ⅱ的因果关系图。

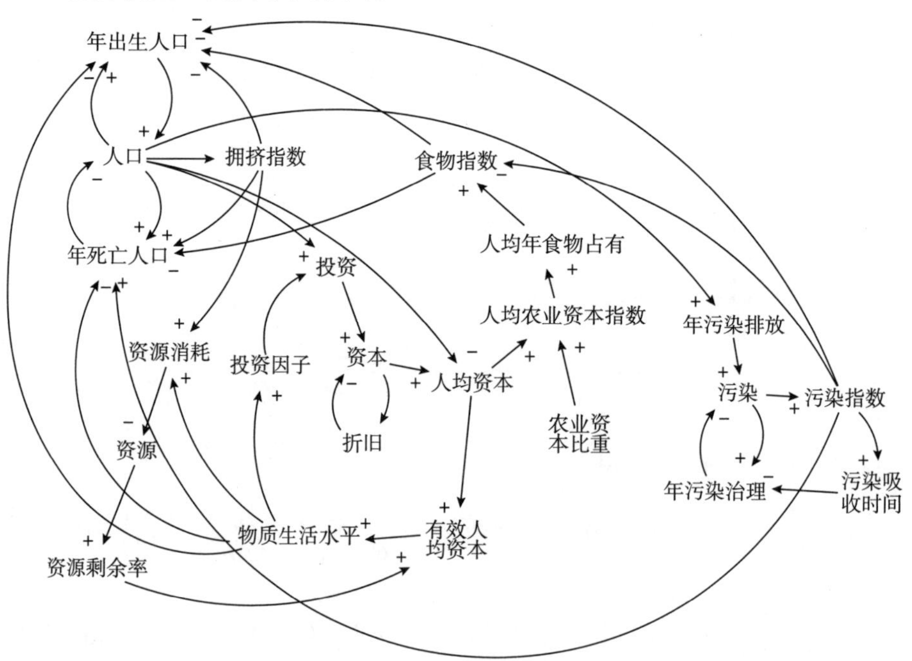

2. 世界模型Ⅱ的流位流率系（用表格的形式，注意变量用斜体）。

序号	流位	流率
1	人口数 P（t）（人）	年出生人口数 BR（t）（人/年） 年死亡人数 DR（t）（人/年）
2	非再生自然资源量 NR（t）（自然资源单位）	年非再生性自然资源消耗量 RUR（t）（自然资源单位/年）
3	资本量 CI（t）（资本单位）	资本投入 CIG（t）（资本单位/年） 资本折旧 CID（t）（资本单位/年）
4	农业资本比重 CIAF（t）（无量纲）	农业资本比重变化量 RAT（t）（1/年）
5	污染量 POL（t）（污染单位）	年污染排放量 POLG（t）（污染单位/年） 年污染治理量 POLA（t）（污染单位/年）

3. 世界模型Ⅱ流率基本入树 T_1 的建流位控制污染流率二部分图；结构模型：入树 T_1（t）仿真方程；污染量 POL、年污染排放量 POLG、年污染治理量 POLA 曲线；仿真检验结果分析。

（1）流率基本入树 T_1 的建流位控制污染流率二部分图。

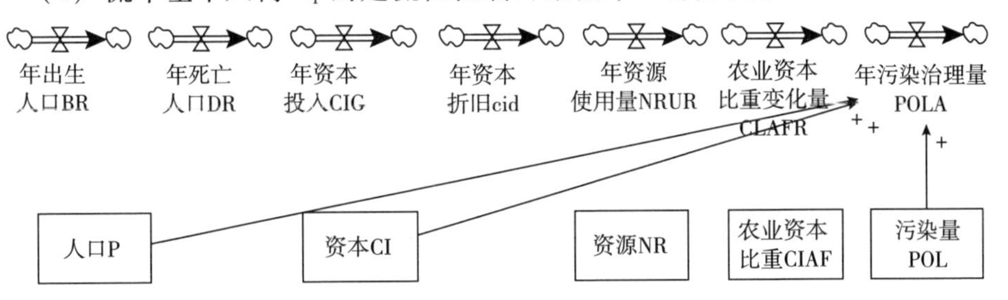

（2）结构模型。

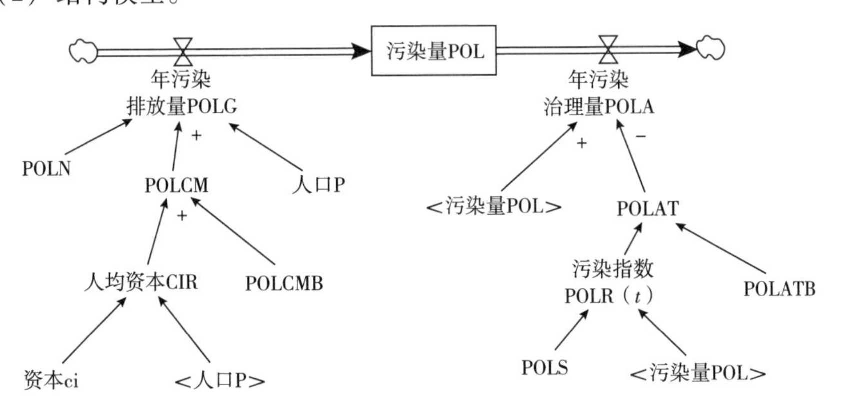

（3）入树 T_1（t）仿真方程。

1）污染量 POL(t+DT)= 污染量 POL(t)+DT×(年污染排放量 POLG（t+DT)-年污染治理量 POLA（t+DT))。

2）污染量 POL(1900)=0.2×10^9(污染单位)。

3）年污染排放量 POLG=人口 P×POLN×POLCM。

4）POLN=1。

5）Polcm=POLCMB(人均资本 CIR)。

6）Polcm=POLCMB(人均资本 CIR)表函数(由 1970 年的数据分析建立)。

人均资本 cir	0	1	2	3	4	5
年污染排放影响因子 Polcm	0.05	1	3	5.4	7.4	8

7）人均资本 CIR=资本 CI/人口 P。

8）设人口 P=3.65307×10^9(1970 年值)。

9）设资本 CI=3.77945×10^9(1970 年值)。

10）年污染治理量 POLA=污染量 POL/POLAT。

11）POLAT=POLATB（污染指数 POLR（t））。

12）表函数。

污染指数 POLR（t）	0	10	20	30	40	50	60
年污染治理需时间 POLAT	0.6	2.5	5	8	11.5	15.5	20

13）污染指数 POLR（t）=污染量 POL/POLS。

14）POLS=3.6×10^9。

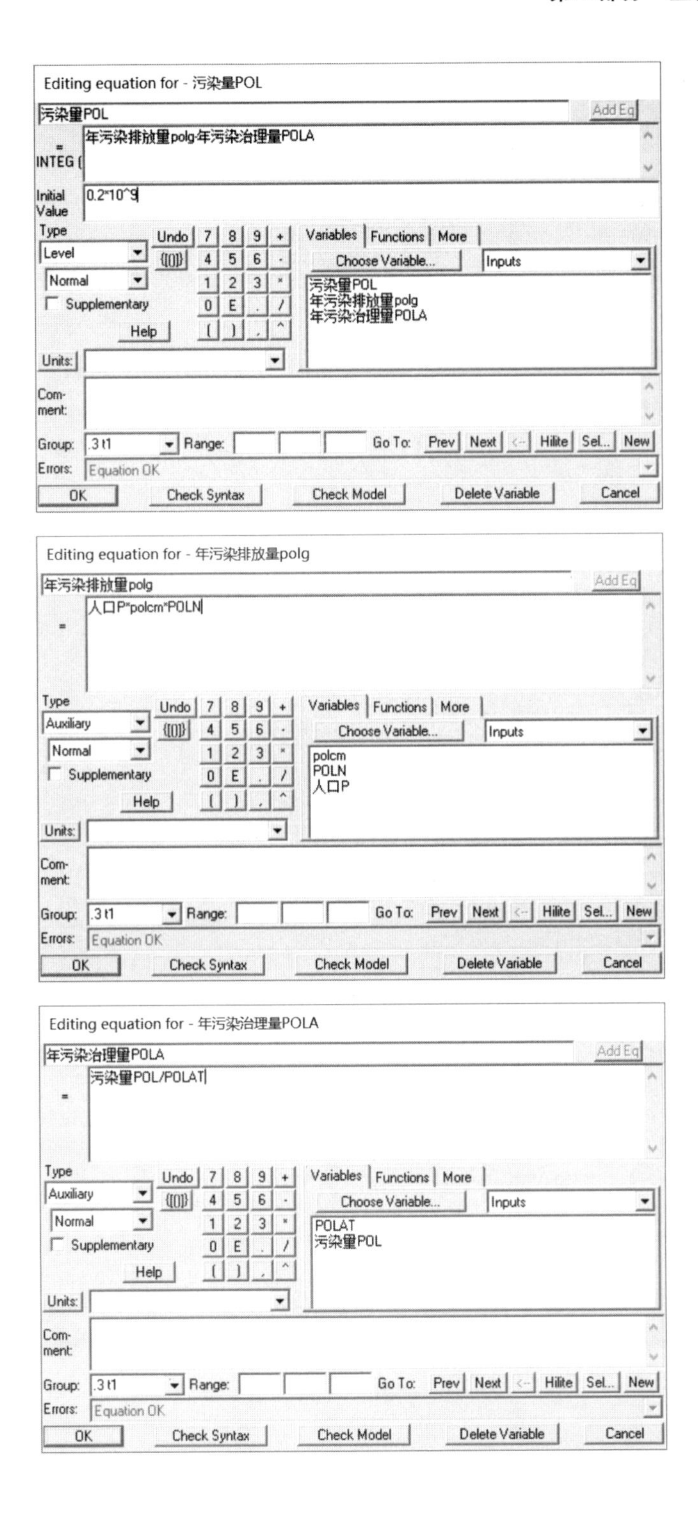
Editing equation for - 污染量POL
污染量POL
INTEG (
年污染排放量polg-年污染治理量POLA
Initial Value
0.2*10^9
Type
Level
Normal
Supplementary
Units:
Com-ment:
Group: .3 t1
Range:
Errors: Equation OK
Variables
Functions
More
Choose Variable...
Inputs
污染量POL
年污染排放量polg
年污染治理量POLA
OK
Check Syntax
Check Model
Delete Variable
Cancel
Editing equation for - 年污染排放量polg
年污染排放量polg
人口P*polcm*POLN
Type
Auxiliary
Normal
Group: .3 t1
Errors: Equation OK
polcm
POLN
人口P
Editing equation for - 年污染治理量POLA
年污染治理量POLA
污染量POL/POLAT
Type
Auxiliary
Normal
Group: .3 t1
Errors: Equation OK
POLAT
污染量POL

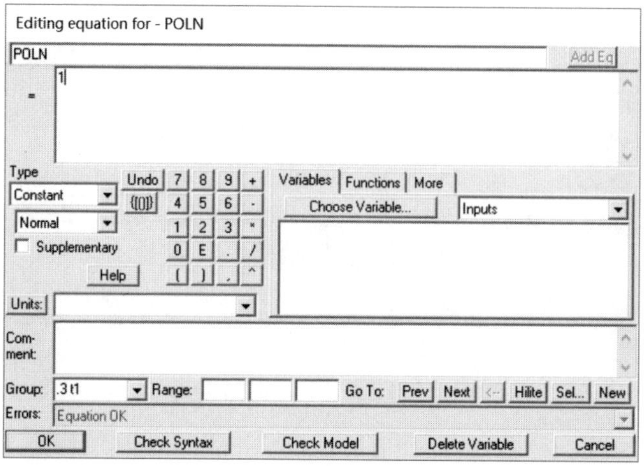

Editing equation for - POLN

POLN | Add Eq

= 1

Type: Constant | Normal | Supplementary | Help
Undo 7 8 9 + | {[()]} 4 5 6 - | 1 2 3 * | 0 E . / | ( ) , ^
Variables | Functions | More | Choose Variable... | Inputs
Units:
Comment:
Group: .3 t1 | Range: | Go To: | Prev | Next | <-- | Hilite | Sel... | New
Errors: Equation OK
OK | Check Syntax | Check Model | Delete Variable | Cancel

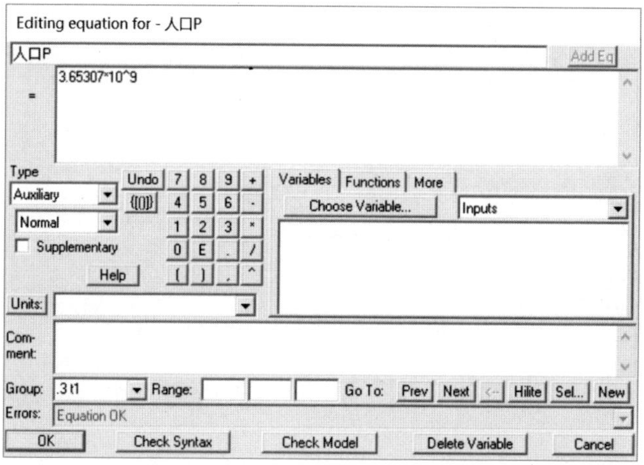

Editing equation for - 人口P

人口P | Add Eq

= 3.65307*10^9

Type: Auxiliary | Normal | Supplementary | Help
Undo 7 8 9 + | {[()]} 4 5 6 - | 1 2 3 * | 0 E . / | ( ) , ^
Variables | Functions | More | Choose Variable... | Inputs
Units:
Comment:
Group: .3 t1 | Range: | Go To: | Prev | Next | <-- | Hilite | Sel... | New
Errors: Equation OK
OK | Check Syntax | Check Model | Delete Variable | Cancel

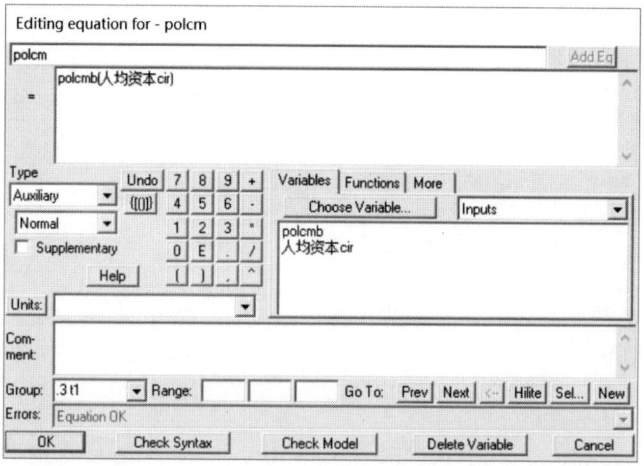

Editing equation for - polcm

polcm | Add Eq

= polcmb(人均资本cir)

Type: Auxiliary | Normal | Supplementary | Help
Undo 7 8 9 + | {[()]} 4 5 6 - | 1 2 3 * | 0 E . / | ( ) , ^
Variables | Functions | More | Choose Variable... | Inputs
polcmb
人均资本cir
Units:
Comment:
Group: .3 t1 | Range: | Go To: | Prev | Next | <-- | Hilite | Sel... | New
Errors: Equation OK
OK | Check Syntax | Check Model | Delete Variable | Cancel

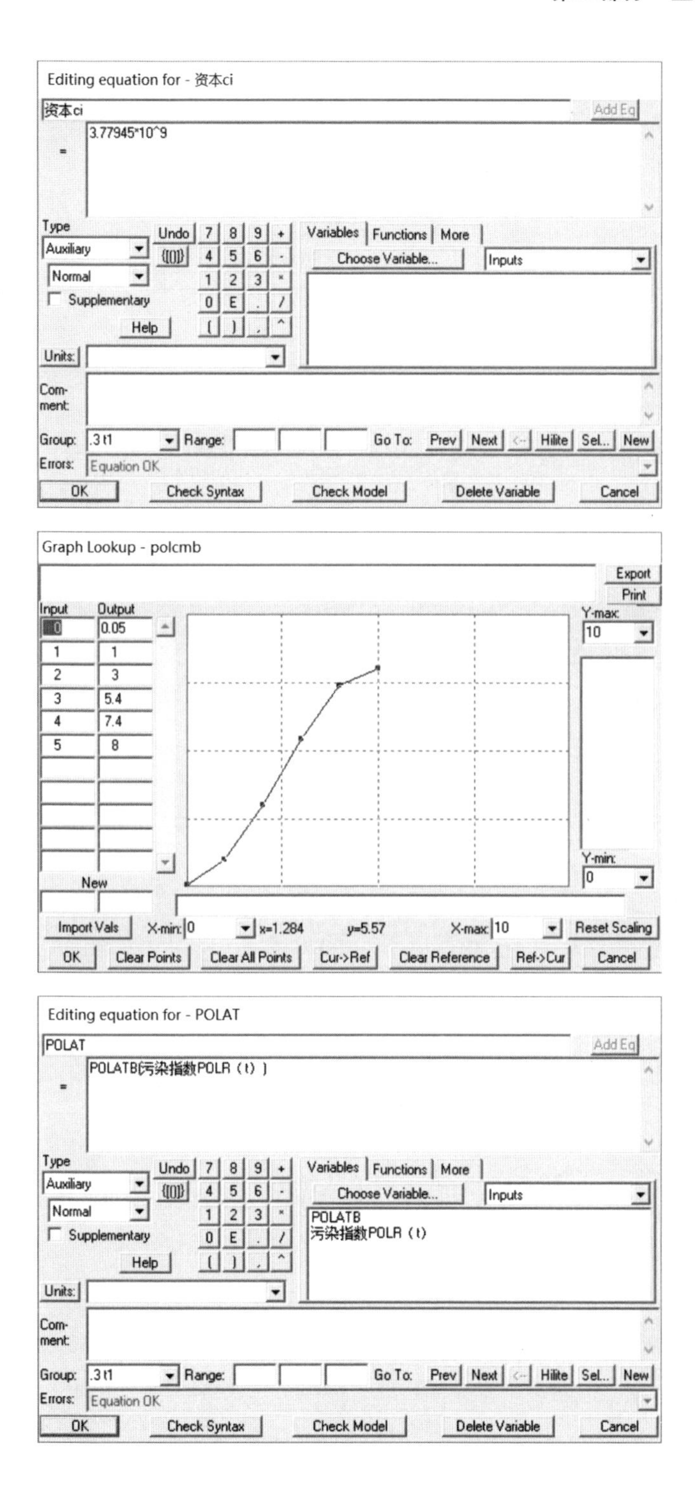

Editing equation for - 资本ci
资本ci
Add Eq
3.77945*10^9
Type
Auxiliary
Normal
Supplementary
Help
Undo
Variables
Functions
More
Choose Variable...
Inputs
Units:
Comment:
Group: .3 t1
Range:
Go To:
Prev
Next
Hilite
Sel...
New
Errors: Equation OK
OK
Check Syntax
Check Model
Delete Variable
Cancel
Graph Lookup - polcmb
Export
Print
Input
Output
0
0.05
1
1
2
3
3
5.4
4
7.4
5
8
New
Y-max: 10
Y-min: 0
Import Vals
X-min: 0
x=1.284
y=5.57
X-max: 10
Reset Scaling
OK
Clear Points
Clear All Points
Cur->Ref
Clear Reference
Ref->Cur
Cancel
Editing equation for - POLAT
POLAT
Add Eq
POLATB(污染指数POLR（t）)
Type
Auxiliary
Normal
Supplementary
Help
Undo
Variables
Functions
More
Choose Variable...
Inputs
POLATB
污染指数POLR（t）
Units:
Comment:
Group: .3 t1
Range:
Go To:
Prev
Next
Hilite
Sel...
New
Errors: Equation OK
OK
Check Syntax
Check Model
Delete Variable
Cancel

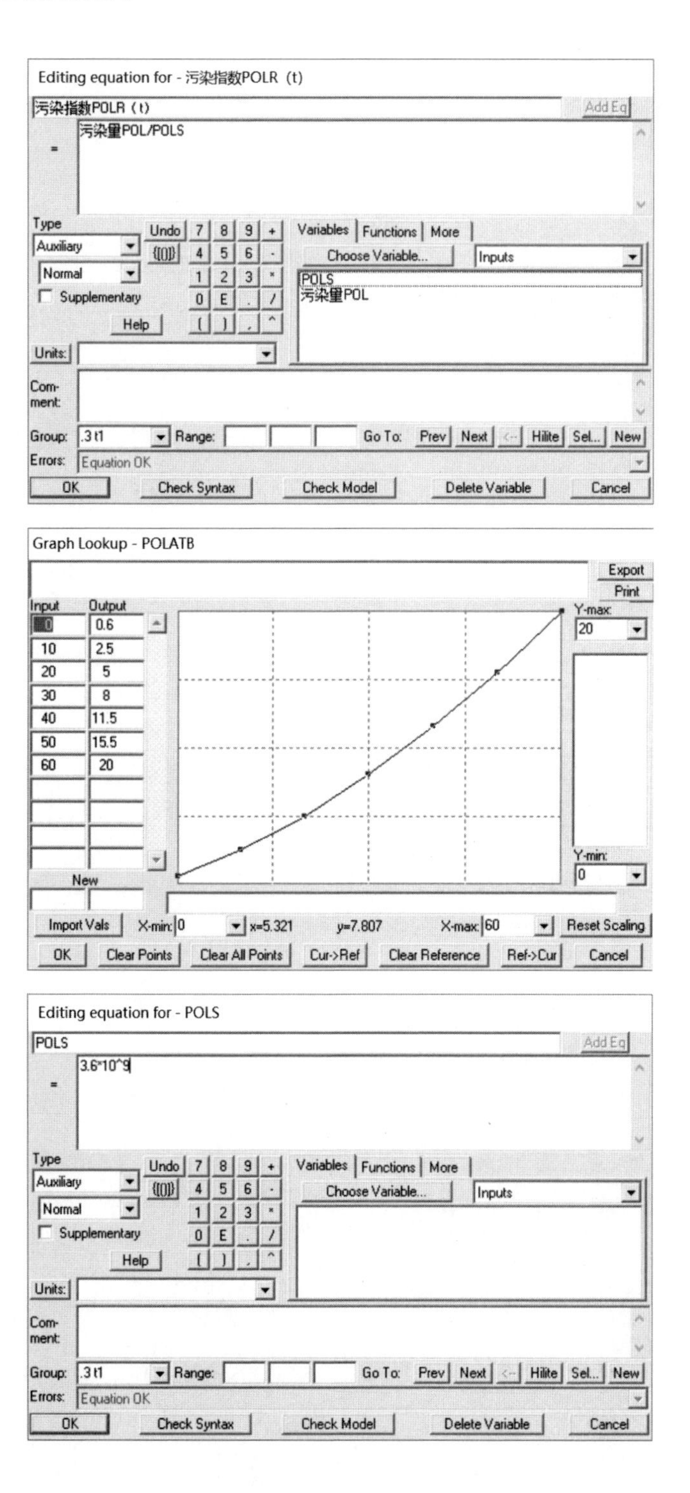
Editing equation for - 污染指数POLR (t)
污染指数POLR (t)
污染量POL/POLS
POLS
污染量POL
Group: .3 t1
Errors: Equation OK
Graph Lookup - POLATB
Input Output
0 0.6
10 2.5
20 5
30 8
40 11.5
50 15.5
60 20
Y-max: 20
Y-min: 0
X-min: 0
x=5.321
y=7.807
X-max: 60
Editing equation for - POLS
POLS
3.6*10^9
Group: .3 t1
Errors: Equation OK

（4）污染量 POL、年污染量 POLG、年污染治理量 POLA 曲线。

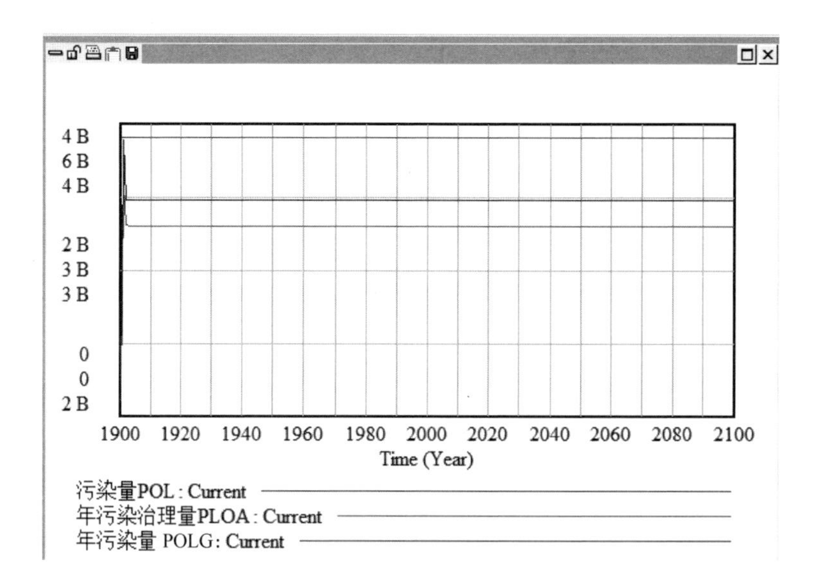

（5）仿真检验结果分析。

1）因为设人口 P=3.65307×10^9（1970 年值），设资本 CI=3.77945×10^9（1970 年值），年污染量 POLG=182803000，不变。

2）年污染治理量 POLA 自 1907 年开始，POLA=182803000，

所以，1907 年开始，污染量 POL=1.10750000，不变。

符合实际，入树 T_1（t）通过仿真检验。

4. 建立表函数的基本步骤及提升规律。

（1）基本步骤。

1）对表函数的有关变量重要关系进行分析。

2）确定自变量的变化和取值范围。

3）确定表函数的增减性。

4）找出表函数的特殊点与特殊线。

5）确定斜率。

6）插值。

重要说明：表函数两自变量之间，按线性插值取值。

（2）提升的规律。

流率 R_i（t）值的主算因果链枝计算的仿真方程可为乘积式仿真方程商式仿真方程。

流入率 R_i（t）值的影响因果链枝计算的仿真方程可为表函数影响因子仿真方程式。

5. 建立世界污染量入树 T_1（t）资本入树 T_2（t）组合模型二部分图；入树 T_1（t）T_2（t）组合结构模型；入树 T_2（t）仿真方程；资本 CI、年资本投入 CIG、年资本折旧 CID 曲线；T_1（t）T_2（t）组合仿真检验结果分析。

（1）世界污染量入树 T_1（t）资本入树 T_2（t）组合模型二部分图。

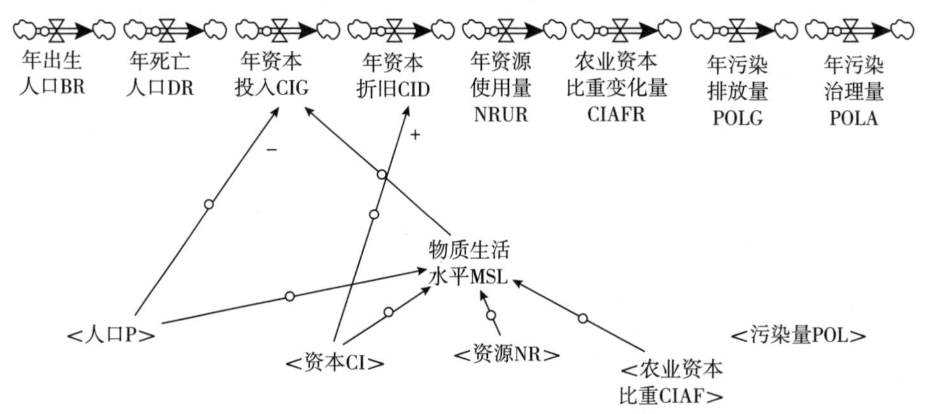

（2）入树 T_1（t）T_2（t）组合结构模型。

先撤销入树 T_1 的资本 CI’，建立完入树 T_2 之后，再恢复入树 T_1 的资本 CI。

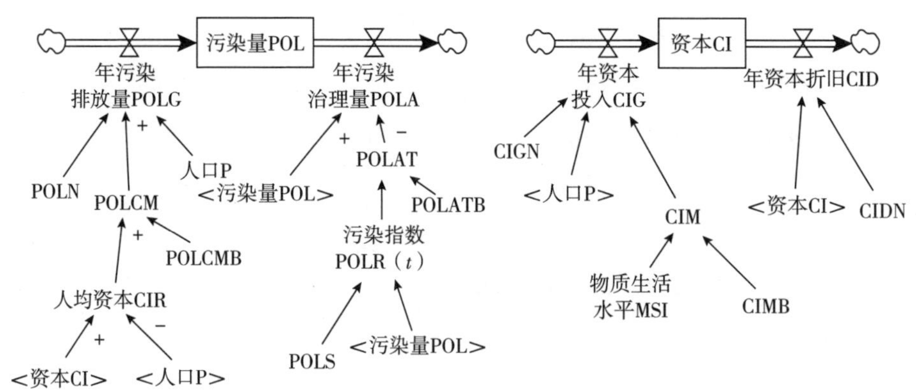

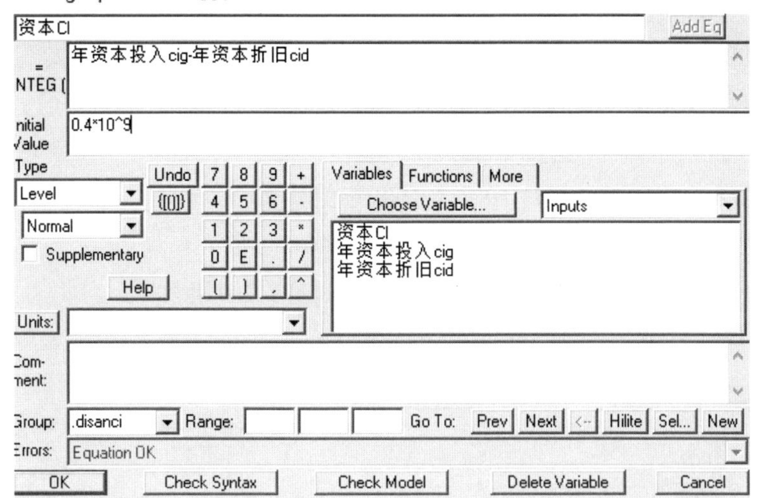

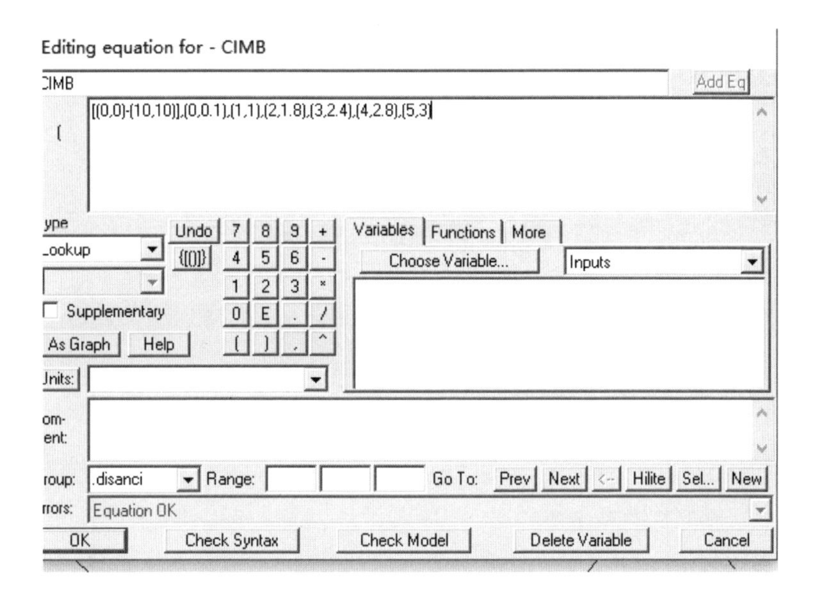

（3）入树 T_2（t）仿真方程。

1）资本 CI（1900）= 0. 4×10^9。

2）年资本投入 CIG = 人口 P×CIGN×CIM。

3）CIGN = 0. 05。

4）CIM = CIMB（物质生活水平 MSL）。

5）CIMB=[(0, 0)-(10, 10)], (0, 0.1), (1, 1), (2, 1.8), (3, 2.4), (4, 2.8), (5, 3)。

6）年资本折旧 CID=资本 CI×CIDN。

7）CIDN=0.025（调控参数）。

8）设人口 P=3.65307×10^9（1970 年值）。

9）设物质生活水平 MSL=0.979299（1970 年值）。

（4）资本 CI、年资本投入 CIG、年资本折旧 CID 曲线。

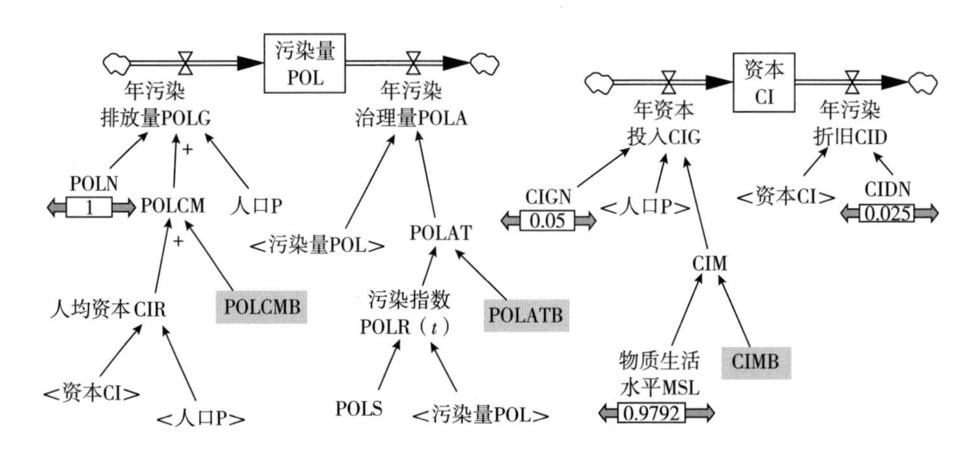

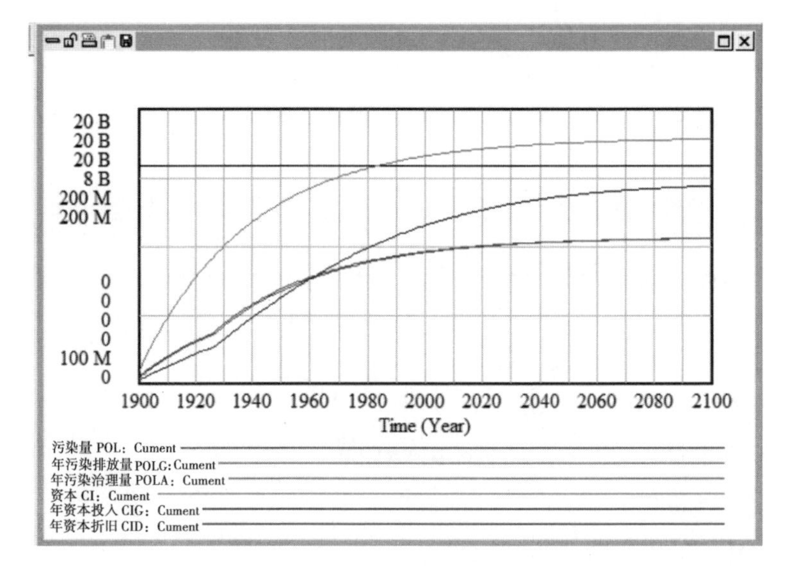

（5）T_1（t）T_2（t）组合仿真检验结果分析。

1）年资本投入 CIG（t）不变，符合实际（因设人口 P＝3. 65307×10^9（1970 年值），设物质生活水平 MSL＝0. 979299（1970 年值））。

2）年资本折旧 CID（t）改变。

3）符合资本 CI(t+DT)＝资本 CI(t)+DT×(年资本投入 CIG（t+DT）−年资本折旧 CID（t+DT））计算公式。

入树 T_1（t）T_2（t）通过仿真检验。

6. 世界污染量入树 T_1（t）资本 CI 入树 T_2（t）资源 NR 入树 T_3（t）组合模型：资源 NR 入树 T_3（t）二部分图；入树 T_3（t）仿真方程；设关联数入树 T_1（t）T_2（t）T_3（t）组合仿真检验；资源 NR 及年资源消耗量 NRUR 曲线；T_1（t）T_2（t）T_3（t）组合仿真检验结果分析。

（1）资源 NR 入树 T_3（t）二部分图。

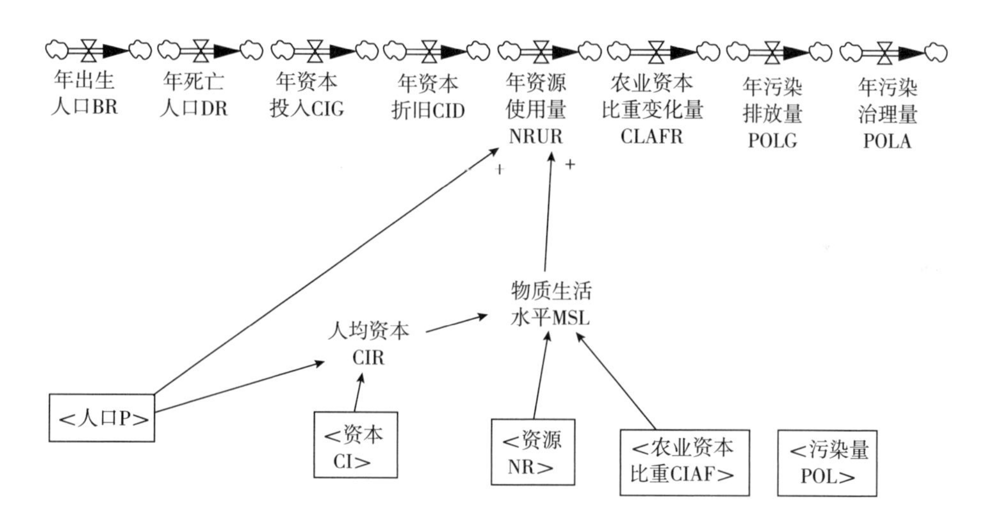

（2）世界污染量入树 T_1（t）资本 CI 入树 T_2（t）资源 NR 入树 T_3（t）组合模型。

先撤销入树 T_2 物质生活水平 MSL，建立完 T_3 之后，再恢复入树 T_2<物质生活水平 MSL>。

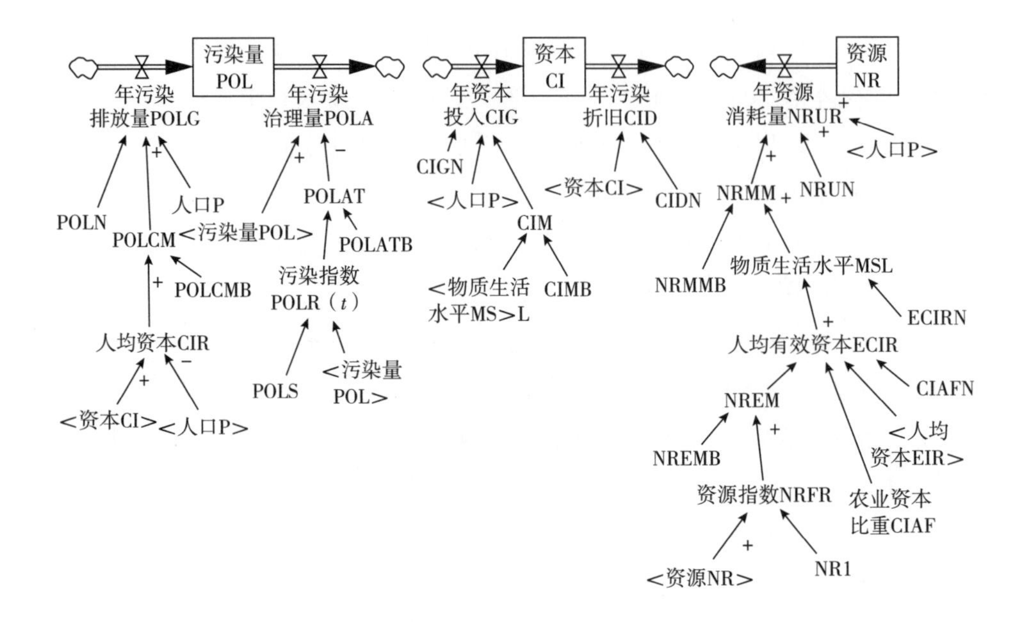

（3）入树 T_3（t）仿真方程。

1）资源 NR（1900）=900×10^9。

2）年资源使用量 NRUR=人口 P×NRUN×NRMM。

3）NRUN=1。

4）NRMM=NRMMB（物质生活水平 MSL）。

5）NRMMB=[（0，0）-（10，10）]，（0，0），（1，1），（2，1.8），（3，2.4），（4，2.9），（5，3.3），（6，3.6），（7，3.8），（8，3.9），（9，3.95），（10，4）。

6）物质生活水平 MSL=人均有效资本 ECIR/ECIRN。

7）ECIRN=1。

8）人均有效资本 ECIR=人均资本 CIR×（1-农业资本比重 CIAF）×NREM/（1-CIAFN）。

9）CIAFN=0.3。

10）NREM=NREMB（资源指数 NRFR）。

11）资源指数 NRFR=资源 NR/NR1。

12）NR1=900×10^9。

13）NREMB=[（0，0）-（10，10）]，（0，0），（0.25，0.15），（0.5，

0.5)，(0.75，0.85)，(1，1)。

14）农业资本比重 CIAF=0.279439（1970 年值）。

（4）设关联数入树 T_1（t）T_2（t）T_3（t）组合仿真检验。

设人口 P=3.65307×10^9（1970 年值）。

设农业资本比重 CIAF=0.279439（1970 年值）。

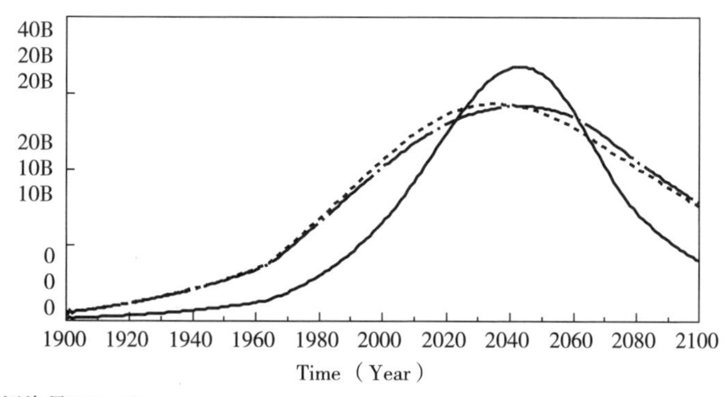

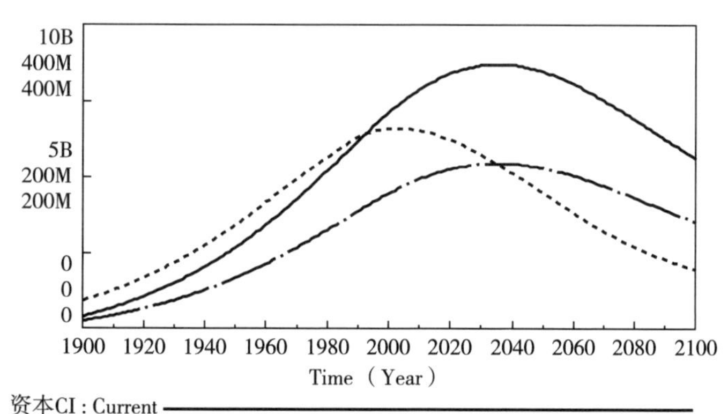

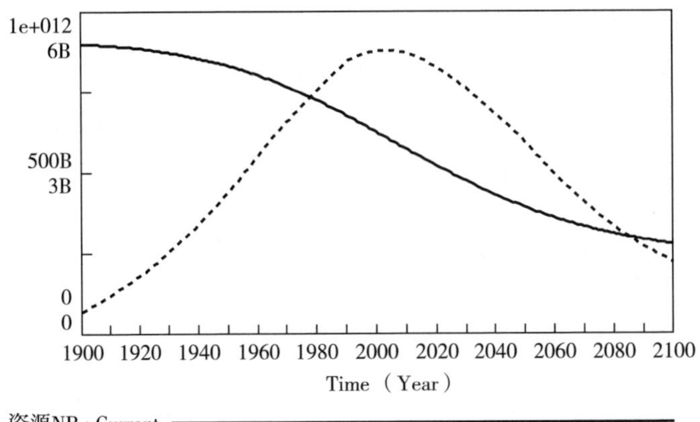
1e+012
6B
500B
3B
0
0
1900 1920 1940 1960 1980 2000 2020 2040 2060 2080 2100
Time （Year）
资源NR : Current
年资源消耗量NRUR : Current

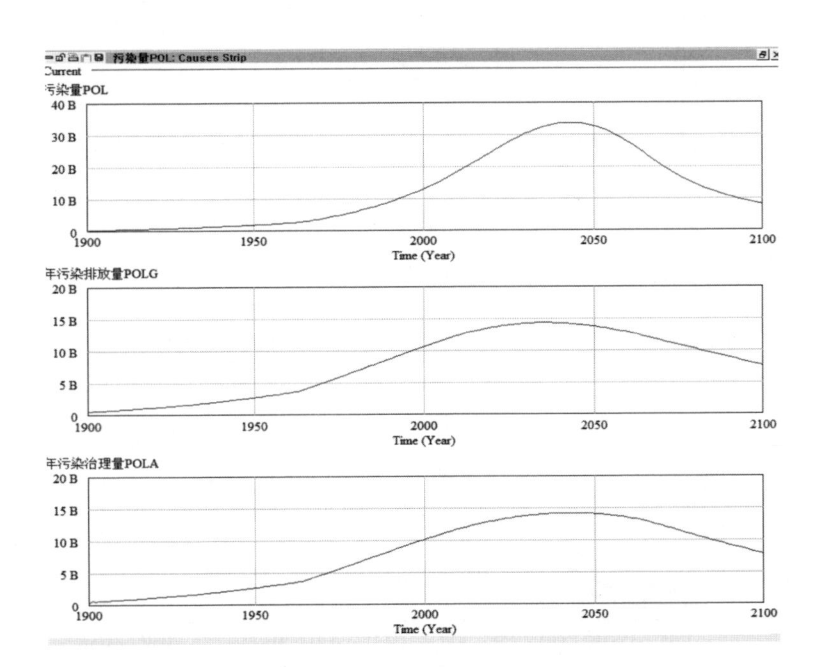
污染量POL: Causes Strip
Current
污染量POL
40 B
30 B
20 B
10 B
0
1900 1950 2000 2050 2100
Time (Year)
年污染排放量POLG
20 B
15 B
10 B
5 B
0
1900 1950 2000 2050 2100
Time (Year)
年污染治理量POLA
20 B
15 B
10 B
5 B
0
1900 1950 2000 2050 2100
Time (Year)

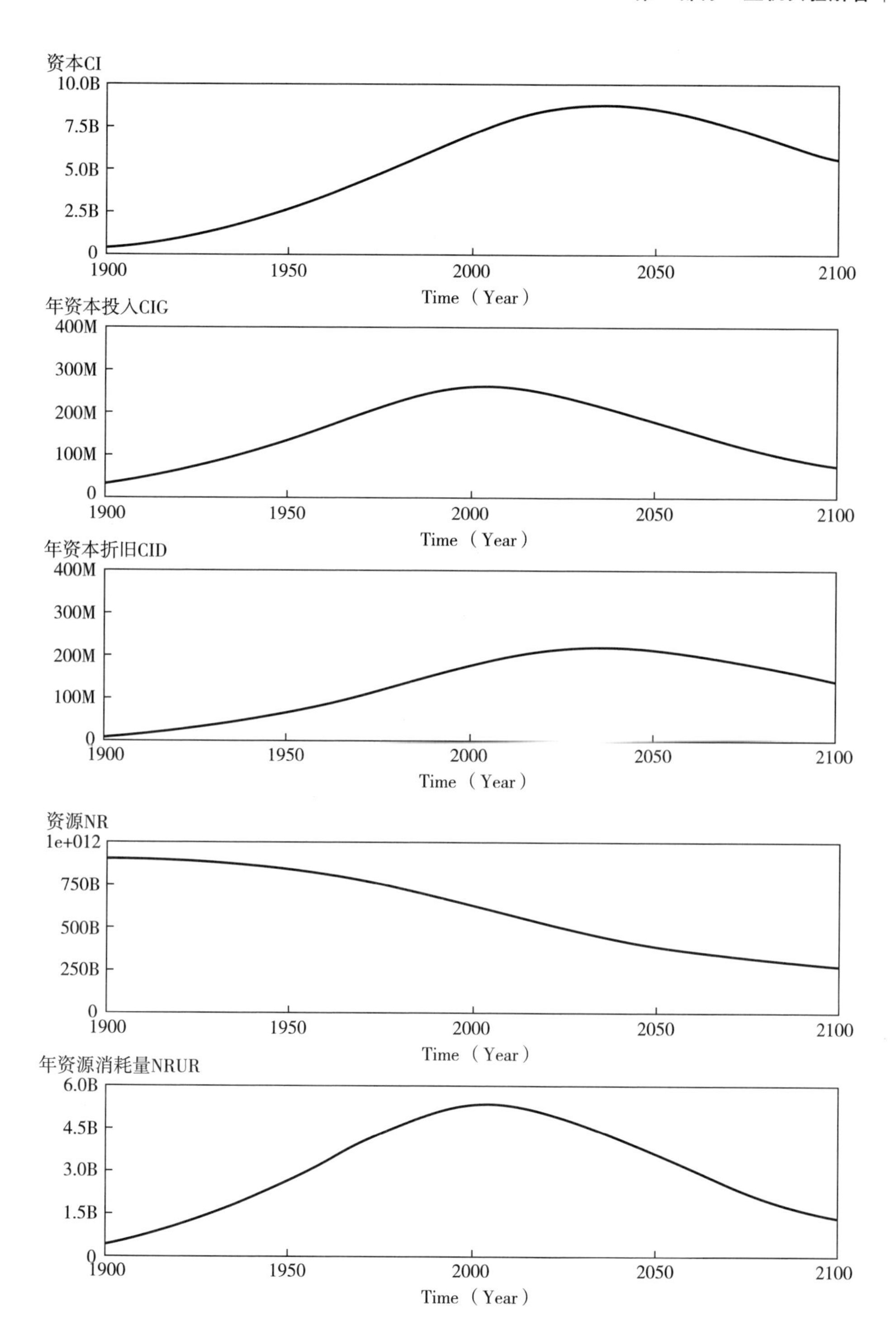
资本CI
10.0B
7.5B
5.0B
2.5B
0
1900
1950
2000
2050
2100
Time（Year）
年资本投入CIG
400M
300M
200M
100M
0
1900
1950
2000
2050
2100
Time（Year）
年资本折旧CID
400M
300M
200M
100M
0
1900
1950
2000
2050
2100
Time（Year）
资源NR
1e+012
750B
500B
250B
0
1900
1950
2000
2050
2100
Time（Year）
年资源消耗量NRUR
6.0B
4.5B
3.0B
1.5B
0
1900
1950
2000
2050
2100
Time（Year）

（5）T_1（t）T_2（t）T_3（t）组合仿真检验结果分析。

1）污染量 POL、资本 CI 相关结果保持符合实际。

2）资源使用量 NRUR 由资本 CI 和人口 P 直接影响，资本 CI 变化使资源 NR 下降，符合实际。资本 CI 后下降减少，资源使用量 NRUR 也减少，符合实际。

入树 T_1（t）T_2（t）T_3（t）组合通过仿真检验。

7. 世界污染量入树 T_1（t）资本 CI 入树 T_2（t）资源 NR 入树 T_3（t）农业资本比重 CIAF 入树 T_4（t）组合模型：农业资本比重 CIAF 入树 T_4（t）二部分图；入树 T_1（t）T_2（t）T_3（t）T_4（t）组合结构模型；入树 T_4（t）仿真方程；T_1（t）T_2（t）T_3（t）T_4（t）组合仿真检验结果分析。

（1）农业资本比重 CIAF 入树 T_4（t）二部分图。

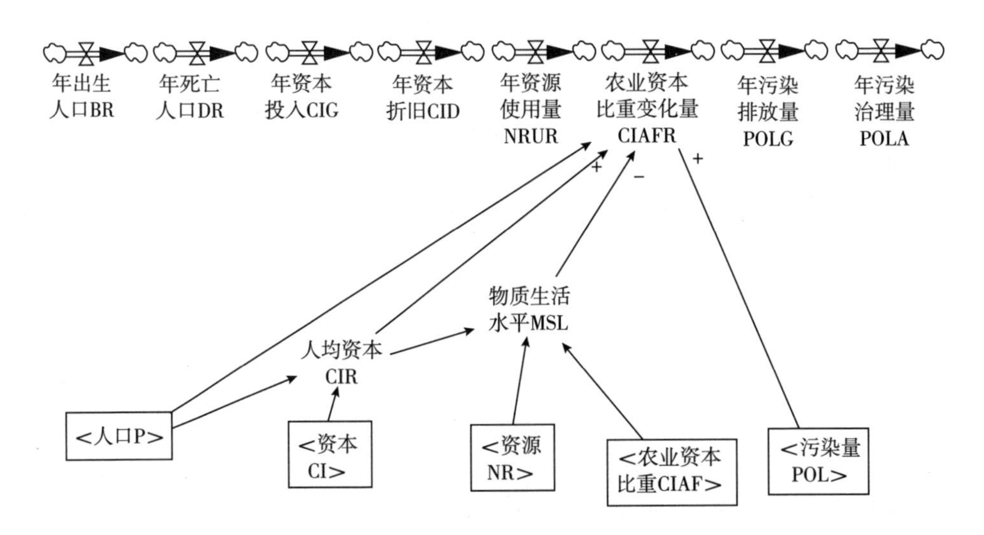

（2）入树 T_1（t）T_2（t）T_3（t）T_4（t）组合结构模型。

先撤销入树 T_3 农业资本比重 CIAF'，建立完入树 T_4 后，恢复入树 T_3 的农业资本比重 CIAF。

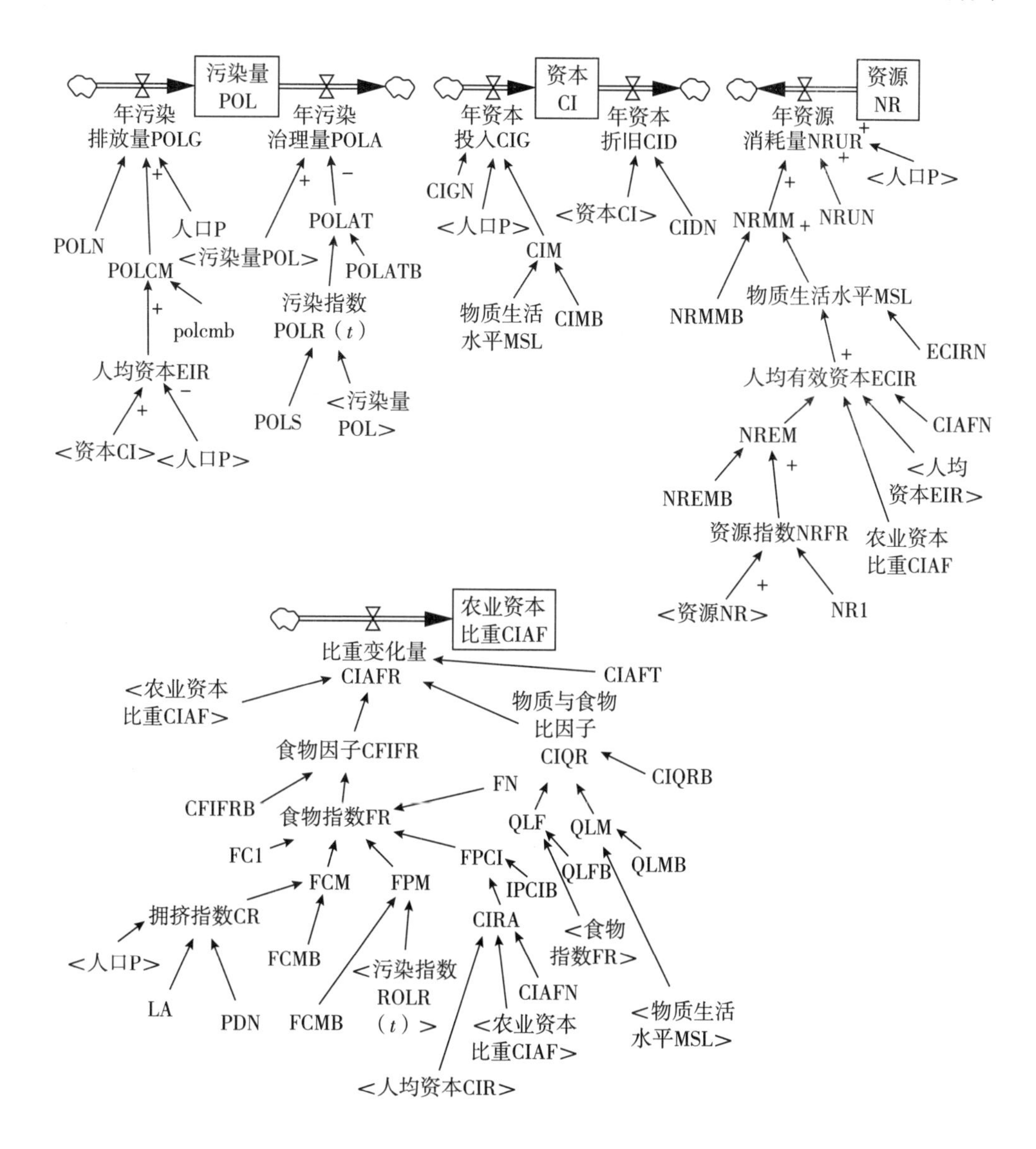

（3）入树 T_4（t）仿真方程。

1）农业资本比重 CIAF（1900）= 0.2。

2）比重变化量 CIAFR =（食物因子 CFIFR×物质与食物比因子 CIQR-农业资本比重 CIAF）/CIAFT。

3）CIAFT = 15。

4）物质与食物比因子 CIQR = CIQRB（QLM/QLF）。

5）食物因子 CFIFR = CFIFRB（食物指数 FR）。

6）CIQRB = [（0，0）-（10，10）]，（0，0.2），（0.5，0.8），（1，1），（1.5，1.5），（2，2）。

7）QLM = QLMB（物质生活水平 MSL）。

8）QLF = QLFB（食物指数 FR）。

9）QLMB = [（0，0）-（10，10）]，（0，0.2），（1，1），（2，1.7），（3，2.3），（4，2.7），（5，2.9）。

10）QLFB = [（0，0）-（10，10）]，（0，0），（1，1），（2，1.8），（3，2.4），（4，2.7）。

11）CFIFRB = [（0，0）-（10，10）]，（0，1），（0.5，0.6），（1，0.3），（1.5，0.15），（2，0.1）。

12）食物指数 FR = 人均食物量 FPCI×FCM×FPM×FC1/FN。

13）FC1 = 1。

14）FN = 1。

15）人均食物量 FPCI = FPCIB（人均农业资本指数 CIRA）。

16）人均农业资本指数 CIRA = 人均资本 CIR×农业资本比重 CIAF/CIAFN。

17）CIAFN = 0.3 人均食物量。

18）FPCIB = [（0，0）-（10，10）]，（0，0.5），（1，1），（2，1.4），（3，1.7），（4，1.9），（5，2.05），（6，2.2）。

19）FPM = FPMB（污染指数 POLR）。

20）FPMB = [（0，0）-（100，2）]，（0，1.02），（10，0.9），（20，0.65），（30，0.35），（40，0.2），（50，0.1），（60，0.05）。

21）FCM = FCMB（拥挤指数 CR）。

22）FCMB = [（0，0）-（10，10）]，（0，2.4），（1，1），（2，0.6），（3，0.4），（4，0.3），（5，0.2）。

23）拥挤指数 CR = 人口 P/（LA×PDN）。

24）LA = 135×10^6。

25）PDN = 26.5。

说明：

世界面积 LA = 135×10^6 平方公里（1.35 亿平方公里）

1958 年人口密度 PDN = 26.5 人/平方公里

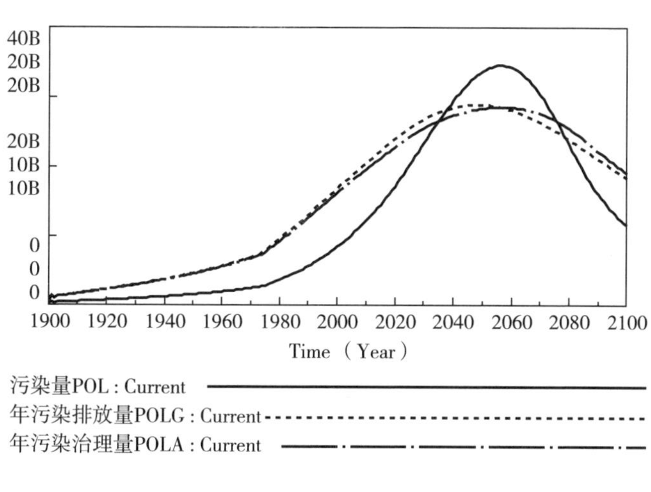

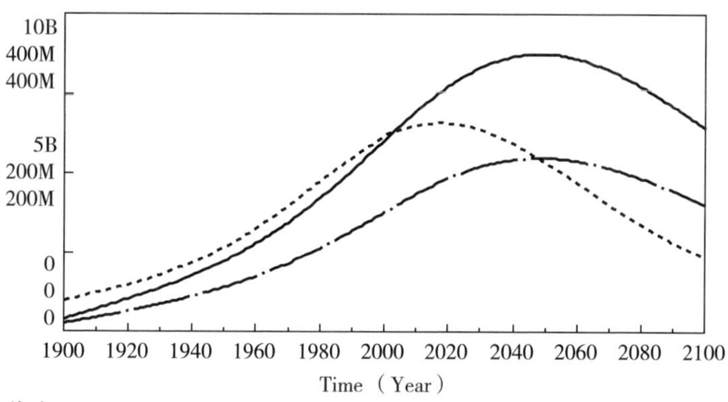

资本CI : Current

年资本投入CIG : Current

年资本折旧CID : Current

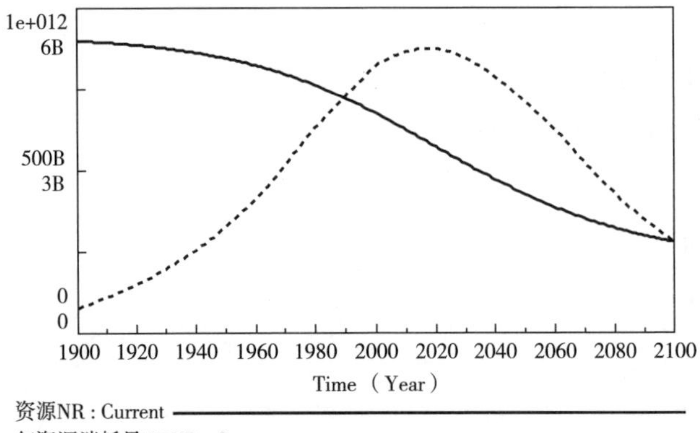
1e+012
6B
500B
3B
0
0
1900 1920 1940 1960 1980 2000 2020 2040 2060 2080 2100
Time（Year）
资源NR : Current
年资源消耗量NRUR : Current

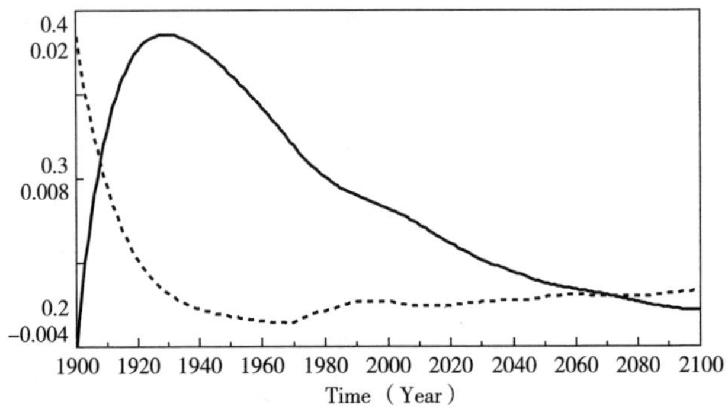
0.4
0.02
0.3
0.008
0.2
-0.004
1900 1920 1940 1960 1980 2000 2020 2040 2060 2080 2100
Time（Year）
农业资本比重CIAF : Current
比重变化量CIAFR : Current

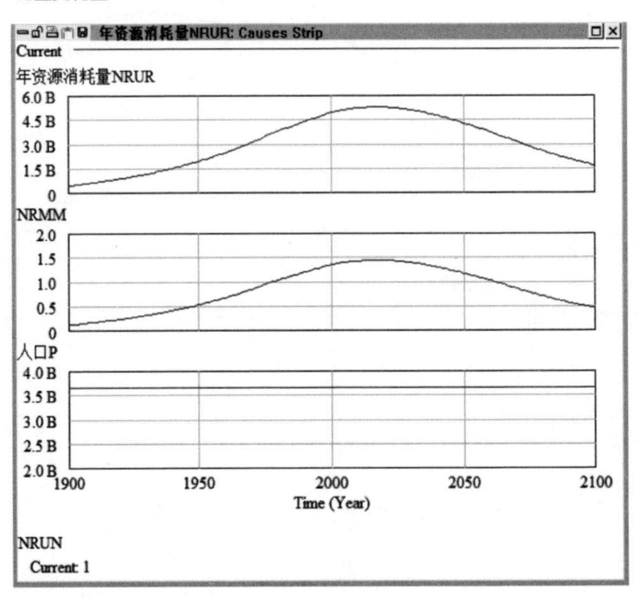
年资源消耗量NRUR: Causes Strip
Current
年资源消耗量NRUR
6.0 B
4.5 B
3.0 B
1.5 B
0
NRMM
2.0
1.5
1.0
0.5
0
人口P
4.0 B
3.5 B
3.0 B
2.5 B
2.0 B
1900 1950 2000 2050 2100
Time (Year)
NRUN
Current: 1

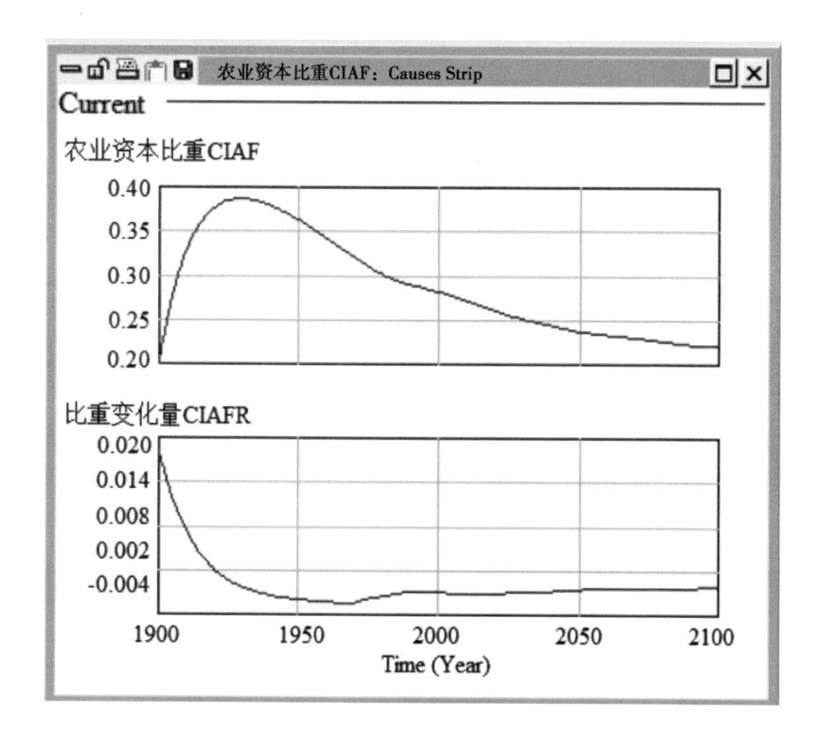

（4）T_1（t）T_2（t）T_3（t）T_4（t）组合仿真检验结果分析。

1）污染量 POL、资本 CI、资源 NR 相关结果保持符合实际。

2）农业资本比重 CIAF 由资本 CI、物质生活水平 MSL 及人口 P 直接影响，资本 CI 变化、物质生活水平 MSL 使农业资本比重 CIAF 下降，符合实际。

入树 T_1（t）T_2（t）T_3（t）T_4（t）组合通过仿真检验。

8. 世界污染量入树 T_1（t）资本 CI 入树 T_2（t）资源 NR 入树 T_3（t）农业资本比重 CIAF 入树 T_4（t）入树 T_5（t）组合模型：入树 T_1（t）T_2（t）T_3（t）T_4（t）T_5（t）组合结构模型：入树 T_5（t）仿真方程；人口 P 及年出生人口 BR、年死亡人口 DR 仿真曲线；T_1（t）T_2（t）T_3（t）T_4（t）T_5（t）组合仿真检验结果分析。

（1）入树 T_1（t）T_2（t）T_3（t）T_4（t）T_5（t）组合结构模型。

撤销各个入树的人口 P'，等到建立完 T_5 之后，再恢复各个入树的影子变量<人口 P>。

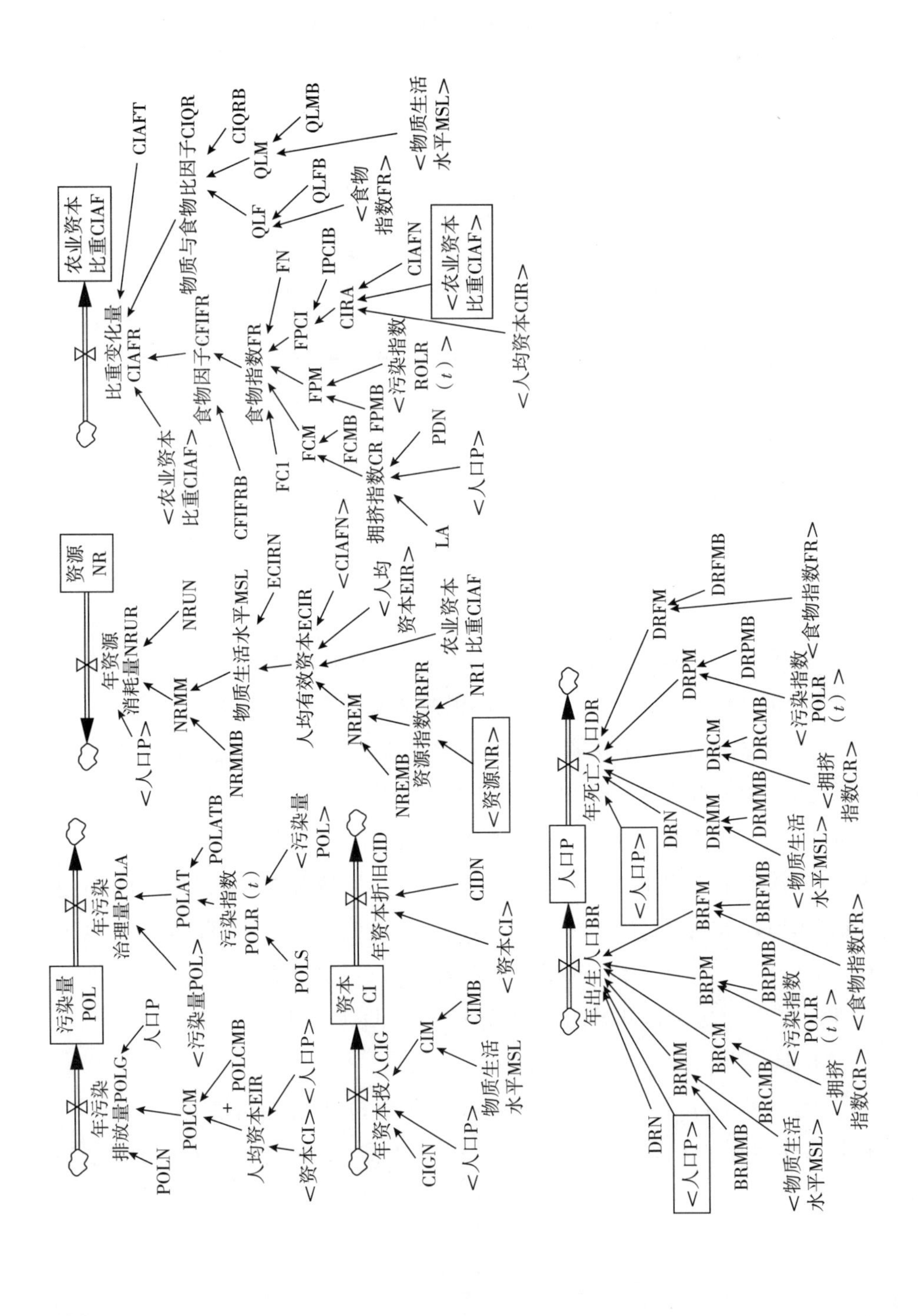

污染量 POL
年污染排放量POLG
年污染治理量POLA
POLN
人口P
POLCM
+
POLCMB
<污染量POL>
POLAT
POLATB
污染指数POLR（t）
<污染量POL>
POLS
人均资本EIR
<资本CI>
<人口P>
资本 CI
年资本投入CIG
年资本折旧CID
CIGN
<人口P>
CIM
CIMB
物质生活水平MSL
CIDN
<资本CI>
资源 NR
年资源消耗量NRUR
NRUN
<人口P>
NRMM
NRMMB
物质生活水平MSL
ECIRN
人均有效资本ECIR
<CIAFN>
<人均资本EIR>
农业资本比重CIAF
NREM
NREMB
资源指数NRFR
NR1
<资源NR>
农业资本比重CIAF
比重变化量CIAFR
CIAFT
<农业资本比重CIAF>
食物因子CFIFR
CFIFRB
物质与食物比因子CIQR
CIQRB
QLM
QLMB
QLF
QLFB
<食物指数FR>
<物质生活水平MSL>
食物指数FR
FN
FC1
FCM
FCMB
FPM
FPMB
FPCI
IPCIB
CIRA
CIAFN
<农业资本比重CIAF>
<人均资本CIR>
<污染指数ROLR（t）>
拥挤指数CR
PDN
<人口P>
LA
人口P
年出生人口BR
年死亡人口DR
DRN
<人口P>
BRMM
BRMMB
BRCM
BRCMB
BRPM
BRPMB
BRFM
BRFMB
<物质生活水平MSL>
<拥挤指数CR>
<污染指数POLR（t）>
<食物指数FR>
DRN
<人口P>
DRMM
DRMMB
DRCM
DRCMB
DRPM
DRPMB
DRFM
DRFMB
<物质生活水平MSL>
<拥挤指数CR>
<污染指数POLR（t）>
<食物指数FR>

（2）人树 T_5（t）仿真方程。

1）人口 P(1900)= 1.65×10^9。

2）年出生人口 BR=人口 P×BRN×BRFM×BRMM×BRCM×BRPM。

3）BRN=0.04。

4）BRMM=BRMMB（物质生活水平 MSL）。

5）BRCM=BRCMB（拥挤指数 CR）。

6）BRPM=BRPMB（污染指数 POLR）。

7）BRFM=BRFMB（食物指数 FR）。

8）BRMMB=[(0，0)-(10，10)]，(0，1.2)，(1，1)，(2，0.85)，(3，0.75)，(4，0.7)，(5，0.7)。

9）BRCMB=[(0，0)-(10，10)]，(0，1.05)，(1，1)，(2，0.9)，(3，0.7)，(4，0.6)，(5，0.55)。

10）BRPMB=[(0，0)-(100，2)]，(0，1.02)，(10，0.9)，(20，0.7)，(30，0.4)，(40，0.25)，(50，0.15)，(60，0.1)。

11）BRFMB=[(0，0)-(10，10)]，(0，0)，(1，1)，(2，1.6)，(3，1.9)，(4，2)。

12）年死亡人口 DR=人口 P * DRN * DRMM * DRPM * DRFM * DRCM。

13）DRN=0.028。

14）DRMM=DRMMB（物质生活水平 MSL）。

15）DRCM=DRCMB（拥挤指数 CR）。

16）DRPM=DRPMB（污染指数 POLR）。

17）DRFM=DRFMB（食物指数 FR）。

18）DRMMB=[(0，0)-(10，10)]，(0，3)，(0.5，1.8)，(1，1)，(1.5，0.8)，(2，0.7)，(2.5，0.6)，(3，0.53)，(3.5，0.5)，(4，0.5)，(4.5，0.5)，(5，0.5)。

19）DRCMB=[(0，0)-(10，10)]，(0，0.9)，(1，1)，(2，1.2)，(3，1.5)，(4，1.9)，(5，3)。

20）DRPMB=[(0，0)-(100，10)]，(0，0.92)，(10，1.3)，(20，2)，(30，3.2)，(40，4.8)，(50，6.8)，(60，9.2)。

21）DRFMB=[(0，0)-(10，40)]，(0，30)，(0.25，3)，(0.5，2)，(0.75，1.4)，(1，1)，(1.25，0.7)，(1.5，0.6)，(1.75，0.5)，(2，0.5)。

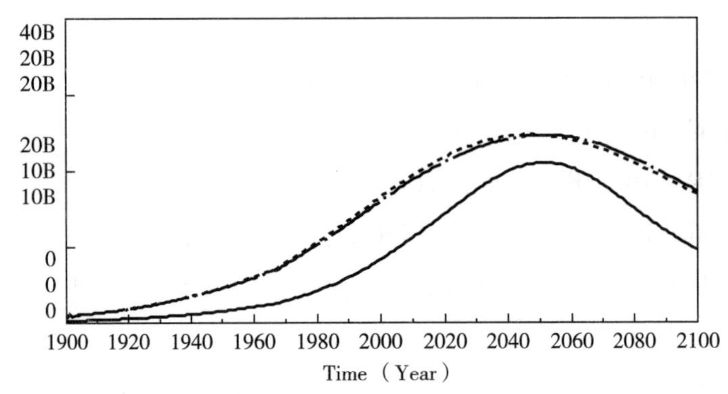

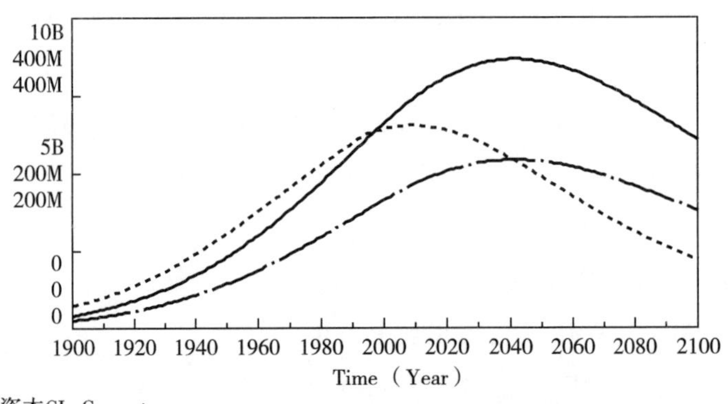

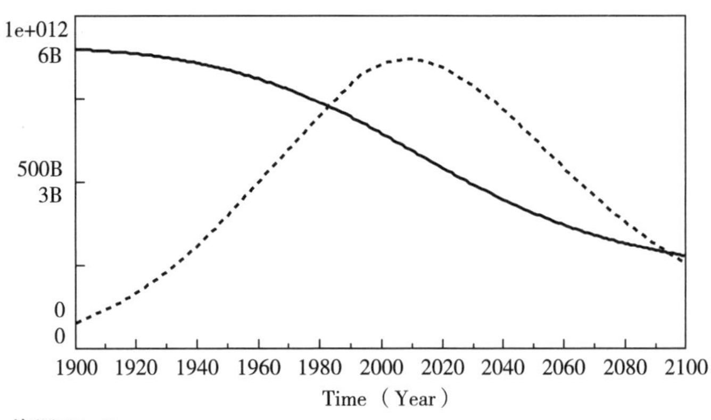
1e+012
6B
500B
3B
0
0
1900 1920 1940 1960 1980 2000 2020 2040 2060 2080 2100
Time（Year）
资源NR : Current
年资源消耗量NRUR : Current

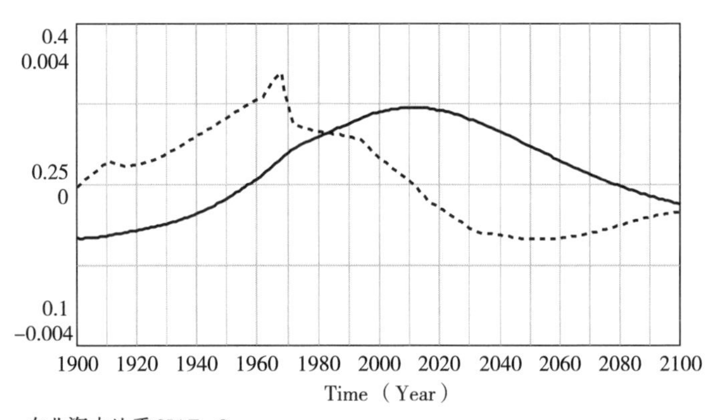
0.4
0.004
0.25
0
0.1
−0.004
1900 1920 1940 1960 1980 2000 2020 2040 2060 2080 2100
Time（Year）
农业资本比重CIAF : Current
比重变化量CIAFR : Current

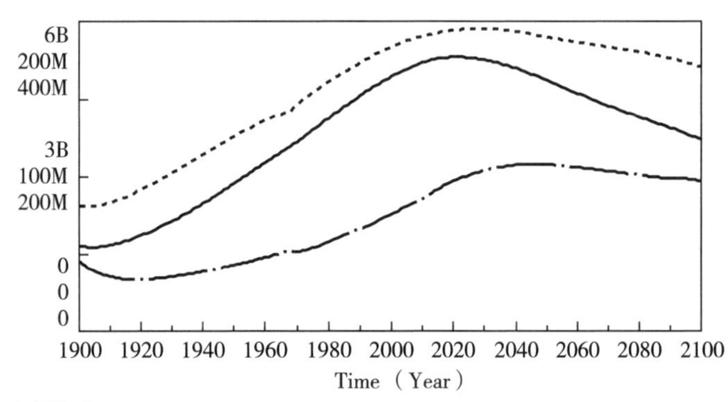
6B
200M
400M
3B
100M
200M
0
0
0
1900 1920 1940 1960 1980 2000 2020 2040 2060 2080 2100
Time（Year）
人口P : Current
年出生人口BR : Current
年死亡人口DR : Current

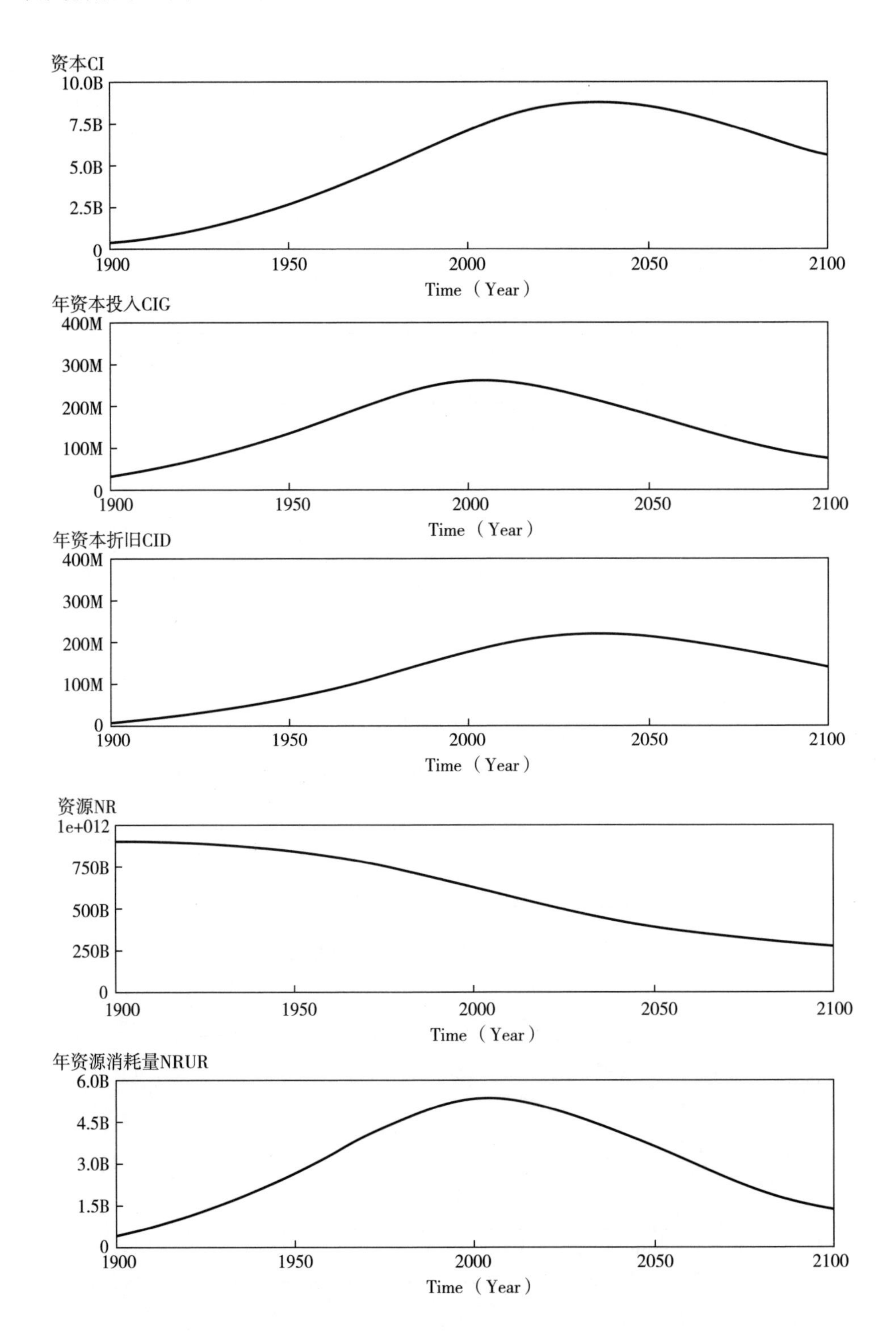

资本CI
10.0B
7.5B
5.0B
2.5B
0
1900
1950
2000
2050
2100
Time（Year）
年资本投入CIG
400M
300M
200M
100M
0
1900
1950
2000
2050
2100
Time（Year）
年资本折旧CID
400M
300M
200M
100M
0
1900
1950
2000
2050
2100
Time（Year）
资源NR
1e+012
750B
500B
250B
0
1900
1950
2000
2050
2100
Time（Year）
年资源消耗量NRUR
6.0B
4.5B
3.0B
1.5B
0
1900
1950
2000
2050
2100
Time（Year）

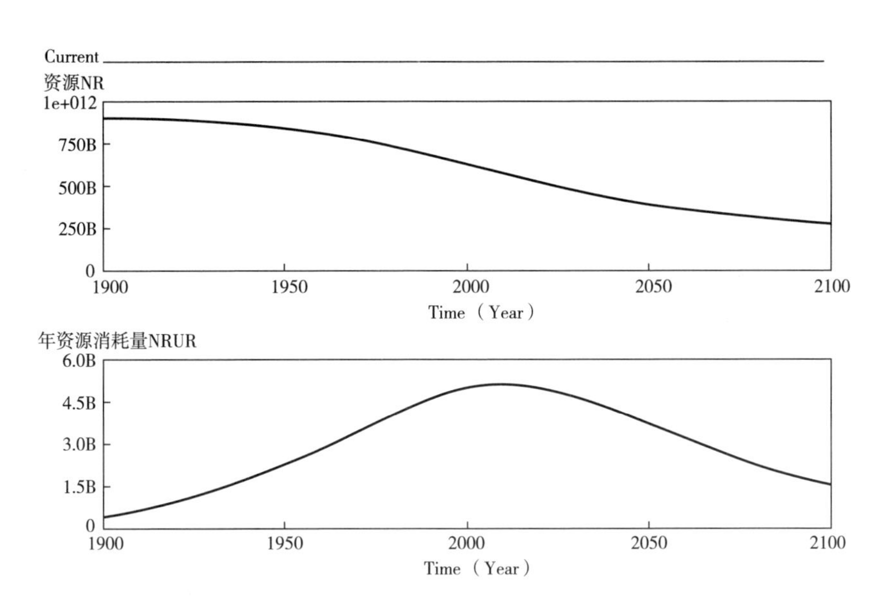

（3）人口 P 及年出生人口 BR、年死亡人口 DR 仿真曲线。

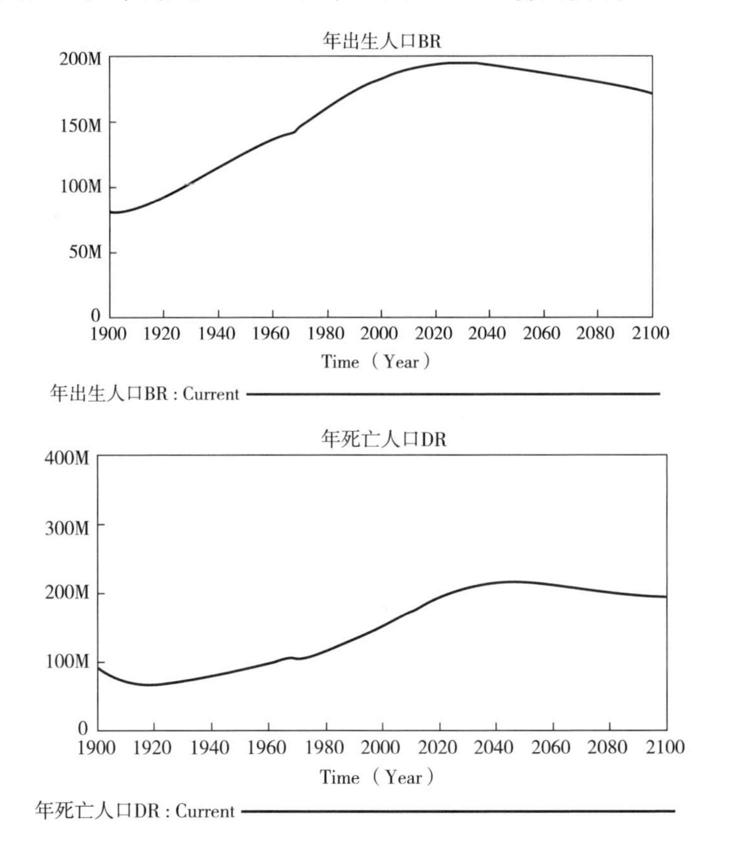

（4）T_1（t）T_2（t）T_3（t）T_4（t）T_5（t）组合仿真检验结果分析。符合实际。

入树 T_1（t）T_2（t）T_3（t）T_4（t）T_5（t）组合通过仿真检验。

上机实验解答 5

1. T_1 入树模型及仿真方程，写出仿真图形 T_1（t）和 R_1（t）。

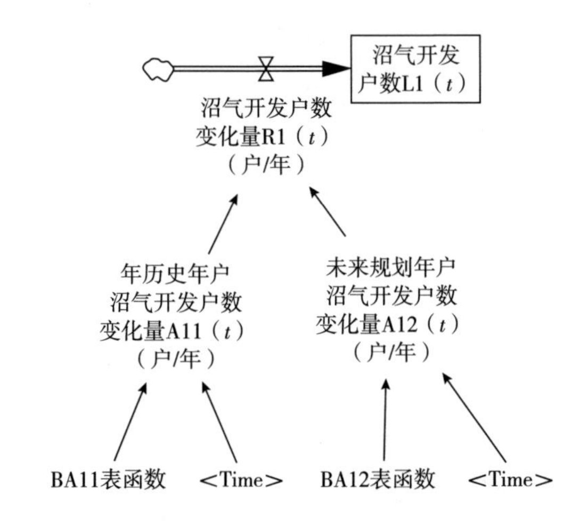

Model Settings - use Sketch to set initial causes

Time Bounds | Info/Pswd | Sketch | Units Equiv | XLS Files | Ref Modes

Time Bounds for Model

INITIAL TIME = 1

FINAL TIME = 15

TIME STEP = 1

☑ Save results every TIME STEP

or use SAVEPER =

Units for Time Year

NOTE: To change later use Model>Settings or edit the equations for the above parameters.

OK　Cancel

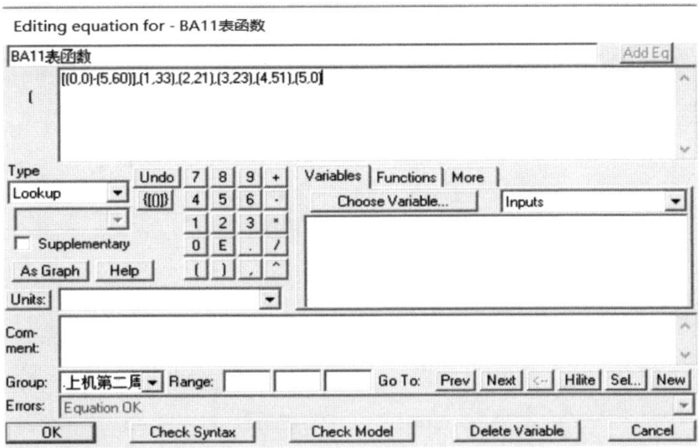

Editing equation for - BA11表函数

BA11表函数

([(0,0)-(5,60)],(1,33),(2,21),(3,23),(4,51),(5,0)

Type: Lookup | Supplementary | As Graph | Help | Units: | Com-ment: | Group: 上机第二届 | Range: | Go To: Prev | Next | Hilite | Sel... | New

Errors: Equation OK

OK | Check Syntax | Check Model | Delete Variable | Cancel

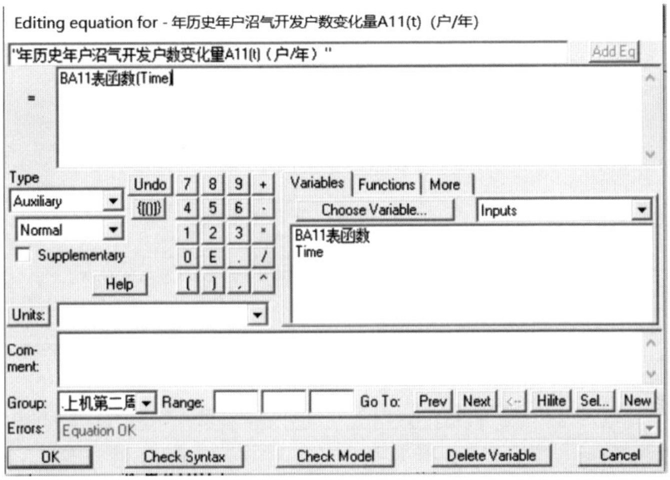

Editing equation for - 年历史年户沼气开发户数变化量A11(t)（户/年）

"年历史年户沼气开发户数变化量A11(t)（户/年）"

= BA11表函数(Time)

Type: Auxiliary | Normal | Supplementary | Help | Variables | Functions | More | Choose Variable... | Inputs

BA11表函数
Time

Units: | Com-ment: | Group: 上机第二届 | Range: | Go To: Prev | Next | Hilite | Sel... | New

Errors: Equation OK

OK | Check Syntax | Check Model | Delete Variable | Cancel

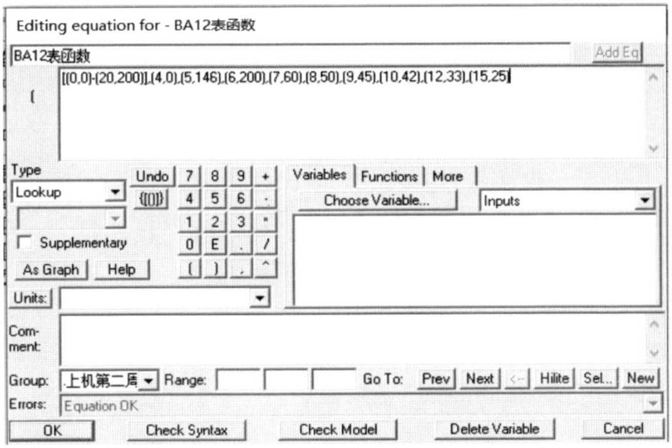

Editing equation for - BA12表函数

BA12表函数

([(0,0)-(20,200)],(4,0),(5,146),(6,200),(7,60),(8,50),(9,45),(10,42),(12,33),(15,25)

Type: Lookup | Supplementary | As Graph | Help | Units: | Com-ment: | Group: 上机第二届 | Range: | Go To: Prev | Next | Hilite | Sel... | New

Errors: Equation OK

OK | Check Syntax | Check Model | Delete Variable | Cancel

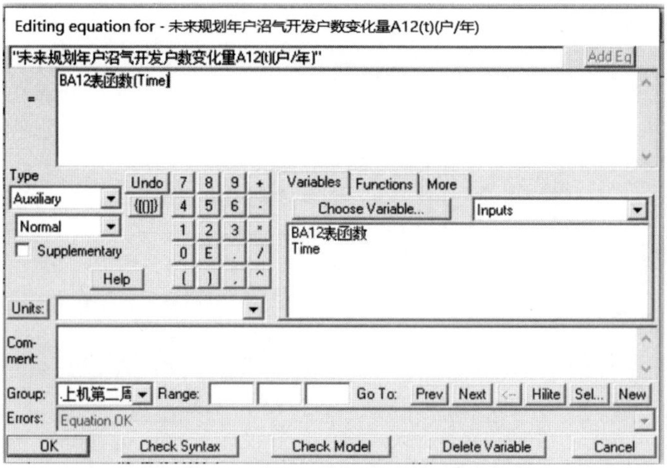

Editing equation for - 未来规划年户沼气开发户数变化量A12(t)(户/年)

"未来规划年户沼气开发户数变化量A12(t)(户/年)"

= BA12表函数(Time)

Type: Auxiliary, Normal, Supplementary

Variables | Functions | More

Choose Variable... Inputs

BA12表函数
Time

Units:

Comment:

Group: 上机第二周 Range: Go To: Prev Next Hilite Sel... New

Errors: Equation OK

OK | Check Syntax | Check Model | Delete Variable | Cancel

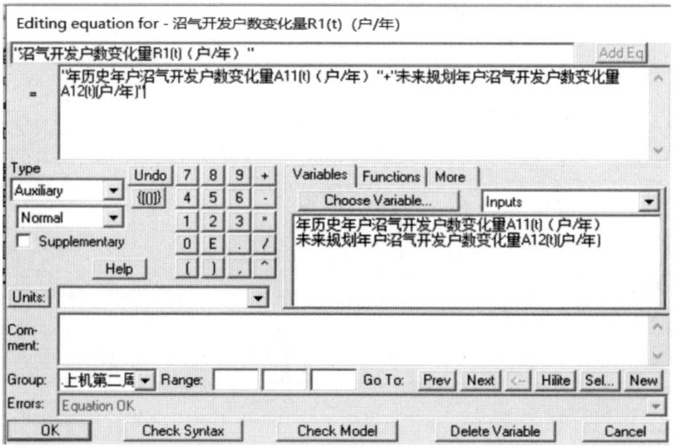

Editing equation for - 沼气开发户数变化量R1(t) (户/年)

"沼气开发户数变化量R1(t)（户/年）"

= "年历史年户沼气开发户数变化量A11(t)（户/年）"+"未来规划年户沼气开发户数变化量A12(t)(户/年)"

Type: Auxiliary, Normal, Supplementary

Variables | Functions | More

Choose Variable... Inputs

年历史年户沼气开发户数变化量A11(t)（户/年）
未来规划年户沼气开发户数变化量A12(t)(户/年)

Units:

Comment:

Group: 上机第二周 Range: Go To: Prev Next Hilite Sel... New

Errors: Equation OK

OK | Check Syntax | Check Model | Delete Variable | Cancel

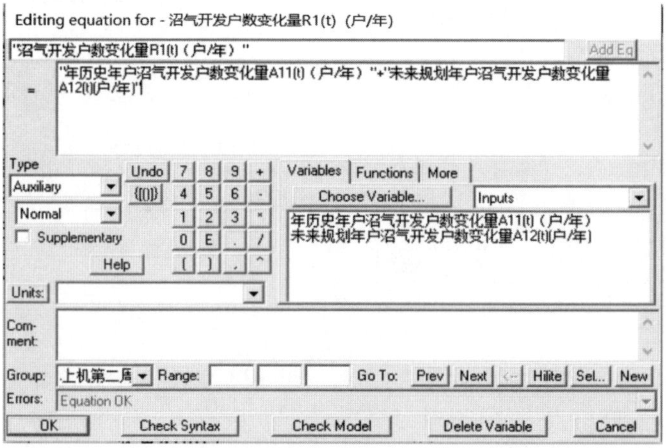

Editing equation for - 沼气开发户数变化量R1(t) (户/年)

"沼气开发户数变化量R1(t)（户/年）"

= "年历史年户沼气开发户数变化量A11(t)（户/年）"+"未来规划年户沼气开发户数变化量A12(t)(户/年)"

Type: Auxiliary, Normal, Supplementary

Variables | Functions | More

Choose Variable... Inputs

年历史年户沼气开发户数变化量A11(t)（户/年）
未来规划年户沼气开发户数变化量A12(t)(户/年)

Units:

Comment:

Group: 上机第二周 Range: Go To: Prev Next Hilite Sel... New

Errors: Equation OK

OK | Check Syntax | Check Model | Delete Variable | Cancel

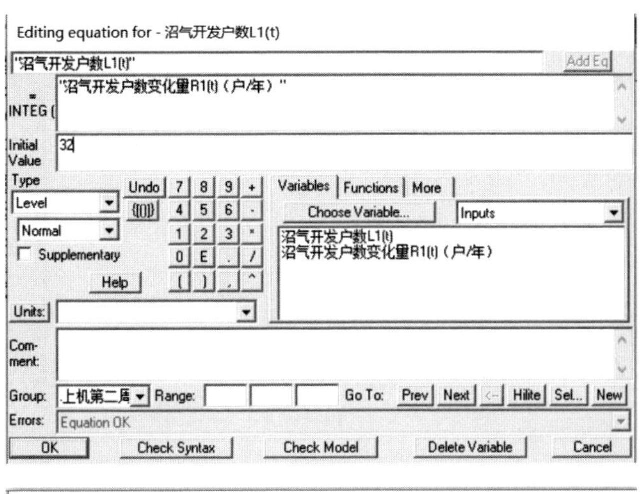
Editing equation for - 沼气开发户数L1(t)
"沼气开发户数L1(t)"
Add Eq
=
INTEG (
"沼气开发户数变化量R1(t)（户/年）"
Initial Value
32
Type
Level
Normal
Supplementary
Undo
Help
Variables
Functions
More
Choose Variable...
Inputs
沼气开发户数L1(t)
沼气开发户数变化量R1(t)（户/年）
Units:
Comment:
Group:
上机第二周
Range:
Go To:
Prev
Next
Hilite
Sel...
New
Errors:
Equation OK
OK
Check Syntax
Check Model
Delete Variable
Cancel

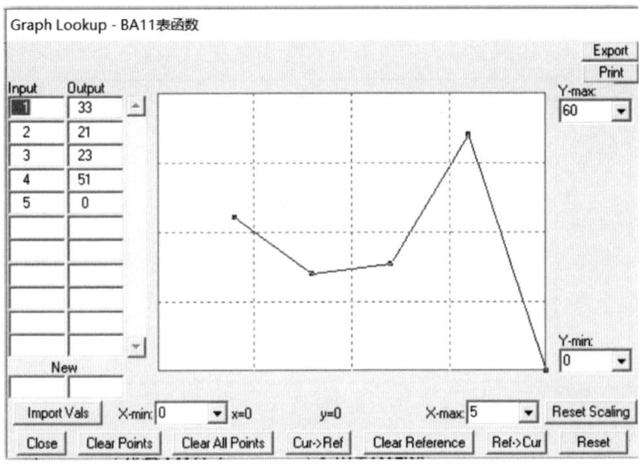
Graph Lookup - BA11表函数
Export
Print
Input
Output
1
33
2
21
3
23
4
51
5
0
New
Y-max:
60
Y-min:
0
Import Vals
X-min:
0
x=0
y=0
X-max:
5
Reset Scaling
Close
Clear Points
Clear All Points
Cur->Ref
Clear Reference
Ref->Cur
Reset

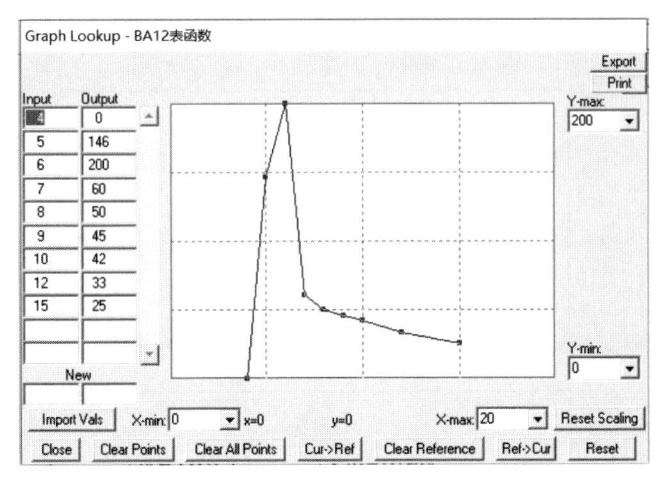
Graph Lookup - BA12表函数
Export
Print
Input
Output
4
0
5
146
6
200
7
60
8
50
9
45
10
42
12
33
15
25
New
Y-max:
200
Y-min:
0
Import Vals
X-min:
0
x=0
y=0
X-max:
20
Reset Scaling
Close
Clear Points
Clear All Points
Cur->Ref
Clear Reference
Ref->Cur
Reset

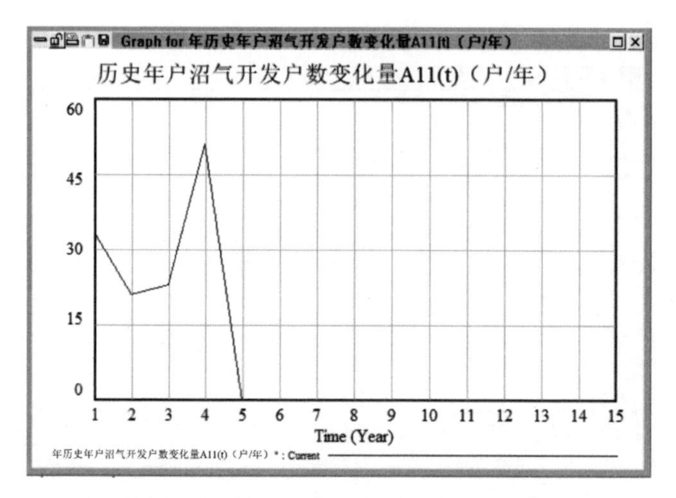
Graph for 年历史年户沼气开发户数变化量A11(t)（户/年）
历史年户沼气开发户数变化量A11(t)（户/年）
60
45
30
15
0
1 2 3 4 5 6 7 8 9 10 11 12 13 14 15
Time (Year)
年历史年户沼气开发户数变化量A11(t)（户/年）*：Current

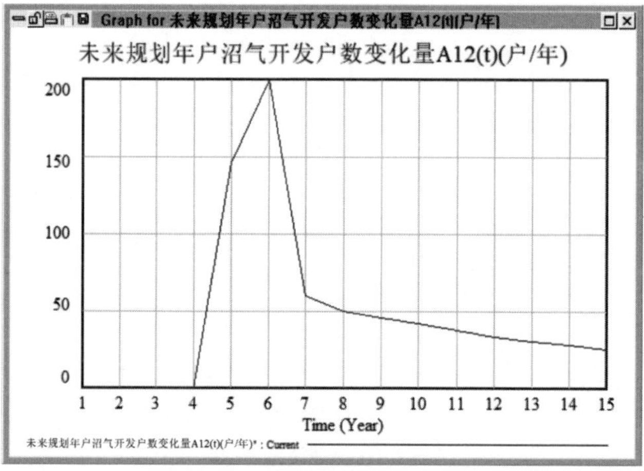
Graph for 未来规划年户沼气开发户数变化量A12(t)(户/年)
未来规划年户沼气开发户数变化量A12(t)(户/年)
200
150
100
50
0
1 2 3 4 5 6 7 8 9 10 11 12 13 14 15
Time (Year)
未来规划年户沼气开发户数变化量A12(t)(户/年)*：Current

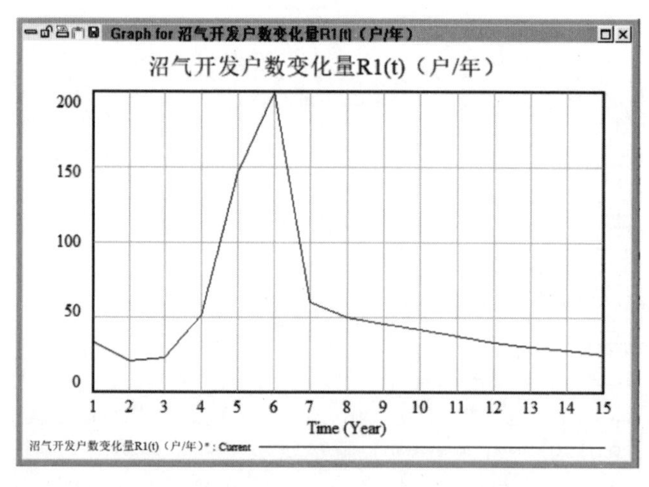
Graph for 沼气开发户数变化量R1(t)（户/年）
沼气开发户数变化量R1(t)（户/年）
200
150
100
50
0
1 2 3 4 5 6 7 8 9 10 11 12 13 14 15
Time (Year)
沼气开发户数变化量R1(t)（户/年）*：Current

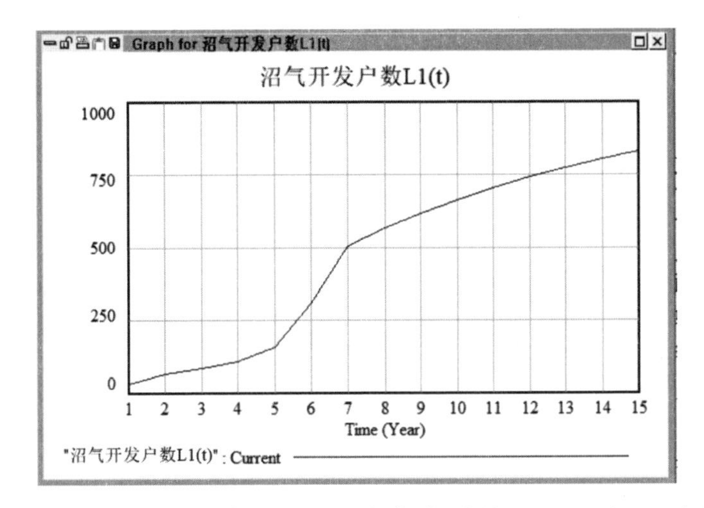

2. $T_1 \sim T_2$ 入树模型及仿真方程，写出仿真图形 T_2（t）和 R_2（t）。

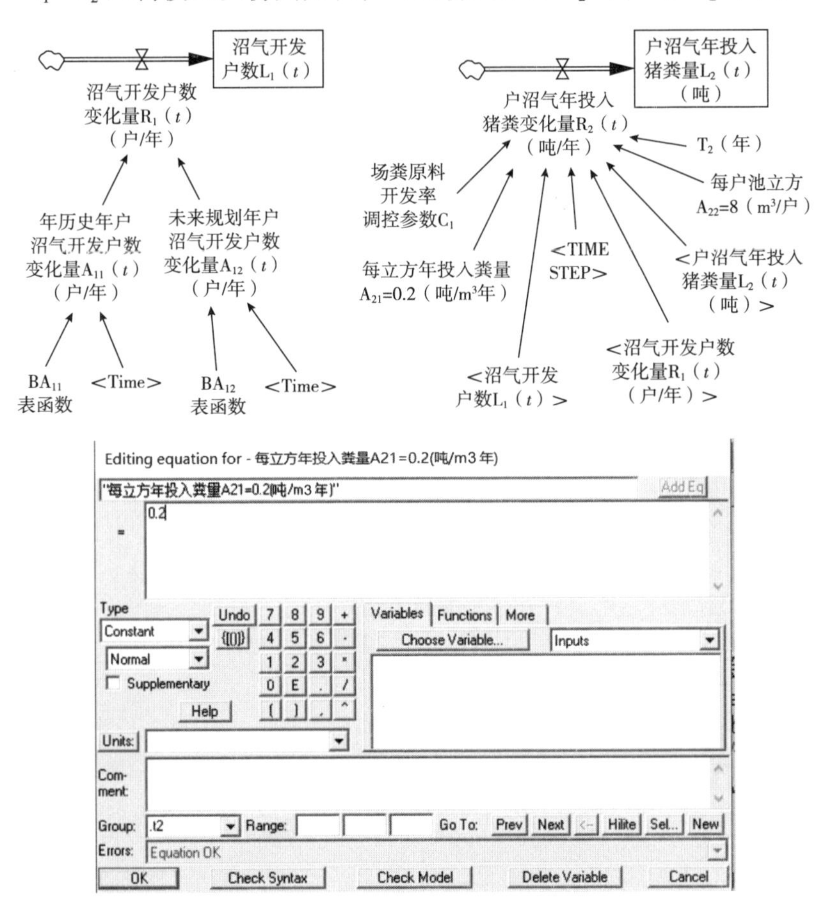

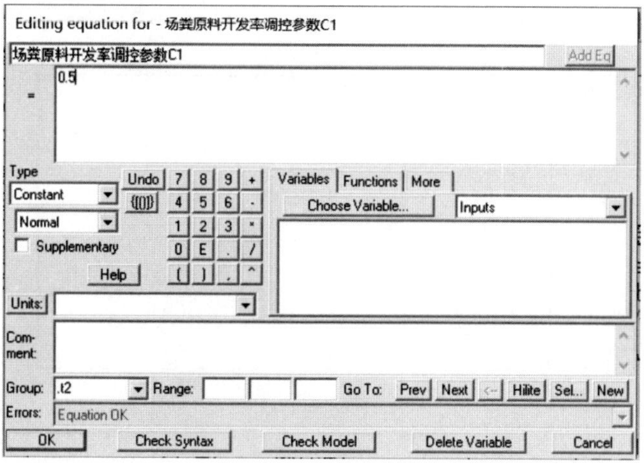

Editing equation for - 场粪原料开发率调控参数C1

场粪原料开发率调控参数C1 Add Eq

= 0.5

Type: Constant, Normal, Supplementary, Help

Undo 7 8 9 + {[()]} 4 5 6 - 1 2 3 * 0 E . / ( ) , ^

Variables Functions More

Choose Variable... Inputs

Units:

Comment:

Group: .t2 Range: Go To: Prev Next <- Hilite Sel... New

Errors: Equation OK

OK Check Syntax Check Model Delete Variable Cancel

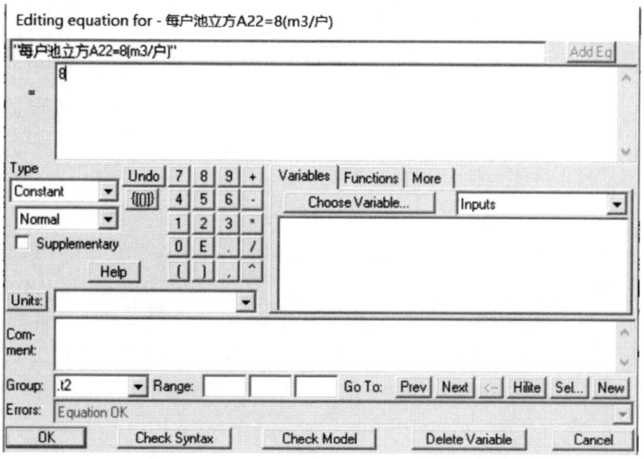

Editing equation for - 每户池立方A22=8(m3/户)

"每户池立方A22=8(m3/户)" Add Eq

= 8

Type: Constant, Normal, Supplementary, Help

Undo 7 8 9 + {[()]} 4 5 6 - 1 2 3 * 0 E . / ( ) , ^

Variables Functions More

Choose Variable... Inputs

Units:

Comment:

Group: .t2 Range: Go To: Prev Next <- Hilite Sel... New

Errors: Equation OK

OK Check Syntax Check Model Delete Variable Cancel

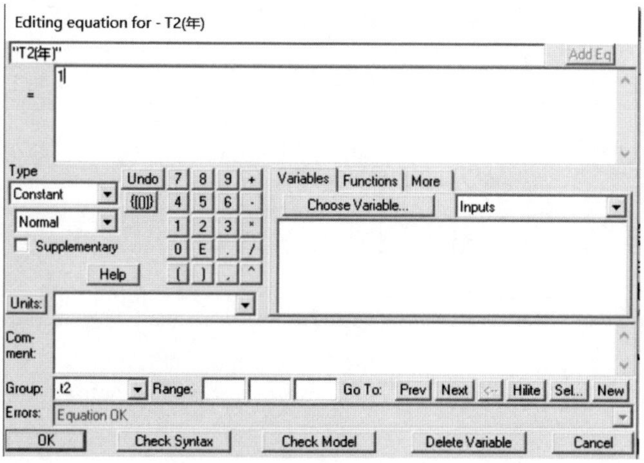

Editing equation for - T2(年)

"T2(年)" Add Eq

= 1

Type: Constant, Normal, Supplementary, Help

Undo 7 8 9 + {[()]} 4 5 6 - 1 2 3 * 0 E . / ( ) , ^

Variables Functions More

Choose Variable... Inputs

Units:

Comment:

Group: .t2 Range: Go To: Prev Next <- Hilite Sel... New

Errors: Equation OK

OK Check Syntax Check Model Delete Variable Cancel

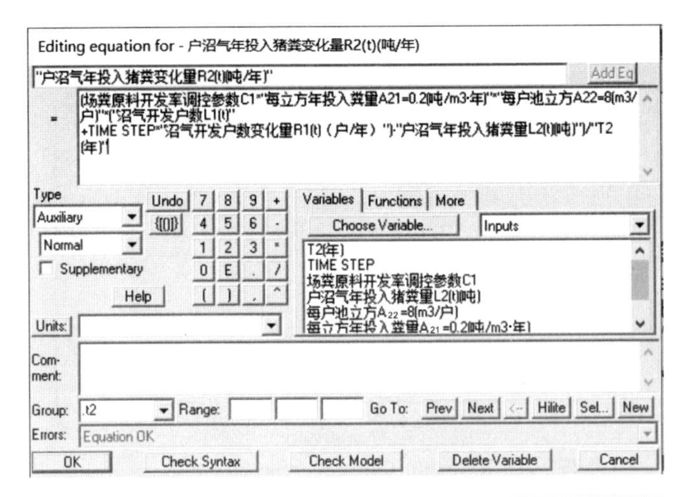
Editing equation for - 户沼气年投入猪粪变化量R2(t)(吨/年)
"户沼气年投入猪粪变化量R2(t)(吨/年)"
Add Eq
(场粪原料开发率调控参数C1*"每立方年投入粪量A21=0.2(吨/m3·年)"*"每户池立方A22=8(m3/户)"*("沼气开发户数L1(t)"
+TIME STEP*"沼气开发户数变化量R1(t)（户/年）")-"户沼气年投入猪粪量L2(t)(吨)")/"T2(年)"
Type
Auxiliary
Normal
Supplementary
Help
Undo
Variables
Functions
More
Choose Variable...
Inputs
T2(年)
TIME STEP
场粪原料开发率调控参数C1
户沼气年投入猪粪量L2(t)(吨)
Units:
Com-
ment:
Group:
.t2
Range:
Go To:
Prev
Next
Hilite
Sel...
New
Errors:
Equation OK
OK
Check Syntax
Check Model
Delete Variable
Cancel

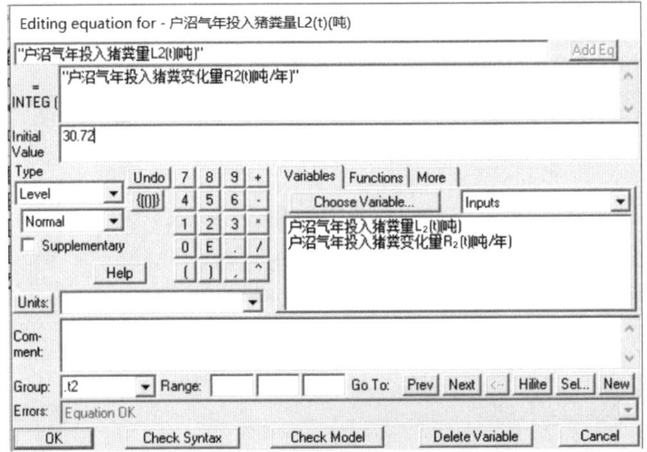
Editing equation for - 户沼气年投入猪粪量L2(t)(吨)
"户沼气年投入猪粪量L2(t)(吨)"
Add Eq
"户沼气年投入猪粪变化量R2(t)(吨/年)"
= INTEG (
Initial Value
30.72
Type
Level
Normal
Supplementary
Help
Undo
Variables
Functions
More
Choose Variable...
Inputs
户沼气年投入猪粪量L2(t)(吨)
户沼气年投入猪粪变化量R2(t)(吨/年)
Units:
Com-
ment:
Group:
.t2
Range:
Go To:
Prev
Next
Hilite
Sel...
New
Errors:
Equation OK
OK
Check Syntax
Check Model
Delete Variable
Cancel

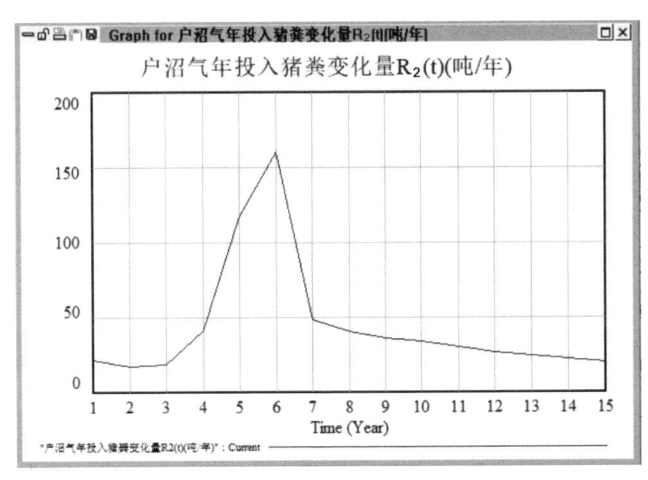
Graph for 户沼气年投入猪粪变化量R2(t)(吨/年)
户沼气年投入猪粪变化量R2(t)(吨/年)
200
150
100
50
0
1
2
3
4
5
6
7
8
9
10
11
12
13
14
15
Time (Year)
"户沼气年投入猪粪变化量R2(t)(吨/年)" : Current

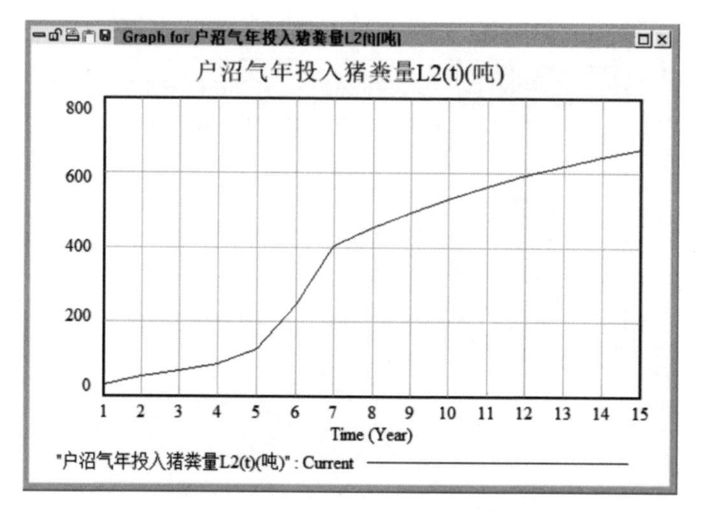

3. $T_1 \sim T_3$ 入树模型及仿真方程，写出仿真图形 T_3（t）和 R_3（t）。

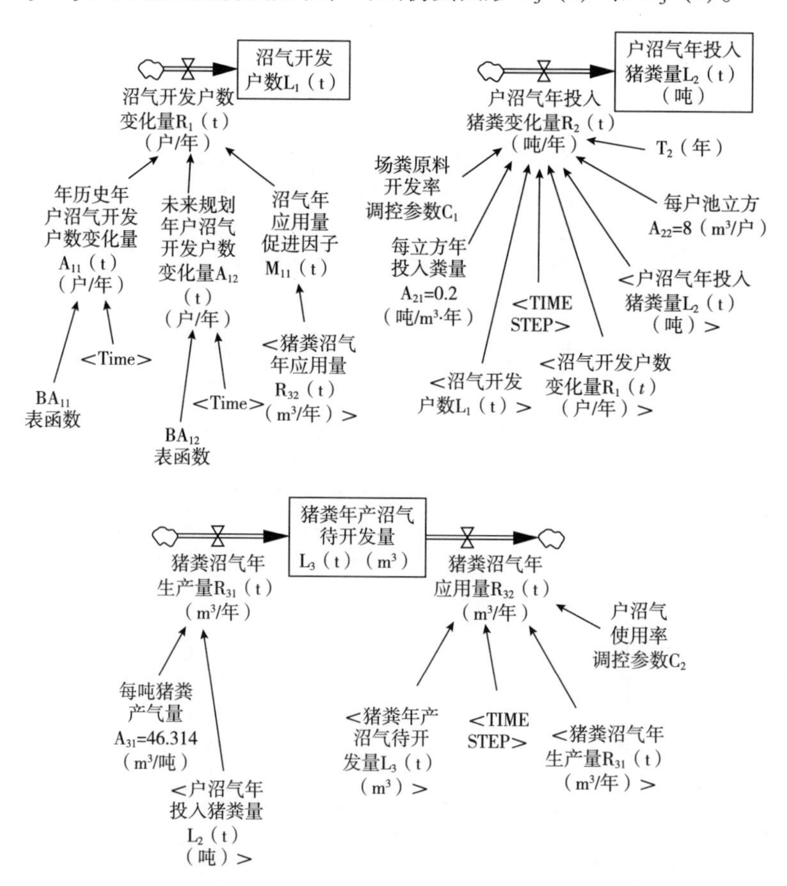

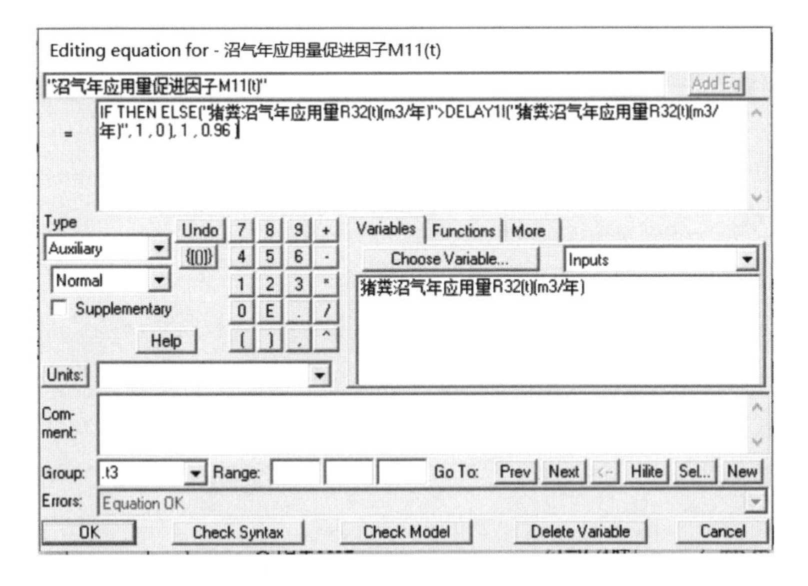
Editing equation for - 沼气年应用量促进因子M11(t)
"沼气年应用量促进因子M11(t)"
Add Eq
IF THEN ELSE("猪粪沼气年应用量R32(t)(m3/年)">DELAY1I("猪粪沼气年应用量R32(t)(m3/年)", 1 , 0), 1 , 0.96)
Type
Auxiliary
Normal
Supplementary
Help
Undo
Variables
Functions
More
Choose Variable...
Inputs
猪粪沼气年应用量R32(t)(m3/年)
Units:
Com-ment:
Group:
.t3
Range:
Go To:
Prev
Next
Hilite
Sel...
New
Errors:
Equation OK
OK
Check Syntax
Check Model
Delete Variable
Cancel

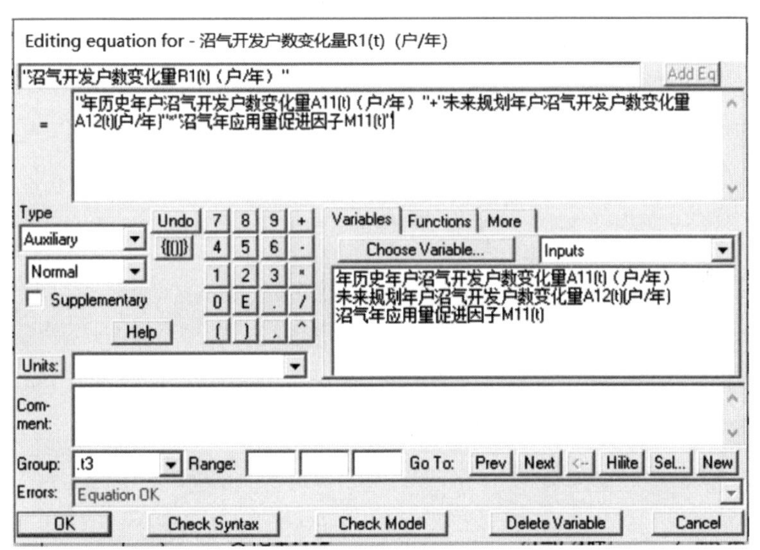
Editing equation for - 沼气开发户数变化量R1(t)（户/年）
"沼气开发户数变化量R1(t)（户/年）"
Add Eq
"年历史年户沼气开发户数变化量A11(t)（户/年）"+"未来规划年户沼气开发户数变化量A12(t)(户/年)"*"沼气年应用量促进因子M11(t)"
Type
Auxiliary
Normal
Supplementary
Help
Undo
Variables
Functions
More
Choose Variable...
Inputs
年历史年户沼气开发户数变化量A11(t)（户/年）
未来规划年户沼气开发户数变化量A12(t)(户/年)
沼气年应用量促进因子M11(t)
Units:
Com-ment:
Group:
.t3
Range:
Go To:
Prev
Next
Hilite
Sel...
New
Errors:
Equation OK
OK
Check Syntax
Check Model
Delete Variable
Cancel

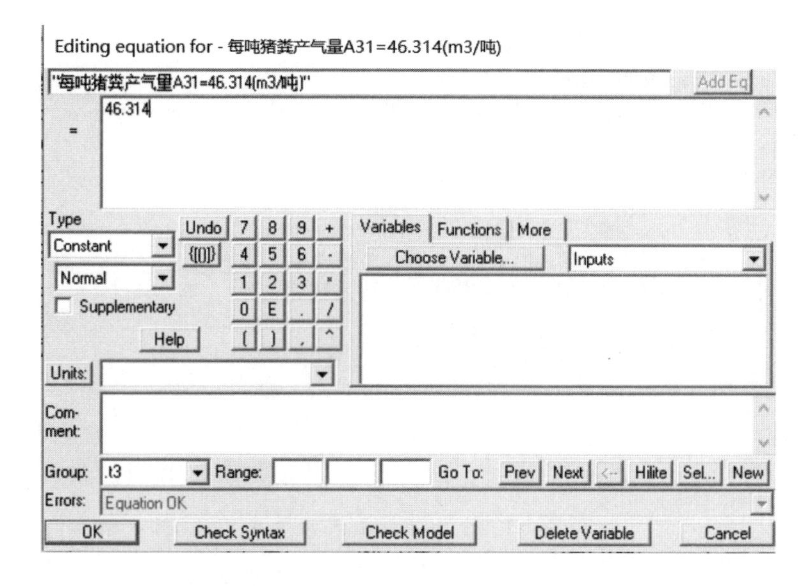
Editing equation for - 每吨猪粪产气量A31=46.314(m3/吨)
"每吨猪粪产气量A31=46.314(m3/吨)"
Add Eq
=
46.314
Type
Constant
Normal
Supplementary
Undo
7 8 9 +
4 5 6 -
1 2 3 *
0 E . /
( ) , ^
Help
Variables
Functions
More
Choose Variable...
Inputs
Units:
Com-
ment:
Group:
.t3
Range:
Go To:
Prev
Next
Hilite
Sel...
New
Errors:
Equation OK
OK
Check Syntax
Check Model
Delete Variable
Cancel

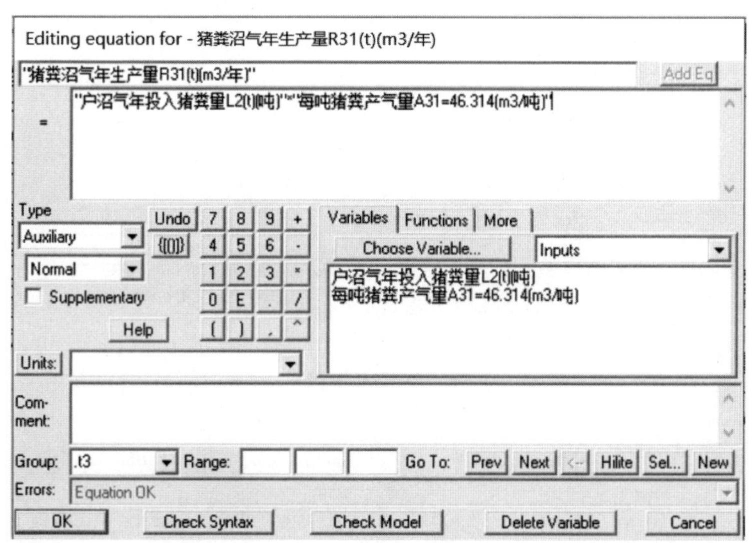
Editing equation for - 猪粪沼气年生产量R31(t)(m3/年)
"猪粪沼气年生产量R31(t)(m3/年)"
Add Eq
=
"户沼气年投入猪粪量L2(t)(吨)"*"每吨猪粪产气量A31=46.314(m3/吨)"
Type
Auxiliary
Normal
Supplementary
Undo
7 8 9 +
4 5 6 -
1 2 3 *
0 E . /
( ) , ^
Help
Variables
Functions
More
Choose Variable...
Inputs
户沼气年投入猪粪量L2(t)(吨)
每吨猪粪产气量A31=46.314(m3/吨)
Units:
Com-
ment:
Group:
.t3
Range:
Go To:
Prev
Next
Hilite
Sel...
New
Errors:
Equation OK
OK
Check Syntax
Check Model
Delete Variable
Cancel

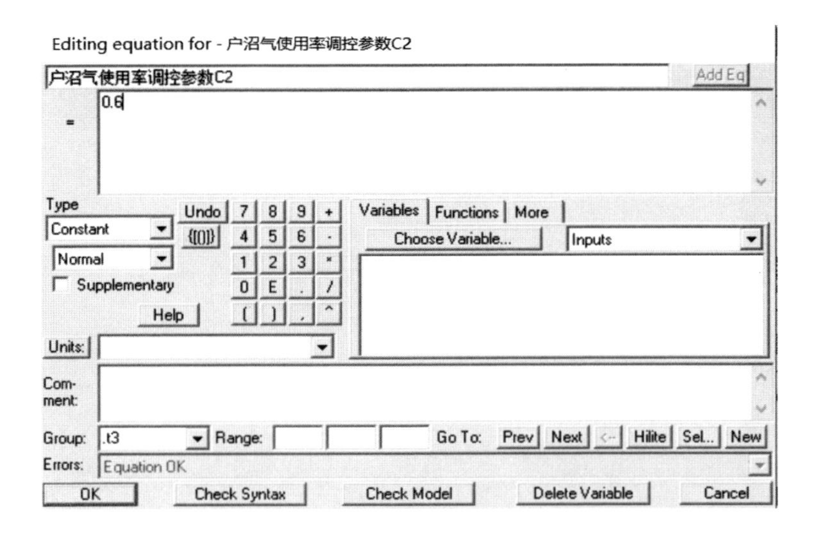

Editing equation for - 户沼气使用率调控参数C2
户沼气使用率调控参数C2
Add Eq
=
0.6
Type
Constant
Normal
Supplementary
Undo
Help
Variables
Functions
More
Choose Variable...
Inputs
Units:
Com-
ment:
Group:
.t3
Range:
Go To:
Prev
Next
Hilite
Sel...
New
Errors:
Equation OK
OK
Check Syntax
Check Model
Delete Variable
Cancel

Editing equation for - 猪粪沼气年应用量R32(t)(m3/年)
"猪粪沼气年应用量R32(t)(m3/年)"
Add Eq
=
("猪粪年产沼气待开发量L3(t)(m3)"+TIME STEP*"猪粪沼气年生产量R31(t)(m3/年)")*户沼气使用率调控参数C2
Type
Auxiliary
Normal
Supplementary
Undo
Help
Variables
Functions
More
Choose Variable...
Inputs
TIME STEP
户沼气使用率调控参数C2
猪粪年产沼气待开发量L3(t)(m3)
猪粪沼气年生产量R31(t)(m3/年)
Units:
Com-
ment:
Group:
.t3
Range:
Go To:
Prev
Next
Hilite
Sel...
New
Errors:
Equation OK
OK
Check Syntax
Check Model
Delete Variable
Cancel

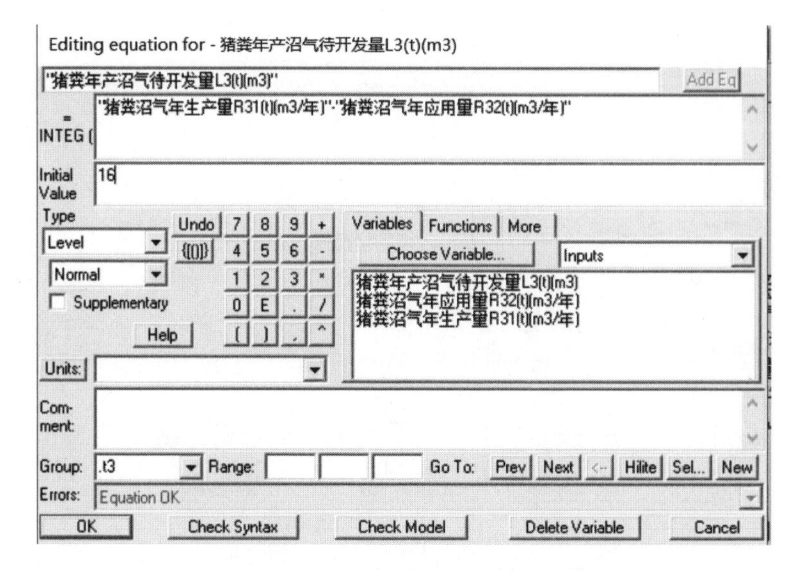
Editing equation for - 猪粪年产沼气待开发量L3(t)(m3)
"猪粪年产沼气待开发量L3(t)(m3)"
Add Eq
= INTEG (
"猪粪沼气年生产量R31(t)(m3/年)"-"猪粪沼气年应用量R32(t)(m3/年)"
Initial Value
16
Type
Level
Normal
Supplementary
Undo
Help
Variables
Functions
More
Choose Variable...
Inputs
猪粪年产沼气待开发量L3(t)(m3)
猪粪沼气年应用量R32(t)(m3/年)
猪粪沼气年生产量R31(t)(m3/年)
Units:
Com-ment:
Group:
.t3
Range:
Go To:
Prev
Next
Hilite
Sel...
New
Errors:
Equation OK
OK
Check Syntax
Check Model
Delete Variable
Cancel

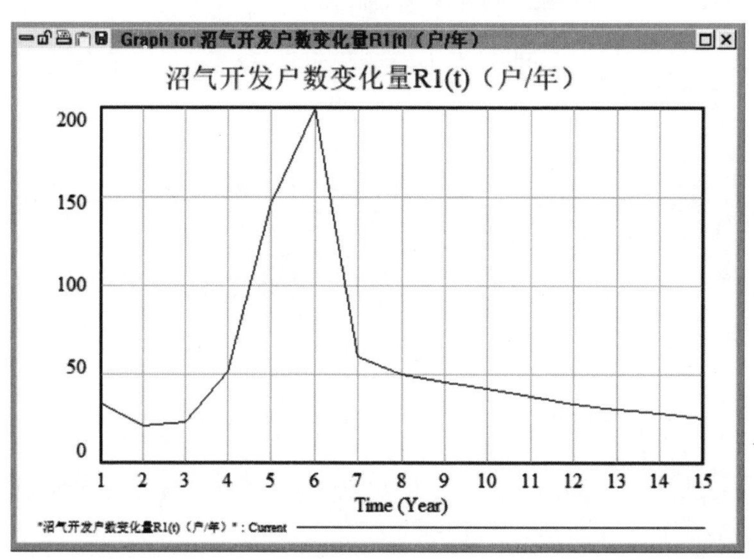
Graph for 沼气开发户数变化量R1(t)（户/年）
沼气开发户数变化量R1(t)（户/年）
200
150
100
50
0
1
2
3
4
5
6
7
8
9
10
11
12
13
14
15
Time (Year)
"沼气开发户数变化量R1(t)（户/年）" : Current

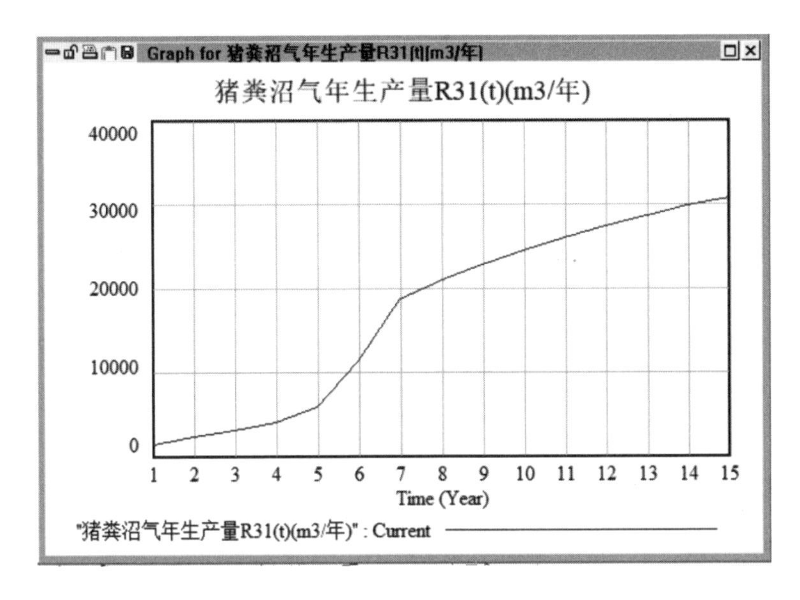
Graph for 猪粪沼气年生产量R31(t)(m3/年)
猪粪沼气年生产量R31(t)(m3/年)
40000
30000
20000
10000
0
1 2 3 4 5 6 7 8 9 10 11 12 13 14 15
Time (Year)
"猪粪沼气年生产量R31(t)(m3/年)" : Current

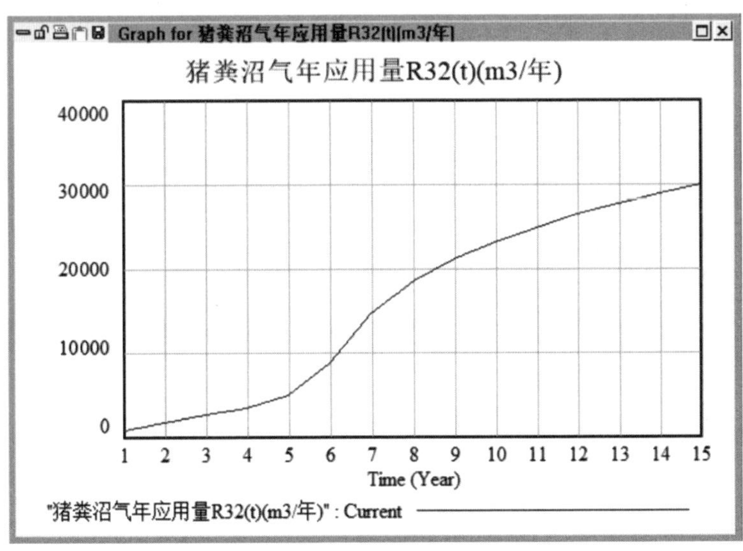
Graph for 猪粪沼气年应用量R32(t)(m3/年)
猪粪沼气年应用量R32(t)(m3/年)
40000
30000
20000
10000
0
1 2 3 4 5 6 7 8 9 10 11 12 13 14 15
Time (Year)
"猪粪沼气年应用量R32(t)(m3/年)" : Current

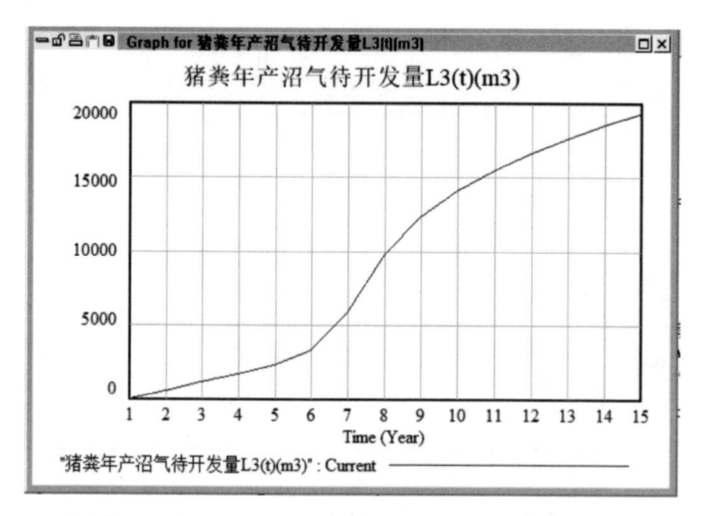

4. $T_1 \sim T_7$ 入树模型及仿真方程，写出仿真图形 T_7（t）和 R_7（t）。

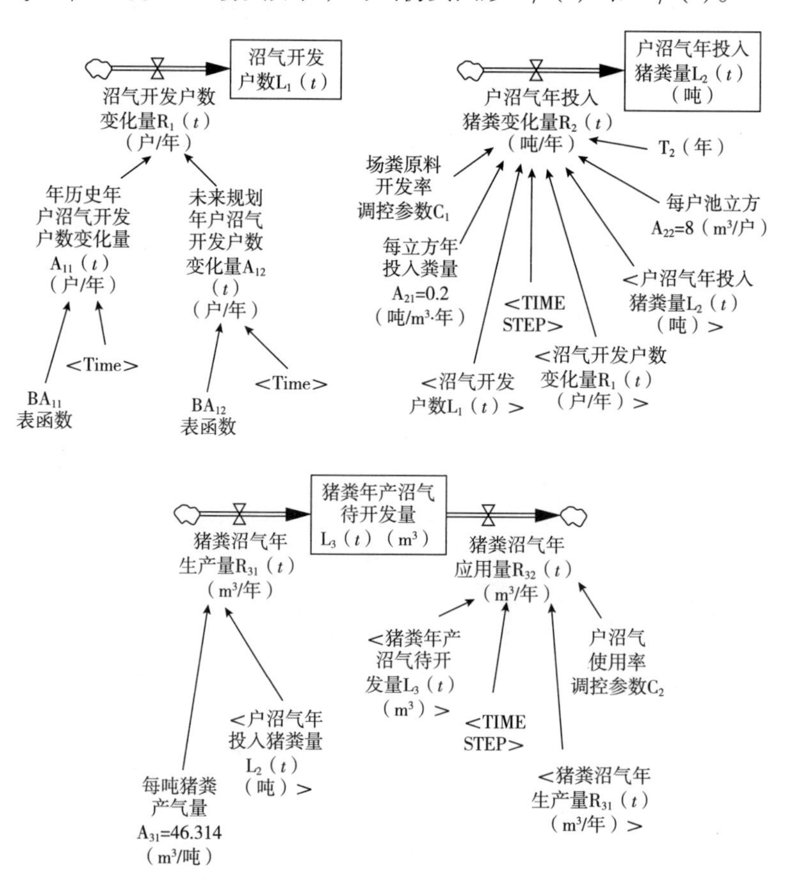

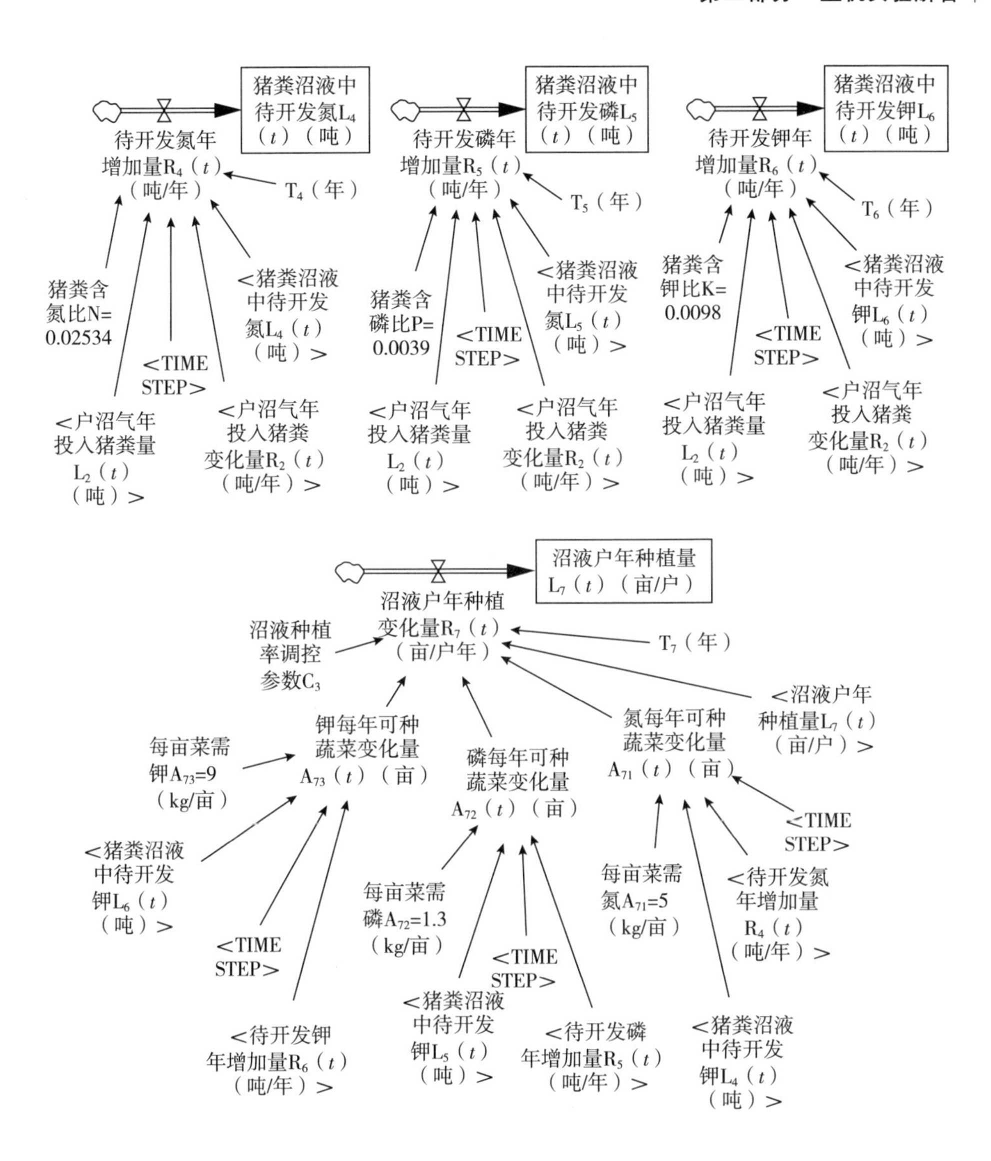

猪粪沼液中待开发氮L_4（t）（吨）
待开发氮年增加量R_4（t）（吨/年）
T_4（年）
猪粪含氮比N=0.02534
<TIME STEP>
<猪粪沼液中待开发氮L_4（t）（吨）>
<户沼气年投入猪粪量L_2（t）（吨）>
<户沼气年投入猪粪变化量R_2（t）（吨/年）>
猪粪沼液中待开发磷L_5（t）（吨）
待开发磷年增加量R_5（t）（吨/年）
T_5（年）
猪粪含磷比P=0.0039
<TIME STEP>
<猪粪沼液中待开发氮L_5（t）（吨）>
<户沼气年投入猪粪量L_2（t）（吨）>
<户沼气年投入猪粪变化量R_2（t）（吨/年）>
猪粪沼液中待开发钾L_6（t）（吨）
待开发钾年增加量R_6（t）（吨/年）
T_6（年）
猪粪含钾比K=0.0098
<TIME STEP>
<猪粪沼液中待开发钾L_6（t）（吨）>
<户沼气年投入猪粪量L_2（t）（吨）>
<户沼气年投入猪粪变化量R_2（t）（吨/年）>
沼液户年种植量L_7（t）（亩/户）
沼液户年种植变化量R_7（t）（亩/户年）
沼液种植率调控参数C_3
T_7（年）
<沼液户年种植量L_7（t）（亩/户）>
钾每年可种蔬菜变化量A_{73}（t）（亩）
磷每年可种蔬菜变化量A_{72}（t）（亩）
氮每年可种蔬菜变化量A_{71}（t）（亩）
每亩菜需钾A_{73}=9（kg/亩）
<猪粪沼液中待开发钾L_6（t）（吨）>
<TIME STEP>
<待开发钾年增加量R_6（t）（吨/年）>
每亩菜需磷A_{72}=1.3（kg/亩）
<TIME STEP>
<猪粪沼液中待开发钾L_5（t）（吨）>
<待开发磷年增加量R_5（t）（吨/年）>
每亩菜需氮A_{71}=5（kg/亩）
<TIME STEP>
<待开发氮年增加量R_4（t）（吨/年）>
<猪粪沼液中待开发钾L_4（t）（吨）>

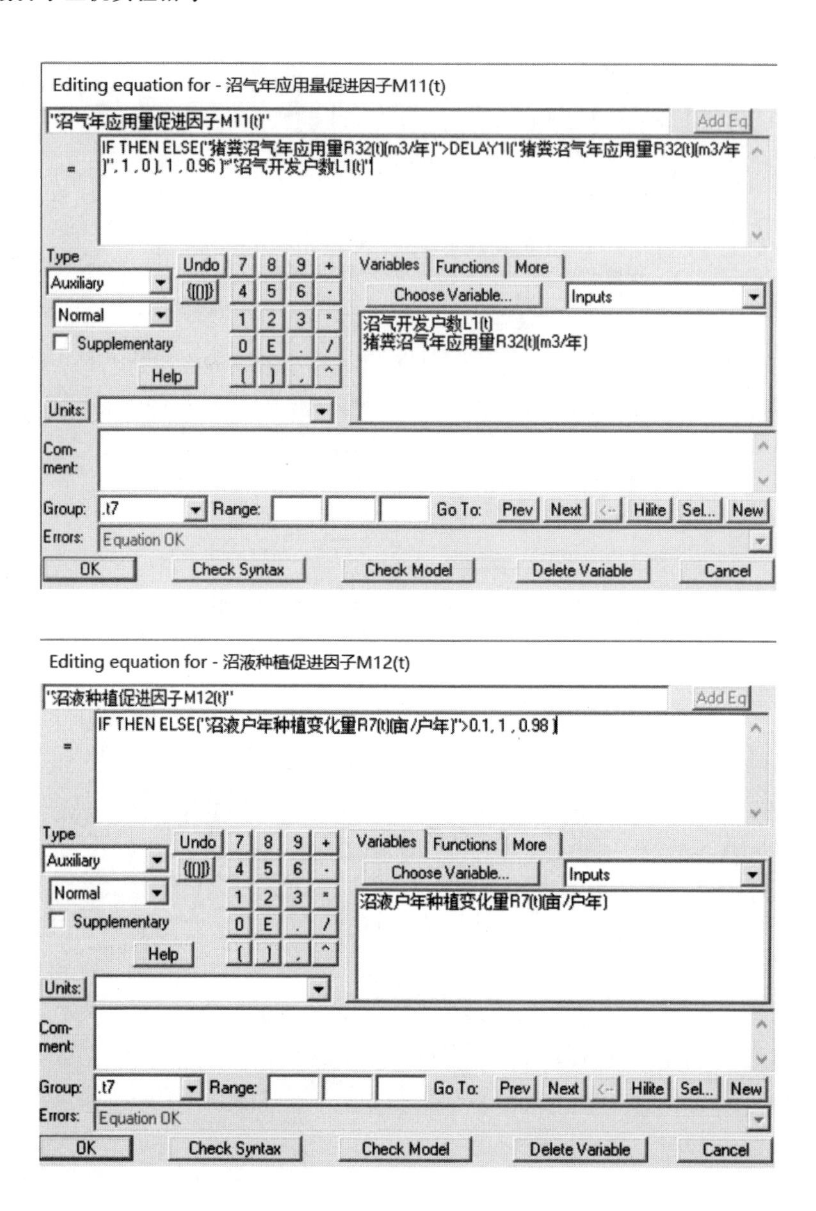
Editing equation for - 沼气年应用量促进因子M11(t)
"沼气年应用量促进因子M11(t)"
Add Eq
IF THEN ELSE("猪粪沼气年应用量R32(t)(m3/年)">DELAY1I("猪粪沼气年应用量R32(t)(m3/年)", 1 , 0), 1 , 0.96)*"沼气开发户数L1(t)"
Type
Auxiliary
Normal
Supplementary
Help
Undo
Variables
Functions
More
Choose Variable...
Inputs
沼气开发户数L1(t)
猪粪沼气年应用量R32(t)(m3/年)
Units:
Com-ment:
Group:
.t7
Range:
Go To:
Prev
Next
Hilite
Sel...
New
Errors:
Equation OK
OK
Check Syntax
Check Model
Delete Variable
Cancel
Editing equation for - 沼液种植促进因子M12(t)
"沼液种植促进因子M12(t)"
Add Eq
IF THEN ELSE("沼液户年种植变化量R7(t)(亩/户年)">0.1, 1 , 0.98)
Type
Auxiliary
Normal
Supplementary
Help
Undo
Variables
Functions
More
Choose Variable...
Inputs
沼液户年种植变化量R7(t)(亩/户年)
Units:
Com-ment:
Group:
.t7
Range:
Go To:
Prev
Next
Hilite
Sel...
New
Errors:
Equation OK
OK
Check Syntax
Check Model
Delete Variable
Cancel

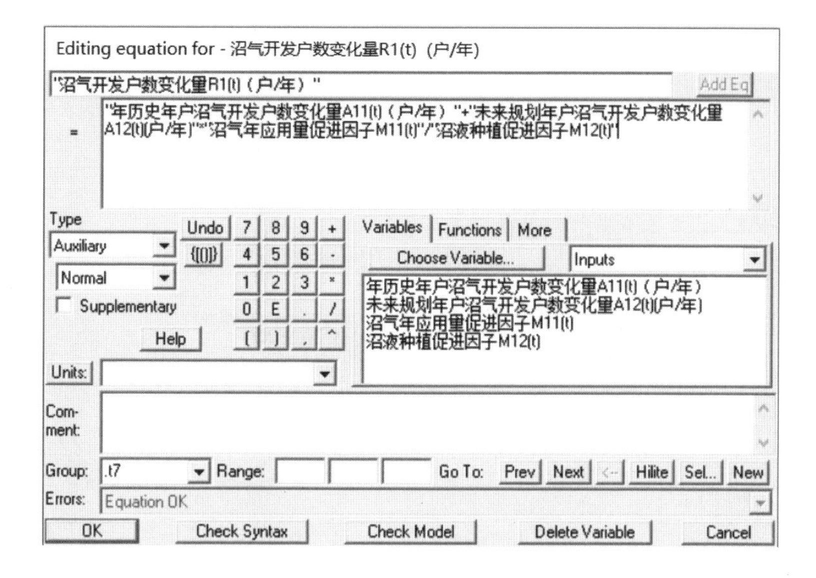
Editing equation for - 沼气开发户数变化量R1(t) (户/年)
"沼气开发户数变化量R1(t)（户/年）"
Add Eq
=
"年历史年户沼气开发户数变化量A11(t)（户/年）"+"未来规划年户沼气开发户数变化量A12(t)(户/年)"*"沼气年应用量促进因子M11(t)"/"沼液种植促进因子M12(t)"
Type
Auxiliary
Normal
Supplementary
Help
Undo
Variables
Functions
More
Choose Variable...
Inputs
年历史年户沼气开发户数变化量A11(t)（户/年）
未来规划年户沼气开发户数变化量A12(t)(户/年)
沼气年应用量促进因子M11(t)
沼液种植促进因子M12(t)
Units:
Com-
ment:
Group:
.t7
Range:
Go To:
Prev
Next
Hilite
Sel...
New
Errors:
Equation OK
OK
Check Syntax
Check Model
Delete Variable
Cancel

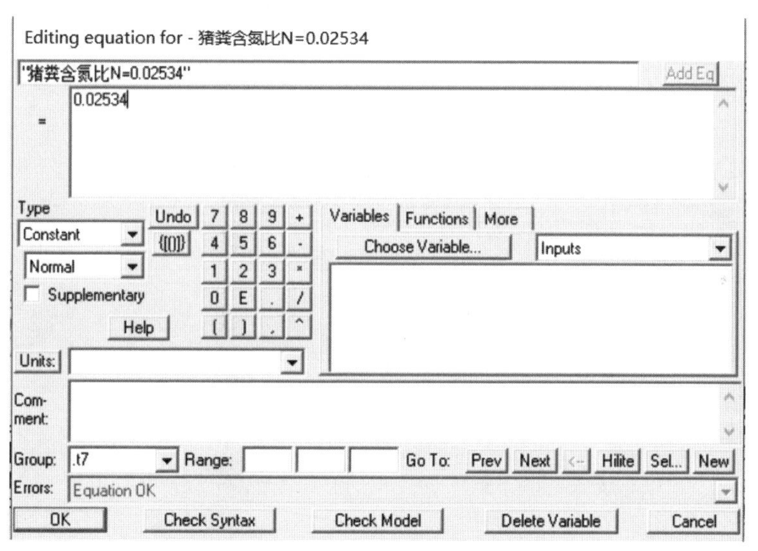
Editing equation for - 猪粪含氮比N=0.02534
"猪粪含氮比N=0.02534"
Add Eq
=
0.02534
Type
Constant
Normal
Supplementary
Help
Undo
Variables
Functions
More
Choose Variable...
Inputs
Units:
Com-
ment:
Group:
.t7
Range:
Go To:
Prev
Next
Hilite
Sel...
New
Errors:
Equation OK
OK
Check Syntax
Check Model
Delete Variable
Cancel

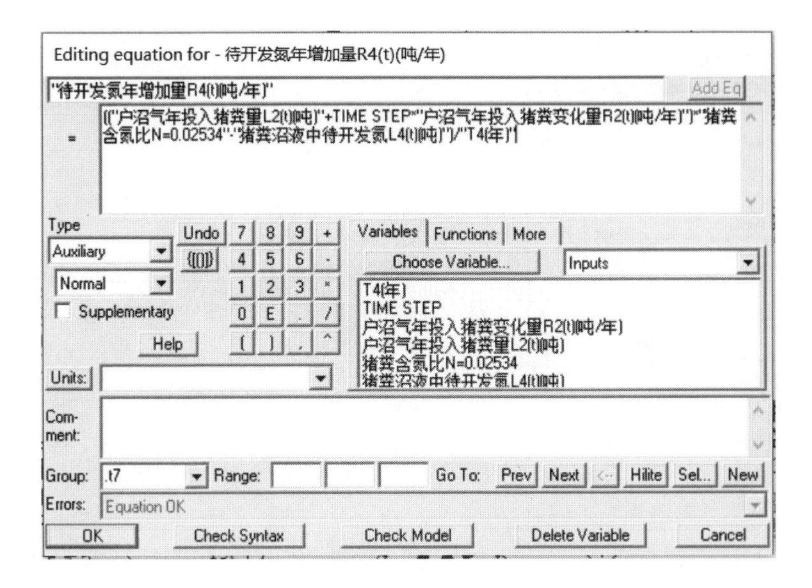
Editing equation for - 待开发氮年增加量R4(t)(吨/年)
"待开发氮年增加量R4(t)(吨/年)"
(("户沼气年投入猪粪量L2(t)(吨)"+TIME STEP*"户沼气年投入猪粪变化量R2(t)(吨/年)")*"猪粪含氮比N=0.02534"-"猪粪沼液中待开发氮L4(t)(吨)")/"T4(年)"
Type
Auxiliary
Normal
Supplementary
Help
Units:
Variables
Functions
More
Choose Variable...
Inputs
T4(年)
TIME STEP
户沼气年投入猪粪变化量R2(t)(吨/年)
户沼气年投入猪粪量L2(t)(吨)
猪粪含氮比N=0.02534
猪粪沼液中待开发氮L4(t)(吨)
Comment:
Group: .t7
Range:
Go To:
Errors: Equation OK
OK
Check Syntax
Check Model
Delete Variable
Cancel

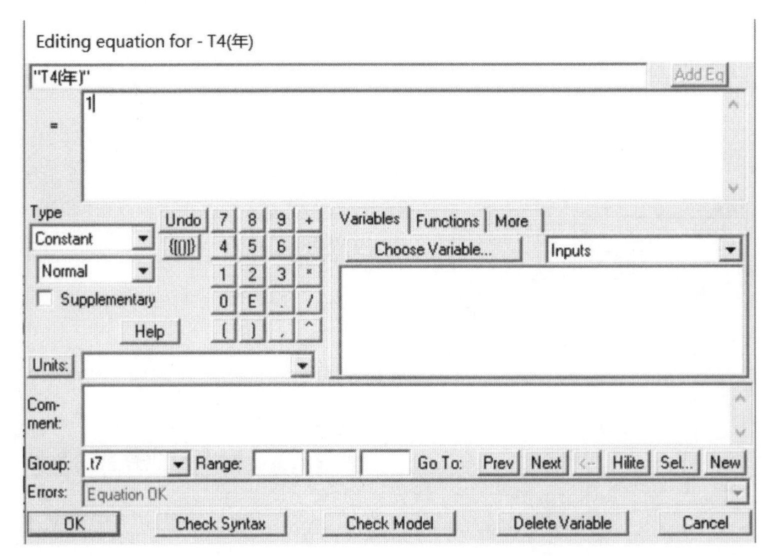
Editing equation for - T4(年)
"T4(年)"
1
Type
Constant
Normal
Supplementary
Help
Units:
Variables
Functions
More
Choose Variable...
Inputs
Comment:
Group: .t7
Range:
Go To:
Errors: Equation OK
OK
Check Syntax
Check Model
Delete Variable
Cancel

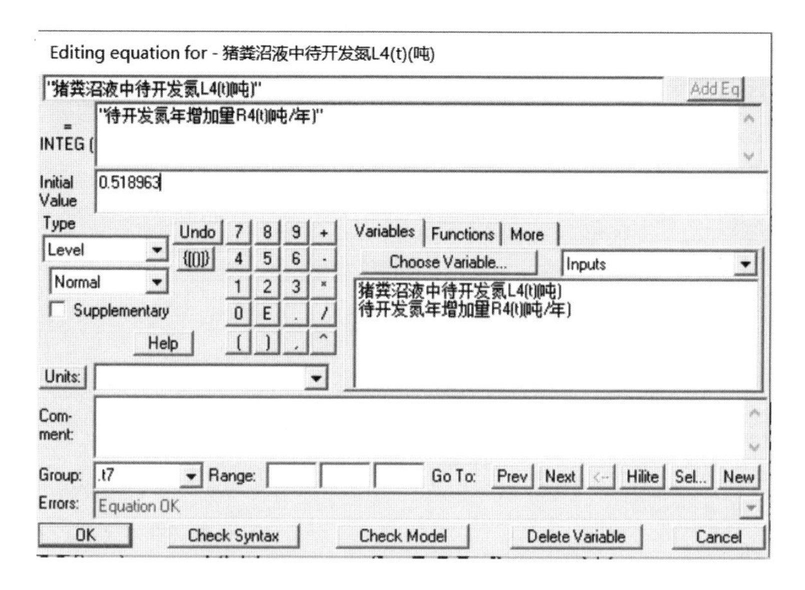
Editing equation for - 猪粪沼液中待开发氮L4(t)(吨)
"猪粪沼液中待开发氮L4(t)(吨)"
Add Eq
=
INTEG (
"待开发氮年增加量R4(t)(吨/年)"
Initial Value
0.518963
Type
Level
Normal
Supplementary
Undo
Help
Variables
Functions
More
Choose Variable...
Inputs
猪粪沼液中待开发氮L4(t)(吨)
待开发氮年增加量R4(t)(吨/年)
Units:
Com-ment:
Group:
.t7
Range:
Go To:
Prev
Next
Hilite
Sel...
New
Errors:
Equation OK
OK
Check Syntax
Check Model
Delete Variable
Cancel

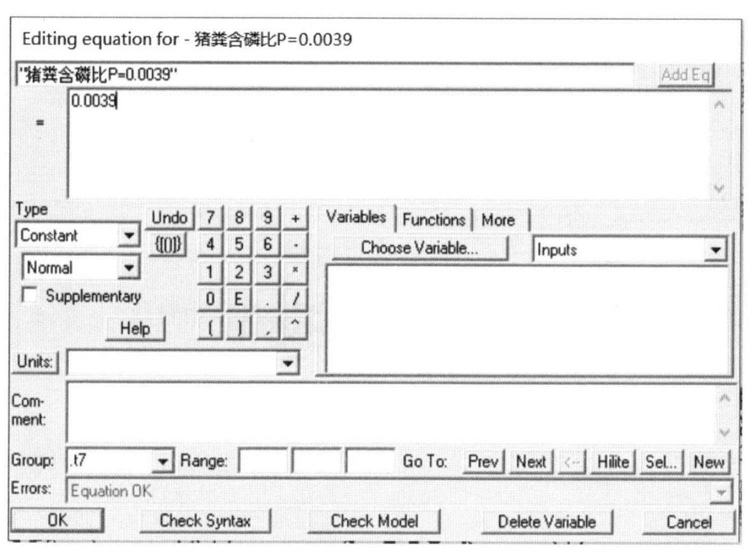
Editing equation for - 猪粪含磷比P=0.0039
"猪粪含磷比P=0.0039"
Add Eq
=
0.0039
Type
Constant
Normal
Supplementary
Undo
Help
Variables
Functions
More
Choose Variable...
Inputs
Units:
Com-ment:
Group:
.t7
Range:
Go To:
Prev
Next
Hilite
Sel...
New
Errors:
Equation OK
OK
Check Syntax
Check Model
Delete Variable
Cancel

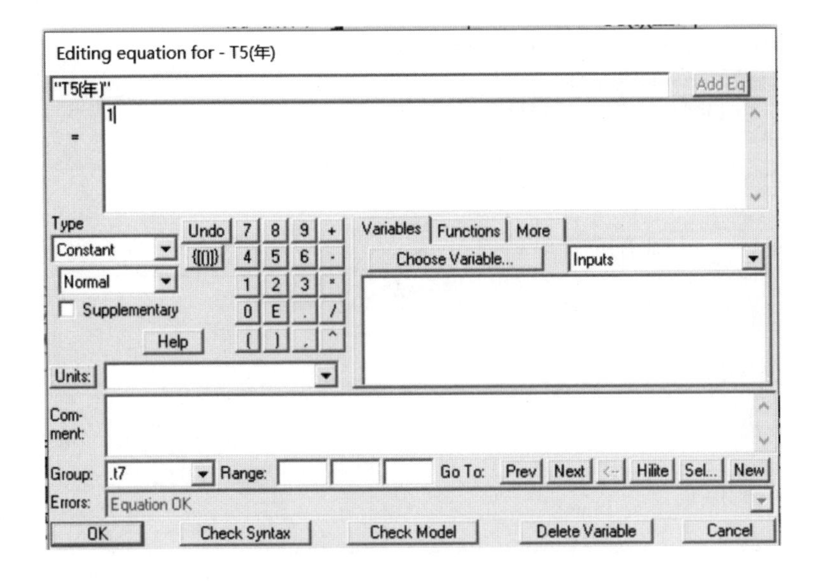
Editing equation for - T5(年)
"T5(年)"
Add Eq
=
1
Type
Constant
Normal
Supplementary
Undo
Help
Variables
Functions
More
Choose Variable...
Inputs
Units:
Com-
ment:
Group:
.t7
Range:
Go To:
Prev
Next
Hilite
Sel...
New
Errors:
Equation OK
OK
Check Syntax
Check Model
Delete Variable
Cancel

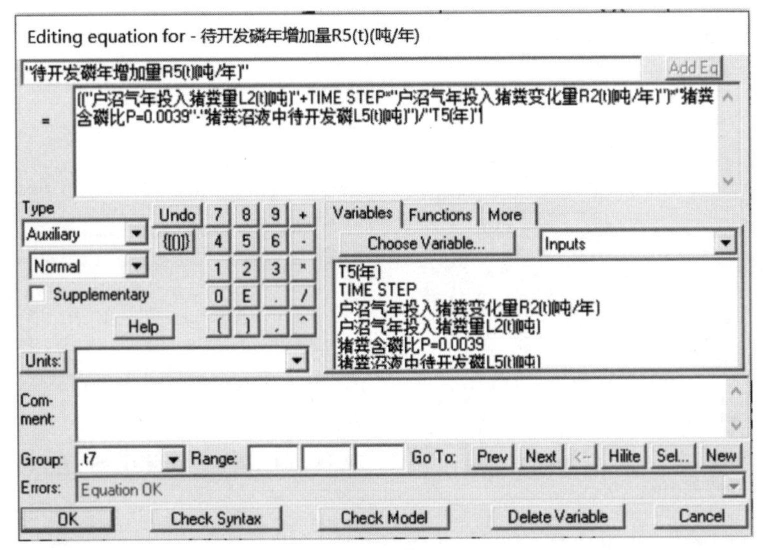
Editing equation for - 待开发磷年增加量R5(t)(吨/年)
"待开发磷年增加量R5(t)(吨/年)"
Add Eq
=
(("户沼气年投入猪粪量L2(t)(吨)"+TIME STEP*"户沼气年投入猪粪变化量R2(t)(吨/年)")*"猪粪含磷比P=0.0039"-"猪粪沼液中待开发磷L5(t)(吨)")/"T5(年)"
Type
Auxiliary
Normal
Supplementary
Undo
Help
Variables
Functions
More
Choose Variable...
Inputs
T5(年)
TIME STEP
户沼气年投入猪粪变化量R2(t)(吨/年)
户沼气年投入猪粪量L2(t)(吨)
猪粪含磷比P=0.0039
猪粪沼液中待开发磷L5(t)(吨)
Units:
Com-
ment:
Group:
.t7
Range:
Go To:
Prev
Next
Hilite
Sel...
New
Errors:
Equation OK
OK
Check Syntax
Check Model
Delete Variable
Cancel

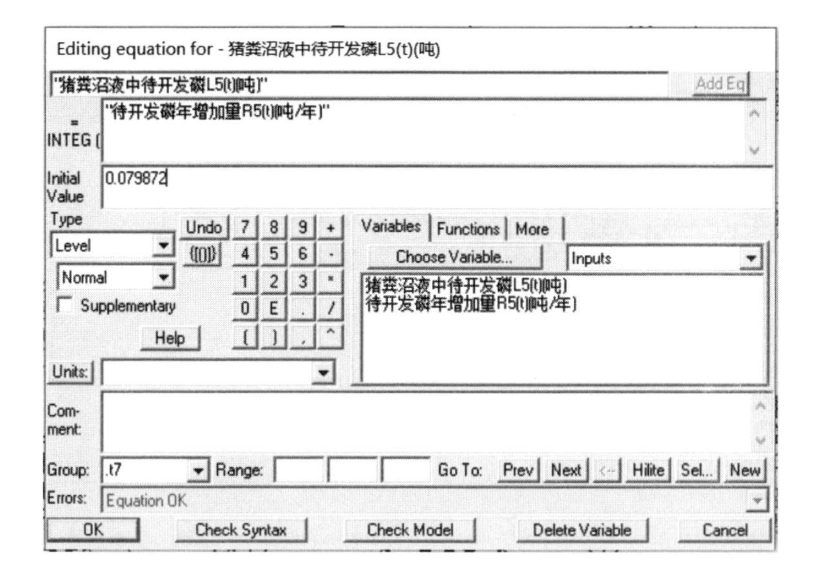
Editing equation for - 猪粪沼液中待开发磷L5(t)(吨)
"猪粪沼液中待开发磷L5(t)(吨)"
Add Eq
=
INTEG (
"待开发磷年增加量R5(t)(吨/年)"
Initial Value
0.079872
Type
Level
Normal
Supplementary
Undo
Help
Variables
Functions
More
Choose Variable...
Inputs
猪粪沼液中待开发磷L5(t)(吨)
待开发磷年增加量R5(t)(吨/年)
Units:
Com-ment:
Group:
.t7
Range:
Go To:
Prev
Next
Hilite
Sel...
New
Errors:
Equation OK
OK
Check Syntax
Check Model
Delete Variable
Cancel

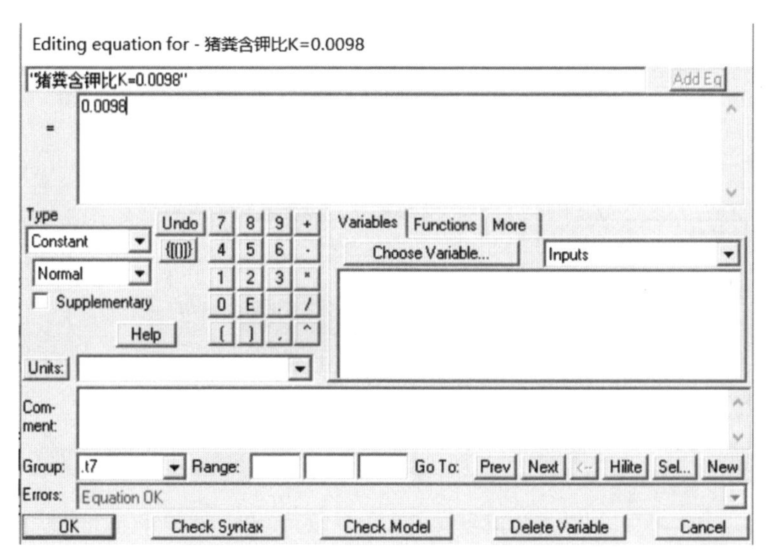
Editing equation for - 猪粪含钾比K=0.0098
"猪粪含钾比K=0.0098"
Add Eq
=
0.0098
Type
Constant
Normal
Supplementary
Undo
Help
Variables
Functions
More
Choose Variable...
Inputs
Units:
Com-ment:
Group:
.t7
Range:
Go To:
Prev
Next
Hilite
Sel...
New
Errors:
Equation OK
OK
Check Syntax
Check Model
Delete Variable
Cancel

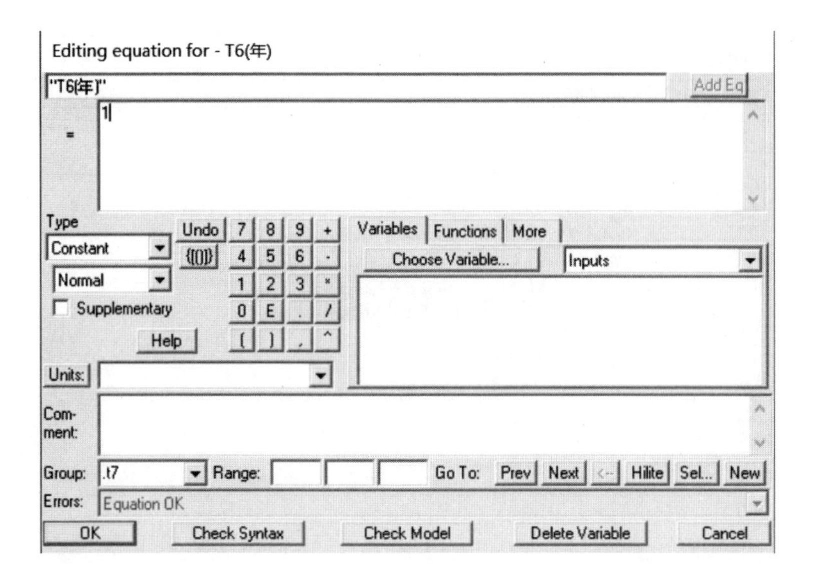
Editing equation for - T6(年)
"T6(年)"
Add Eq
1
Type
Constant
Normal
Supplementary
Undo
Help
Variables
Functions
More
Choose Variable...
Inputs
Units:
Comment:
Group: .t7
Range:
Go To:
Prev
Next
Hilite
Sel...
New
Errors: Equation OK
OK
Check Syntax
Check Model
Delete Variable
Cancel

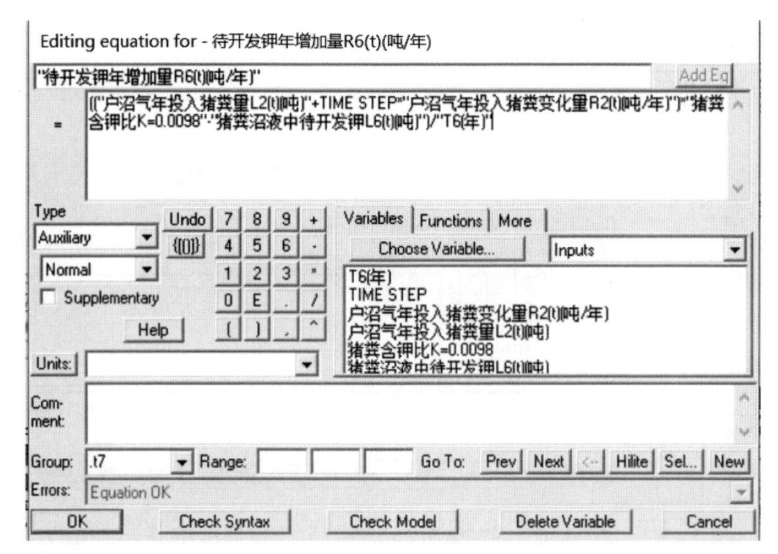
Editing equation for - 待开发钾年增加量R6(t)(吨/年)
"待开发钾年增加量R6(t)(吨/年)"
Add Eq
(("户沼气年投入猪粪量L2(t)(吨)"+TIME STEP*"户沼气年投入猪粪变化量R2(t)(吨/年)")*"猪粪含钾比K=0.0098"-"猪粪沼液中待开发钾L6(t)(吨)")/"T6(年)"
Type
Auxiliary
Normal
Supplementary
Undo
Help
Variables
Functions
More
Choose Variable...
Inputs
T6(年)
TIME STEP
户沼气年投入猪粪变化量R2(t)(吨/年)
户沼气年投入猪粪量L2(t)(吨)
猪粪含钾比K=0.0098
猪粪沼液中待开发钾L6(t)(吨)
Units:
Comment:
Group: .t7
Range:
Go To:
Prev
Next
Hilite
Sel...
New
Errors: Equation OK
OK
Check Syntax
Check Model
Delete Variable
Cancel

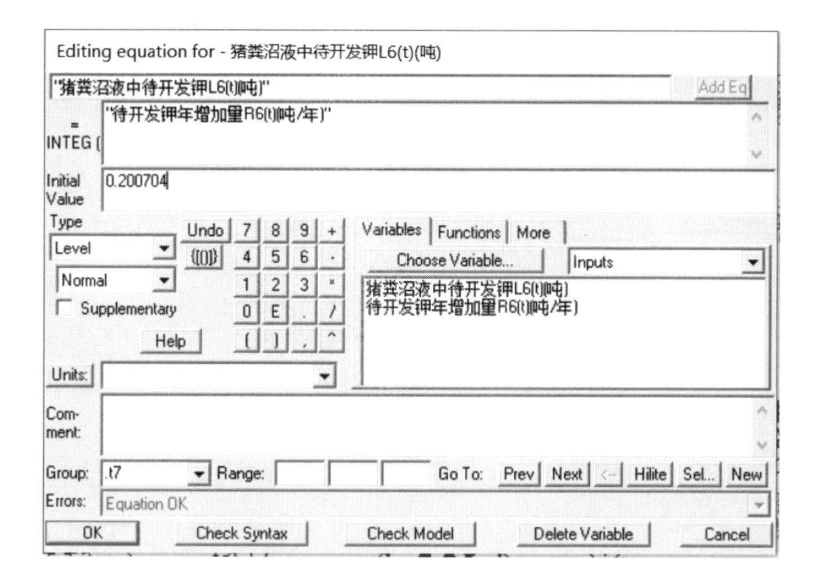
Editing equation for - 猪粪沼液中待开发钾L6(t)(吨)
"猪粪沼液中待开发钾L6(t)(吨)"
Add Eq
=
INTEG (
"待开发钾年增加量R6(t)(吨/年)"
Initial Value
0.200704
Type
Level
Normal
Supplementary
Undo
Help
Variables
Functions
More
Choose Variable...
Inputs
猪粪沼液中待开发钾L6(t)(吨)
待开发钾年增加量R6(t)(吨/年)
Units:
Com-ment:
Group:
.t7
Range:
Go To:
Prev
Next
Hilite
Sel...
New
Errors:
Equation OK
OK
Check Syntax
Check Model
Delete Variable
Cancel

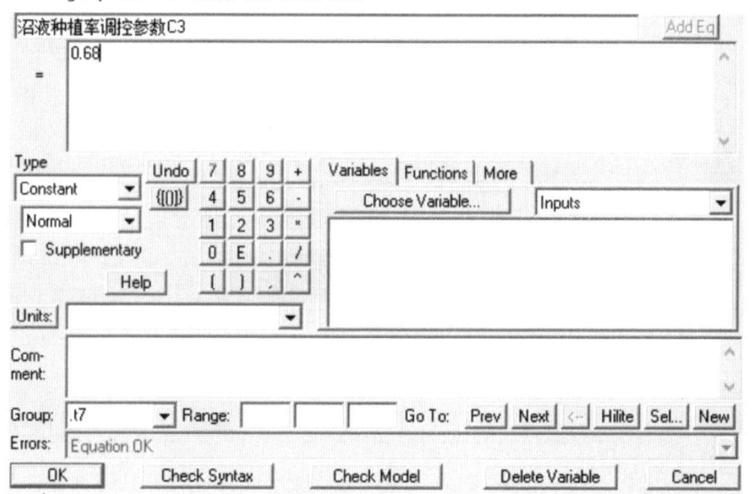
Editing equation for - 沼液种植率调控参数C3
沼液种植率调控参数C3
Add Eq
=
0.68
Type
Constant
Normal
Supplementary
Undo
Help
Variables
Functions
More
Choose Variable...
Inputs
Units:
Com-ment:
Group:
.t7
Range:
Go To:
Prev
Next
Hilite
Sel...
New
Errors:
Equation OK
OK
Check Syntax
Check Model
Delete Variable
Cancel

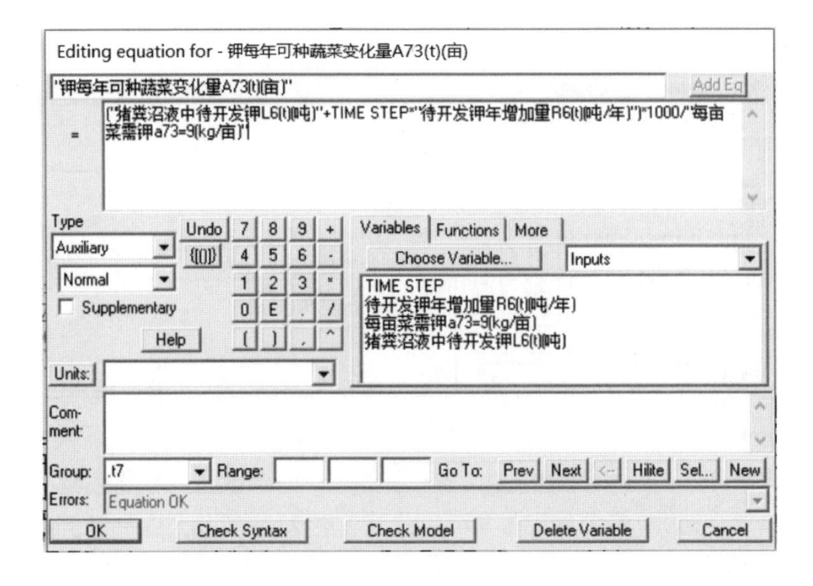
Editing equation for - 钾每年可种蔬菜变化量A73(t)(亩)
"钾每年可种蔬菜变化量A73(t)(亩)"
Add Eq
("猪粪沼液中待开发钾L6(t)(吨)"+TIME STEP*"待开发钾年增加量R6(t)(吨/年)")*1000/"每亩菜需钾a73=9(kg/亩)"
Type
Auxiliary
Normal
Supplementary
Help
Undo
Variables
Functions
More
Choose Variable...
Inputs
TIME STEP
待开发钾年增加量R6(t)(吨/年)
每亩菜需钾a73=9(kg/亩)
猪粪沼液中待开发钾L6(t)(吨)
Units:
Com-
ment:
Group:
.t7
Range:
Go To:
Prev
Next
Hilite
Sel...
New
Errors:
Equation OK
OK
Check Syntax
Check Model
Delete Variable
Cancel

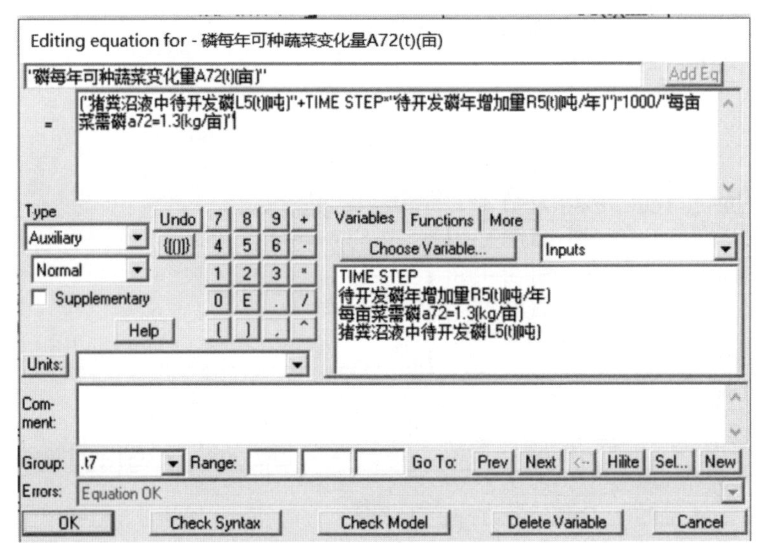
Editing equation for - 磷每年可种蔬菜变化量A72(t)(亩)
"磷每年可种蔬菜变化量A72(t)(亩)"
Add Eq
("猪粪沼液中待开发磷L5(t)(吨)"+TIME STEP*"待开发磷年增加量R5(t)(吨/年)")*1000/"每亩菜需磷a72=1.3(kg/亩)"
Type
Auxiliary
Normal
Supplementary
Help
Undo
Variables
Functions
More
Choose Variable...
Inputs
TIME STEP
待开发磷年增加量R5(t)(吨/年)
每亩菜需磷a72=1.3(kg/亩)
猪粪沼液中待开发磷L5(t)(吨)
Units:
Com-
ment:
Group:
.t7
Range:
Go To:
Prev
Next
Hilite
Sel...
New
Errors:
Equation OK
OK
Check Syntax
Check Model
Delete Variable
Cancel

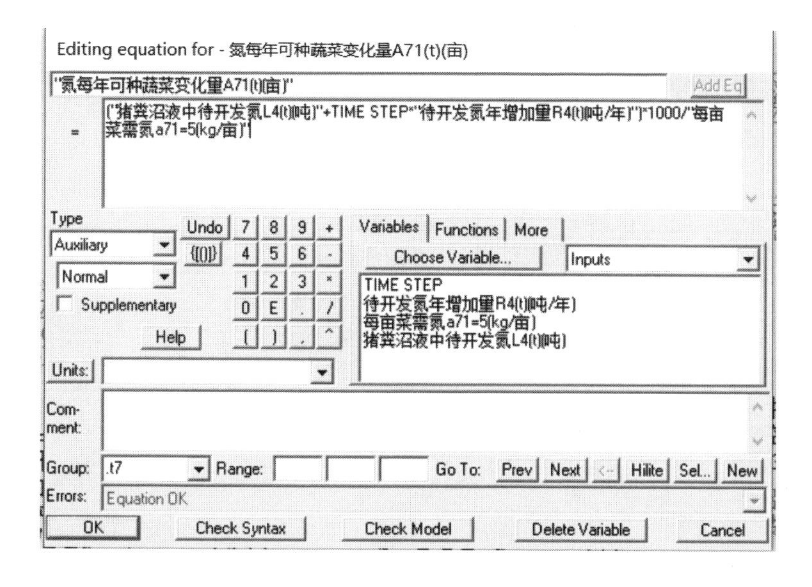
Editing equation for - 氮每年可种蔬菜变化量A71(t)(亩)
"氮每年可种蔬菜变化量A71(t)(亩)"
("猪粪沼液中待开发氮L4(t)吨)"+TIME STEP*"待开发氮年增加量R4(t)吨/年)")*1000/"每亩菜需氮a71=5(kg/亩)"
Type
Auxiliary
Normal
Supplementary
Help
Undo
Variables
Functions
More
Choose Variable...
Inputs
TIME STEP
待开发氮年增加量R4(t)吨/年)
每亩菜需氮a71=5(kg/亩)
猪粪沼液中待开发氮L4(t)吨)
Units:
Com-
ment:
Group: .t7
Range:
Go To:
Prev
Next
Hilite
Sel...
New
Errors: Equation OK
OK
Check Syntax
Check Model
Delete Variable
Cancel

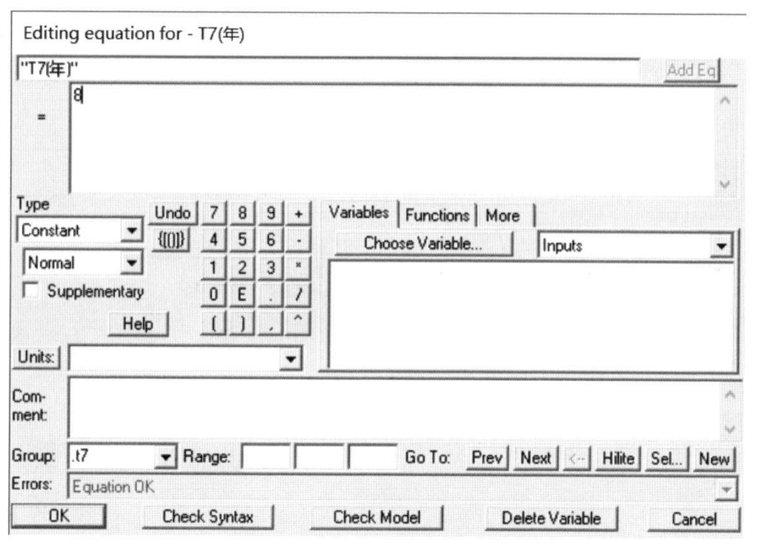
Editing equation for - T7(年)
"T7(年)"
8
Type
Constant
Normal
Supplementary
Help
Undo
Variables
Functions
More
Choose Variable...
Inputs
Units:
Com-
ment:
Group: .t7
Range:
Go To:
Prev
Next
Hilite
Sel...
New
Errors: Equation OK
OK
Check Syntax
Check Model
Delete Variable
Cancel

Editing equation for - 沼液户年种植量L7(t)(亩/户)
"沼液户年种植量L7(t)(亩/户)"
Add Eq
= INTEG ("沼液户年种植变化量R7(t)(亩/户年)"
Initial Value 0.8
Type
Level
Normal
Supplementary
Help
Undo
Variables
Functions
More
Choose Variable...
Inputs
沼液户年种植量L7(t)(亩/户)
沼液户年种植变化量R7(t)(亩/户年)
Units:
Com-ment:
Group: .t7
Range:
Go To:
Prev
Next
Hilite
Sel...
New
Errors: Equation OK
OK
Check Syntax
Check Model
Delete Variable
Cancel

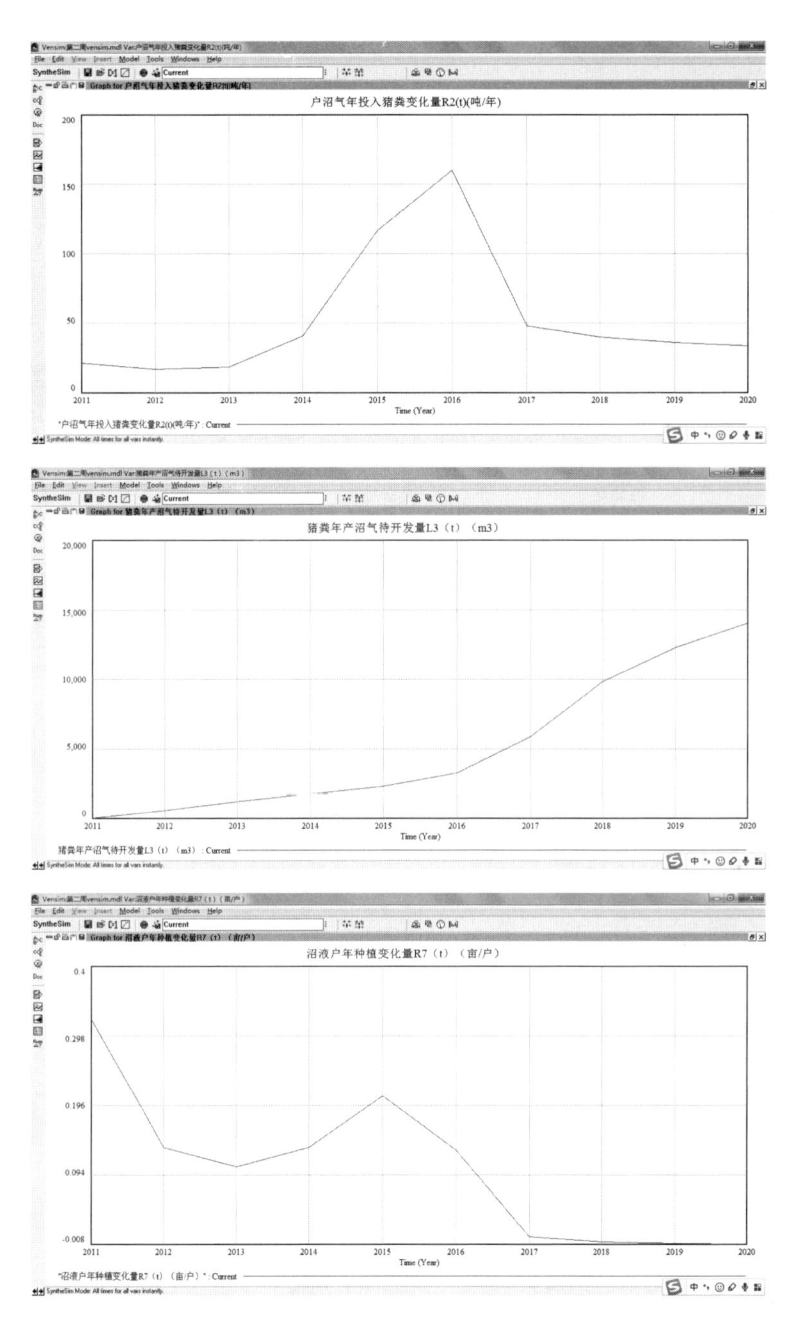

5. T_1 ~ T_9 入树模型及仿真方程，写出仿真图形 T_8（t）和 R_8（t）、T_9（t）和 R_9（t）。

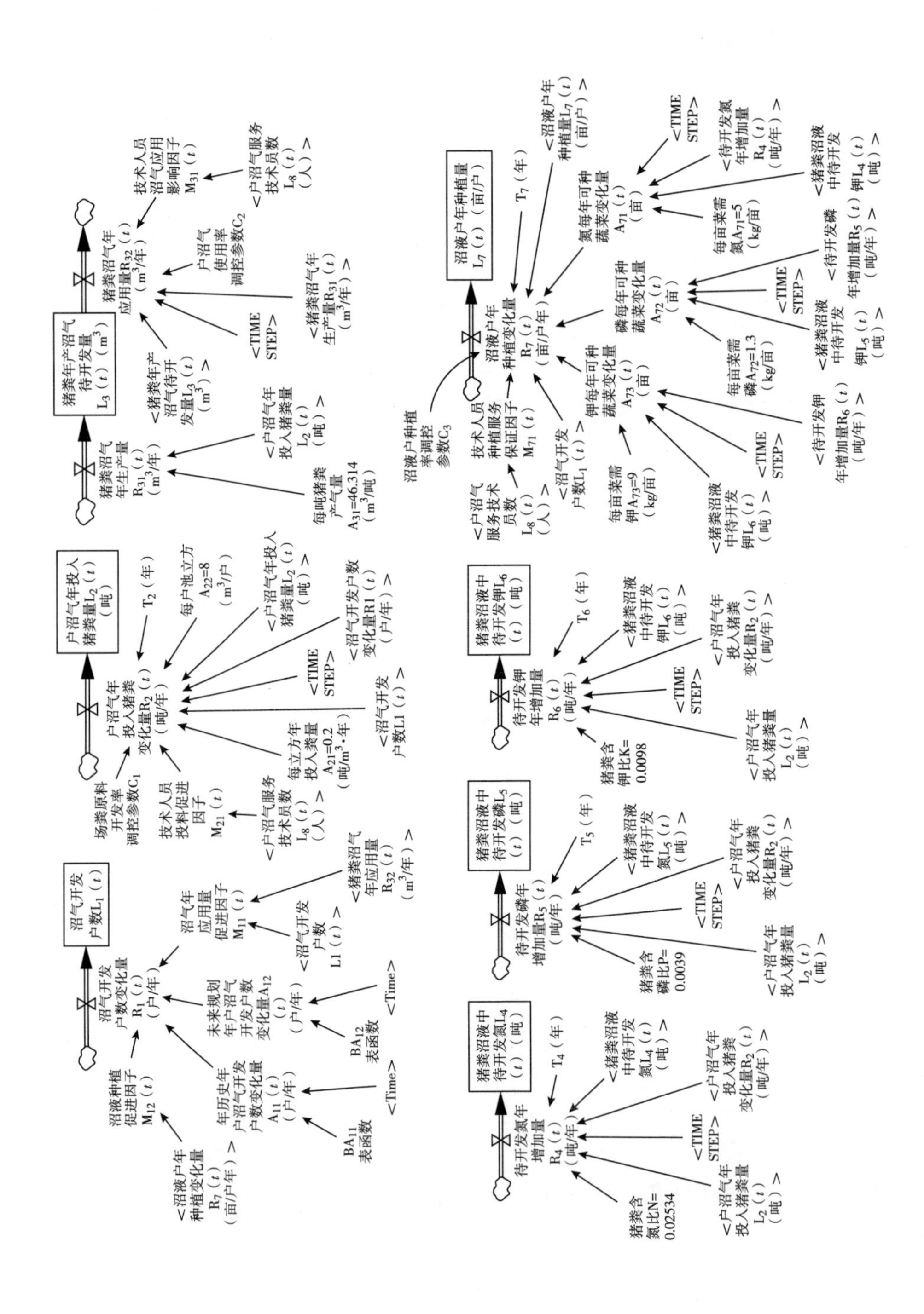
沼气开发户数L1（t）
沼气开发户数变化量R1（t）（户/年）
沼液种植促进因子M12（t）
<沼液户年种植变化量R7（t）（亩/户年）>
年历史年户沼气开发户数变化量A11（t）（户/年）
未来规划年户沼气开发户数变化量A12（t）（户/年）
沼气年应用量促进因子M11（t）
<沼气开发户数L1（t）>
<猪粪沼气年应用量R32（t）（m3/年）>
BA11表函数
BA12表函数
<Time>
<Time>
户沼气年投入猪粪量L2（t）（吨）
户沼气年投入猪粪变化量R2（t）（吨/年）
场粪原料开发率调控参数C1
技术人员投料促进因子M21（t）
<户沼气服务技术员数L8（t）（人）>
每立方年投入粪量A21=0.2（吨/m3·年）
<沼气开发户数L1（t）>
<TIME STEP>
<沼气开发户数变化量R1（t）（户/年）>
<户沼气年投入猪粪量L2（t）（吨）>
每户池立方A22=8（m3/户）
T2（年）
猪粪年产沼气待开发量L3（t）（m3）
猪粪沼气年生产量R31（t）（m3/年）
每吨猪粪产气量A31=46.314（m3/吨）
<户沼气年投入猪粪量L2（t）（吨）>
猪粪沼气年应用量R32（t）（m3/年）
<猪粪年产沼气待开发量L3（t）（m3）>
<TIME STEP>
<猪粪沼气年生产量R31（t）（m3/年）>
户沼气使用率调控参数C2
技术人员沼气应用影响因子M31（t）
<户沼气服务技术员数L8（t）（人）>
猪粪沼液中待开发氮L4（t）（吨）
待开发氮年增加量R4（t）（吨/年）
T4（年）
猪粪含氮比N=0.02534
<户沼气年投入猪粪量L2（t）（吨）>
<TIME STEP>
<户沼气年投入猪粪变化量R2（t）（吨/年）>
<猪粪沼液中待开发氮L4（t）（吨）>
猪粪沼液中待开发磷L5（t）（吨）
待开发磷年增加量R5（t）（吨/年）
T5（年）
猪粪含磷比P=0.0039
<户沼气年投入猪粪量L2（t）（吨）>
<TIME STEP>
<户沼气年投入猪粪变化量R2（t）（吨/年）>
<猪粪沼液中待开发氮L5（t）（吨）>
猪粪沼液中待开发钾L6（t）（吨）
待开发钾年增加量R6（t）（吨/年）
T6（年）
猪粪含钾比K=0.0098
<户沼气年投入猪粪量L2（t）（吨）>
<TIME STEP>
<户沼气年投入猪粪变化量R2（t）（吨/年）>
<猪粪沼液中待开发钾L6（t）（吨）>
沼液户年种植量L7（t）（亩/户）
沼液户种植率调控参数C3
沼液户年种植变化量R7（t）（亩/户年）
<户沼气服务技术员数L8（t）（人）>
技术人员种植服务保证因子M71（t）
<沼气开发户数L1（t）>
T7（年）
<沼液户年种植量L7（t）（亩/户）>
钾每年可种蔬菜变化量A73（t）（亩）
每亩菜需钾A73=9（kg/亩）
<猪粪沼液中待开发钾L6（t）（吨）>
<TIME STEP>
<待开发钾年增加量R6（t）（吨/年）>
磷每年可种蔬菜变化量A72（t）（亩）
每亩菜需磷A72=1.3（kg/亩）
<猪粪沼液中待开发钾L5（t）（吨）>
<TIME STEP>
<待开发磷年增加量R5（t）（吨/年）>
氮每年可种蔬菜变化量A71（t）（亩）
每亩菜需氮A71=5（kg/亩）
<待开发氮年增加量R4（t）（吨/年）>
<猪粪沼液中待开发钾L4（t）（吨）>
<TIME STEP>

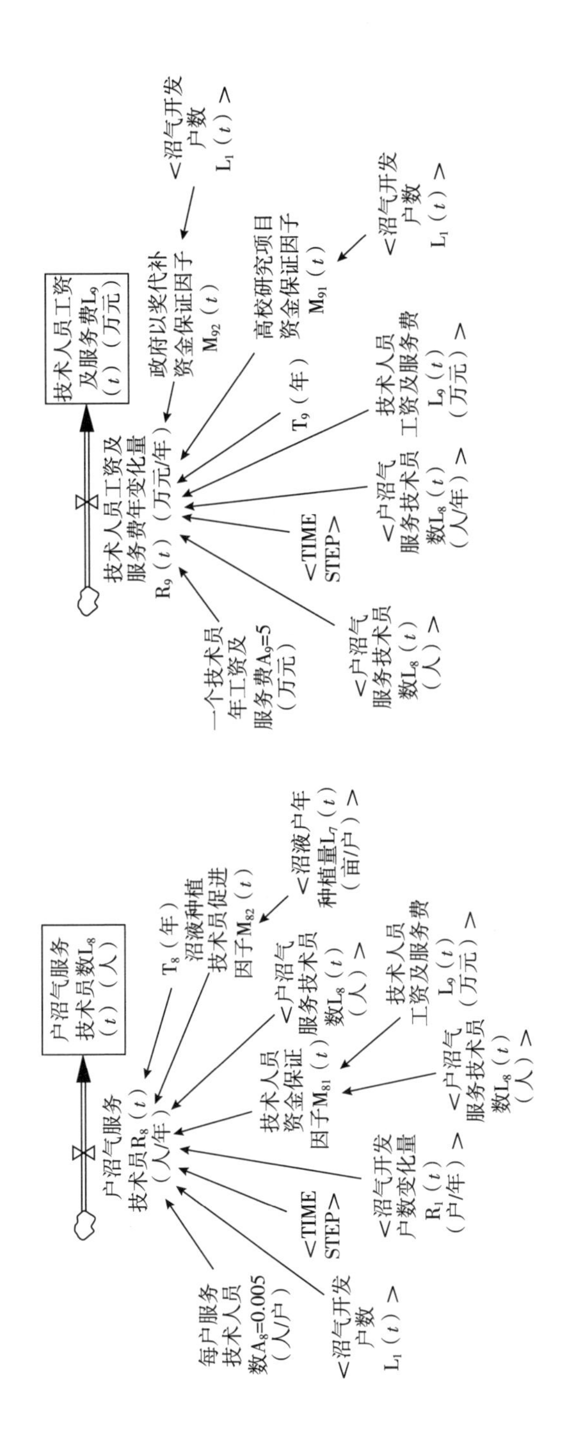
技术人员工资及服务费L9（t）（万元）
技术人员工资及服务费年变化量R9（t）（万元/年）
政府以奖代补资金保证因子M92（t）
<沼气开发户数L1（t）>
高校研究项目资金保证因子M91（t）
<沼气开发户数L1（t）>
T9（年）
技术人员工资及服务费L9（t）（万元）>
<户沼气服务技术员数L8（t）（人/年）>
<TIME STEP>
<户沼气服务技术员数L8（t）（人）>
一个技术员年工资及服务费A9=5（万元）
户沼气服务技术员数L8（t）（人）
户沼气服务技术员R8（t）（人/年）
T8（年）
沼液种植技术员促进因子M82（t）
<沼液户年种植量L7（t）（亩/户）>
<户沼气服务技术员数L8（t）（人）>
技术人员资金保证因子M81（t）
技术人员工资及服务费L9（t）（万元）>
<户沼气服务技术员数L8（t）（人）>
<沼气开发户数变化量R1（t）（户/年）>
<TIME STEP>
<沼气开发户数L1（t）>
每户服务技术人员数A8=0.005（人/户）

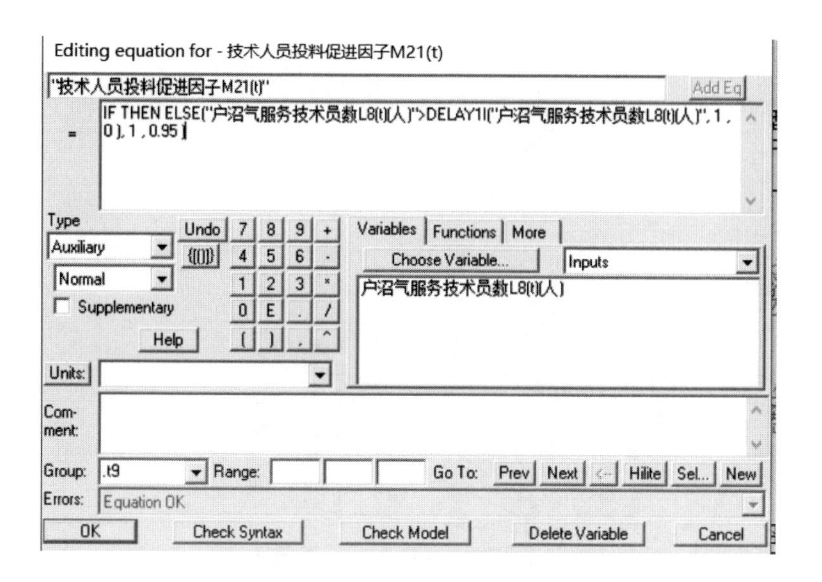

Editing equation for - 技术人员投料促进因子M21(t)
"技术人员投料促进因子M21(t)"
Add Eq
IF THEN ELSE("户沼气服务技术员数L8(t)(人)">DELAY1I("户沼气服务技术员数L8(t)(人)", 1 , 0), 1 , 0.95)
Type
Auxiliary
Normal
Supplementary
Help
Undo
Variables
Functions
More
Choose Variable...
Inputs
户沼气服务技术员数L8(t)(人)
Units:
Com-
ment:
Group:
.t9
Range:
Go To:
Prev
Next
Hilite
Sel...
New
Errors:
Equation OK
OK
Check Syntax
Check Model
Delete Variable
Cancel

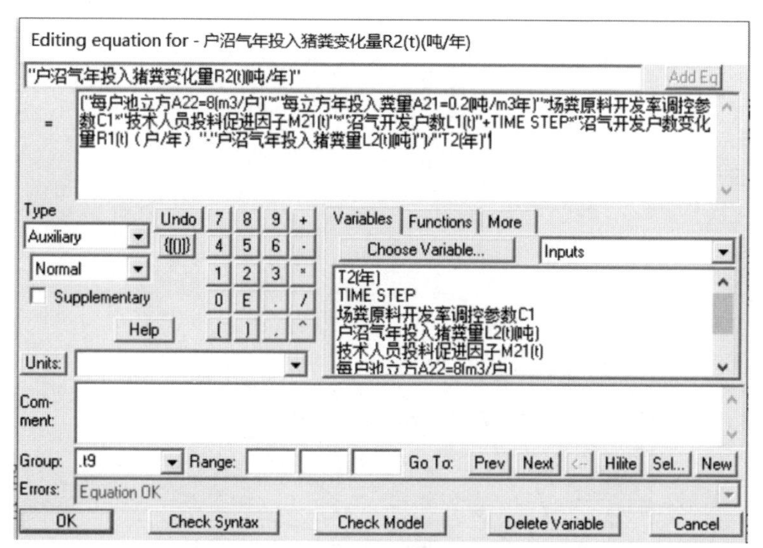

Editing equation for - 户沼气年投入猪粪变化量R2(t)(吨/年)
"户沼气年投入猪粪变化量R2(t)(吨/年)"
Add Eq
("每户池立方A22=8(m3/户)"*"每立方年投入粪量A21=0.2(吨/m3年)"*场粪原料开发率调控参数C1*"技术人员投料促进因子M21(t)"*"沼气开发户数L1(t)"+TIME STEP*"沼气开发户数变化量R1(t)(户/年)"-"户沼气年投入猪粪量L2(t)(吨)")/"T2(年)"
Type
Auxiliary
Normal
Supplementary
Help
Undo
Variables
Functions
More
Choose Variable...
Inputs
T2(年)
TIME STEP
场粪原料开发率调控参数C1
户沼气年投入猪粪量L2(t)(吨)
技术人员投料促进因子M21(t)
每户池立方A22=8(m3/户)
Units:
Com-
ment:
Group:
.t9
Range:
Go To:
Prev
Next
Hilite
Sel...
New
Errors:
Equation OK
OK
Check Syntax
Check Model
Delete Variable
Cancel

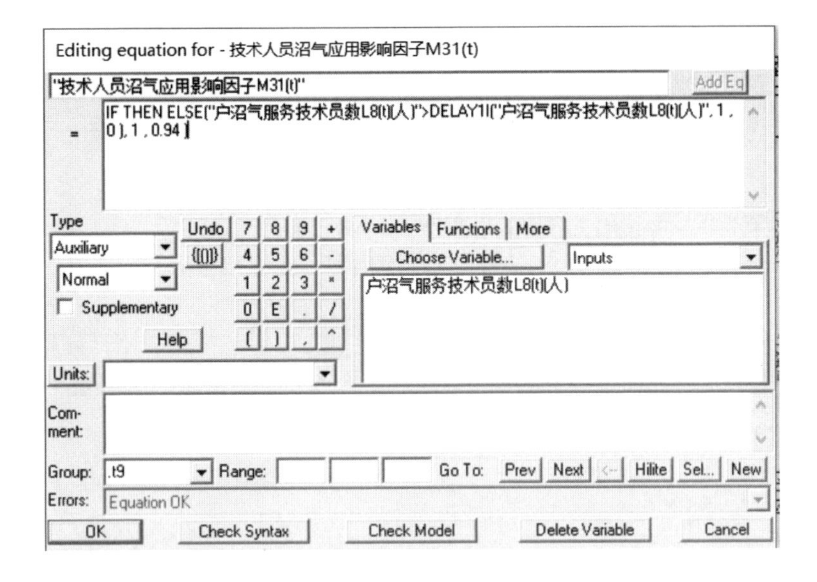

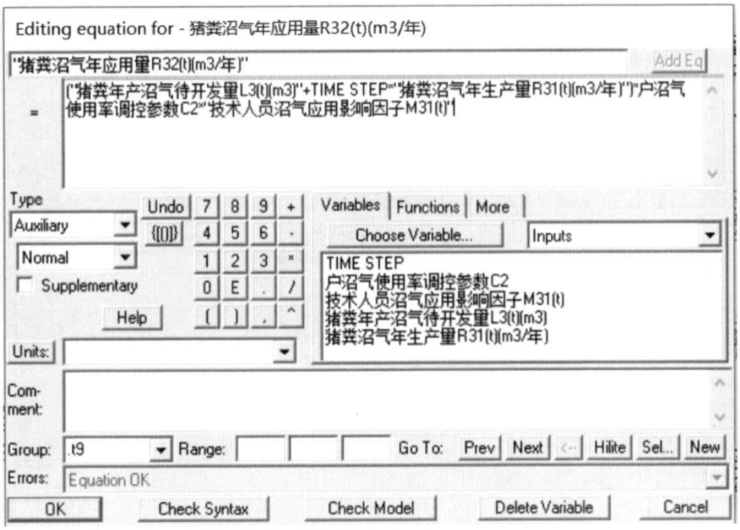

Editing equation for - 猪粪沼气年应用量R32(t)(m3/年)

"猪粪沼气年应用量R32(t)(m3/年)"

Add Eq

= ("猪粪年产沼气待开发量L3(t)(m3)"+TIME STEP*"猪粪沼气年生产量R31(t)(m3/年)")*户沼气使用率调控参数C2*"技术人员沼气应用影响因子M31(t)"

Type: Auxiliary | Normal | Supplementary | Help | Undo

Variables | Functions | More

Choose Variable... | Inputs

TIME STEP
户沼气使用率调控参数C2
技术人员沼气应用影响因子M31(t)
猪粪年产沼气待开发量L3(t)(m3)
猪粪沼气年生产量R31(t)(m3/年)

Units:

Com-
ment:

Group: .t9 | Range: | Go To: | Prev | Next | Hilite | Sel... | New

Errors: Equation OK

OK | Check Syntax | Check Model | Delete Variable | Cancel

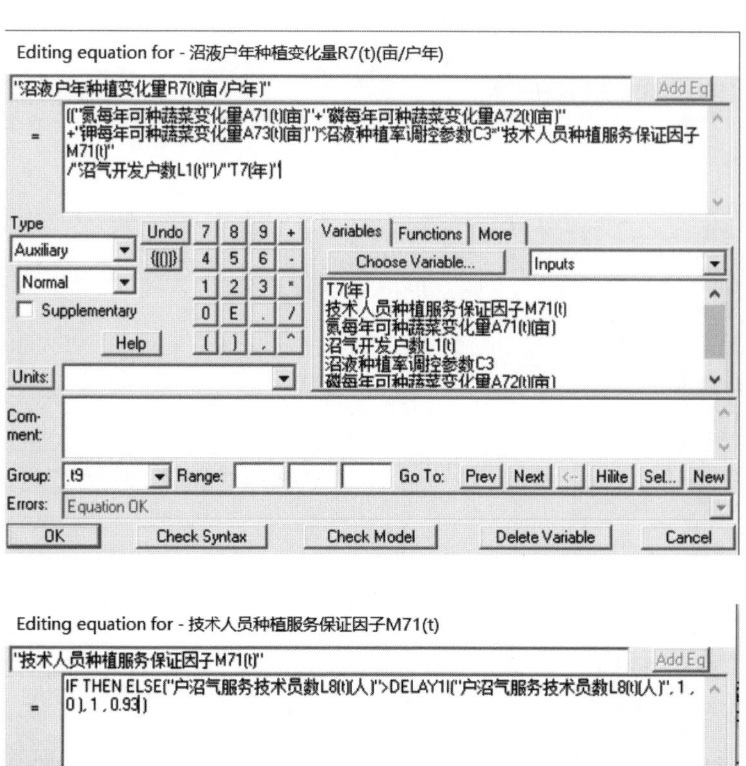
Editing equation for - 沼液户年种植变化量R7(t)(亩/户年)
"沼液户年种植变化量R7(t)(亩/户年)"
Add Eq
=
(("氮每年可种蔬菜变化量A71(t)(亩)"+"磷每年可种蔬菜变化量A72(t)(亩)"
+"钾每年可种蔬菜变化量A73(t)(亩)")*沼液种植率调控参数C3*"技术人员种植服务保证因子M71(t)"
/"沼气开发户数L1(t)")/"T7(年)"
Type
Auxiliary
Normal
Supplementary
Undo
Help
Variables
Functions
More
Choose Variable...
Inputs
T7(年)
技术人员种植服务保证因子M71(t)
氮每年可种蔬菜变化量A71(t)(亩)
沼气开发户数L1(t)
沼液种植率调控参数C3
Units:
Com-
ment:
Group:
.t9
Range:
Go To:
Prev
Next
Hilite
Sel...
New
Errors:
Equation OK
OK
Check Syntax
Check Model
Delete Variable
Cancel

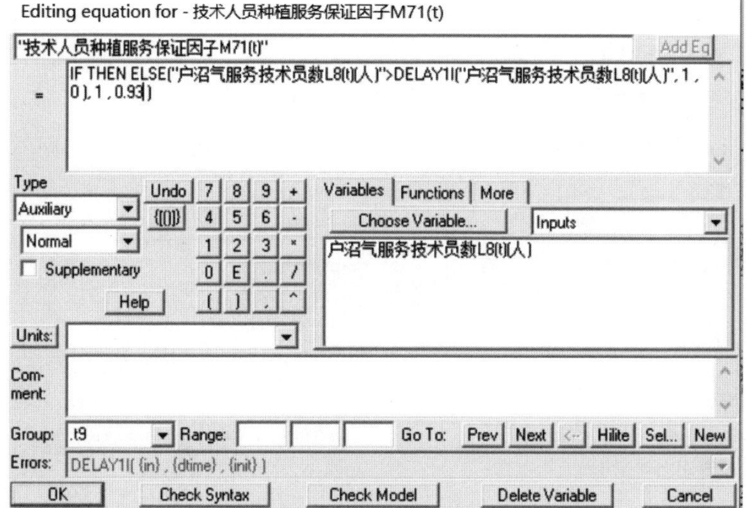
Editing equation for - 技术人员种植服务保证因子M71(t)
"技术人员种植服务保证因子M71(t)"
Add Eq
=
IF THEN ELSE("户沼气服务技术员数L8(t)(人)">DELAY1I("户沼气服务技术员数L8(t)(人)", 1 , 0), 1 , 0.93)
Type
Auxiliary
Normal
Supplementary
Undo
Help
Variables
Functions
More
Choose Variable...
Inputs
户沼气服务技术员数L8(t)(人)
Units:
Com-
ment:
Group:
.t9
Range:
Go To:
Prev
Next
Hilite
Sel...
New
Errors:
DELAY1I({in} , {dtime} , {init})
OK
Check Syntax
Check Model
Delete Variable
Cancel

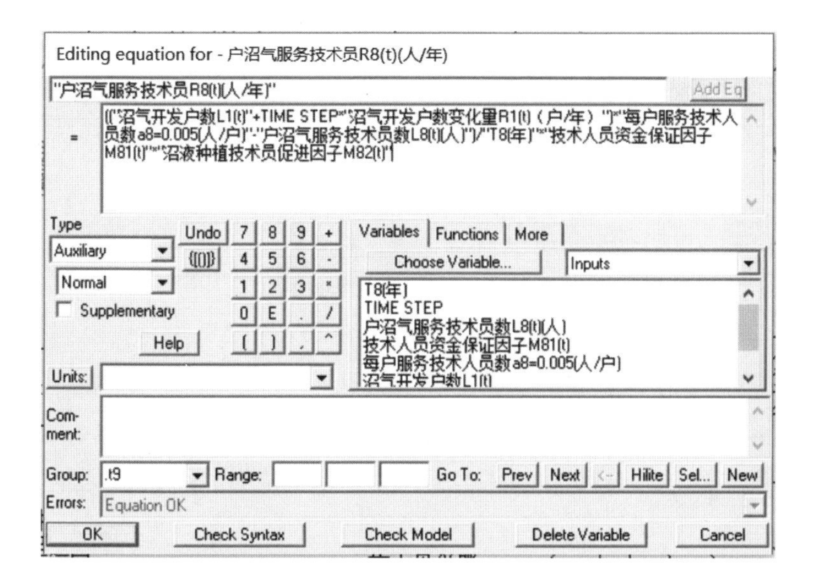
Editing equation for - 户沼气服务技术员R8(t)(人/年)
"户沼气服务技术员R8(t)(人/年)"
Add Eq
(("沼气开发户数L1(t)"+TIME STEP*"沼气开发户数变化量R1(t)（户/年）")*"每户服务技术人员数a8=0.005(人/户)"-"户沼气服务技术员数L8(t)(人)")/"T8(年)"*"技术人员资金保证因子M81(t)"*"沼液种植技术员促进因子M82(t)"
Type
Auxiliary
Normal
Supplementary
Help
Undo
Variables
Functions
More
Choose Variable...
Inputs
T8(年)
TIME STEP
户沼气服务技术员数L8(t)(人)
技术人员资金保证因子M81(t)
每户服务技术人员数a8=0.005(人/户)
沼气开发户数L1(t)
Units:
Comment:
Group: .t9
Range:
Go To:
Prev
Next
Hilite
Sel...
New
Errors: Equation OK
OK
Check Syntax
Check Model
Delete Variable
Cancel

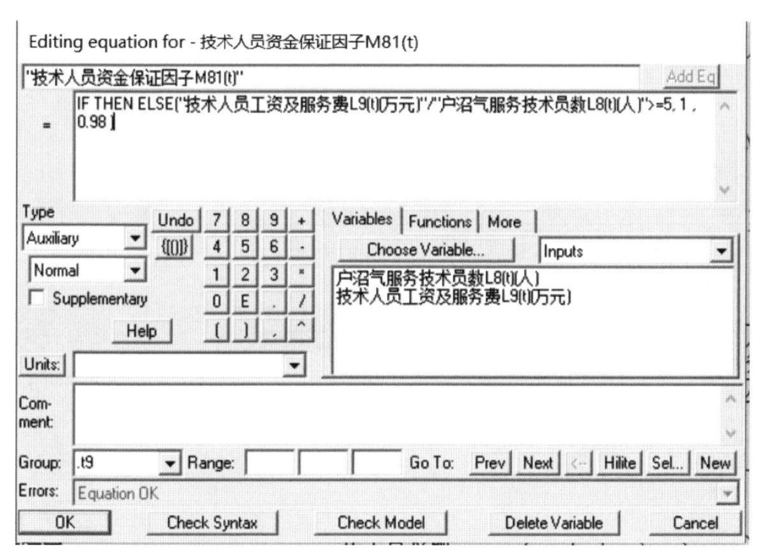
Editing equation for - 技术人员资金保证因子M81(t)
"技术人员资金保证因子M81(t)"
Add Eq
IF THEN ELSE("技术人员工资及服务费L9(t)(万元)"/"户沼气服务技术员数L8(t)(人)">=5, 1 , 0.98)
Type
Auxiliary
Normal
Supplementary
Help
Undo
Variables
Functions
More
Choose Variable...
Inputs
户沼气服务技术员数L8(t)(人)
技术人员工资及服务费L9(t)(万元)
Units:
Comment:
Group: .t9
Range:
Go To:
Prev
Next
Hilite
Sel...
New
Errors: Equation OK
OK
Check Syntax
Check Model
Delete Variable
Cancel

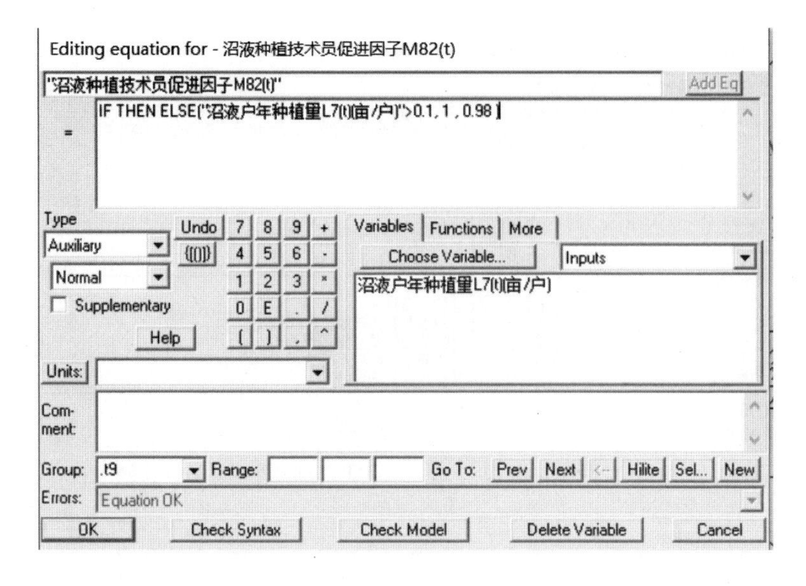
Editing equation for - 沼液种植技术员促进因子M82(t)
"沼液种植技术员促进因子M82(t)"
IF THEN ELSE("沼液户年种植量L7(t)(亩/户)">0.1, 1 , 0.98)
Type
Auxiliary
Normal
Supplementary
Undo
Help
Variables
Functions
More
Choose Variable...
Inputs
沼液户年种植量L7(t)(亩/户)
Units:
Com-ment:
Group:
.t9
Range:
Go To:
Prev
Next
Hilite
Sel...
New
Errors:
Equation OK
OK
Check Syntax
Check Model
Delete Variable
Cancel

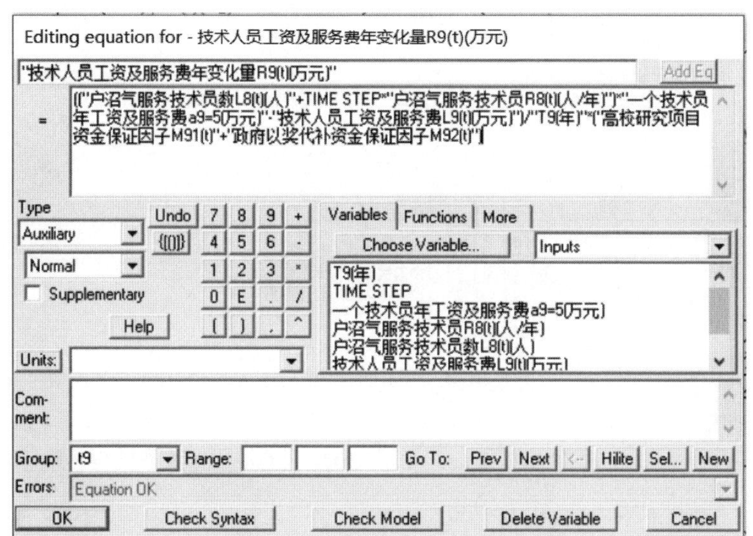
Editing equation for - 技术人员工资及服务费年变化量R9(t)(万元)
"技术人员工资及服务费年变化量R9(t)(万元)"
(("户沼气服务技术员数L8(t)(人)"+TIME STEP*"户沼气服务技术员R8(t)(人/年)")*"一个技术员年工资及服务费a9=5(万元)"-"技术人员工资及服务费L9(t)(万元)")/"T9(年)"*("高校研究项目资金保证因子M91(t)"+"政府以奖代补资金保证因子M92(t)")
Type
Auxiliary
Normal
Supplementary
Undo
Help
Variables
Functions
More
Choose Variable...
Inputs
T9(年)
TIME STEP
一个技术员年工资及服务费a9=5(万元)
户沼气服务技术员R8(t)(人/年)
户沼气服务技术员数L8(t)(人)
Units:
Com-ment:
Group:
.t9
Range:
Go To:
Prev
Next
Hilite
Sel...
New
Errors:
Equation OK
OK
Check Syntax
Check Model
Delete Variable
Cancel

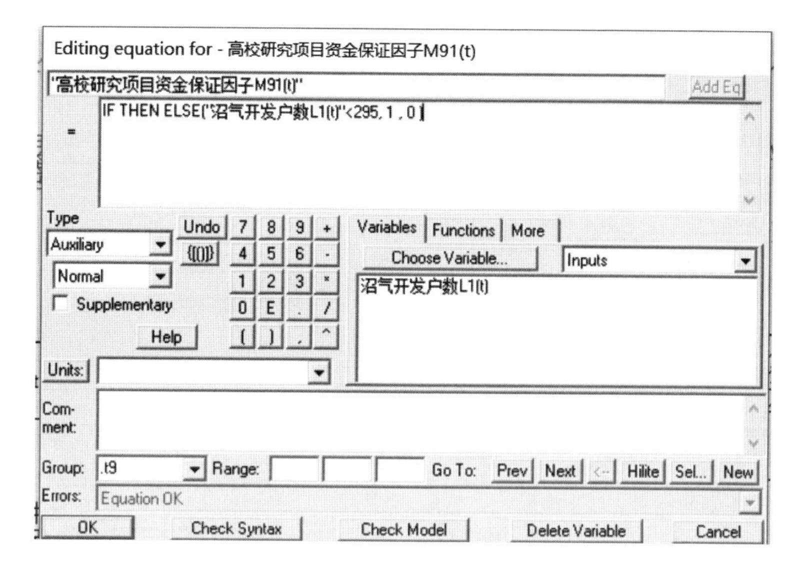
Editing equation for - 高校研究项目资金保证因子M91(t)
"高校研究项目资金保证因子M91(t)"
Add Eq
=
IF THEN ELSE("沼气开发户数L1(t)"<295, 1 , 0)
Type
Auxiliary
Normal
Supplementary
Help
Undo
7 8 9 +
4 5 6 -
1 2 3 *
0 E . /
( ) , ^
Variables
Functions
More
Choose Variable...
Inputs
沼气开发户数L1(t)
Units:
Com-
ment:
Group:
.t9
Range:
Go To:
Prev
Next
Hilite
Sel...
New
Errors:
Equation OK
OK
Check Syntax
Check Model
Delete Variable
Cancel

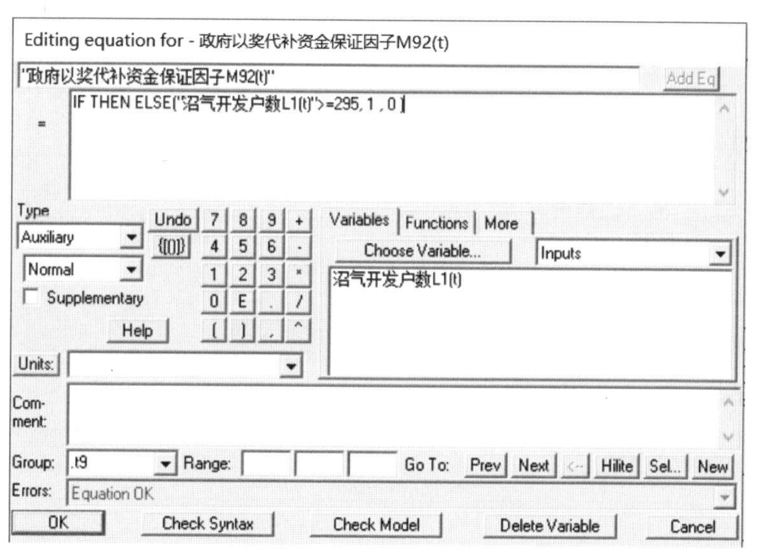
Editing equation for - 政府以奖代补资金保证因子M92(t)
"政府以奖代补资金保证因子M92(t)"
Add Eq
=
IF THEN ELSE("沼气开发户数L1(t)">=295, 1 , 0)
Type
Auxiliary
Normal
Supplementary
Help
Undo
7 8 9 +
4 5 6 -
1 2 3 *
0 E . /
( ) , ^
Variables
Functions
More
Choose Variable...
Inputs
沼气开发户数L1(t)
Units:
Com-
ment:
Group:
.t9
Range:
Go To:
Prev
Next
Hilite
Sel...
New
Errors:
Equation OK
OK
Check Syntax
Check Model
Delete Variable
Cancel

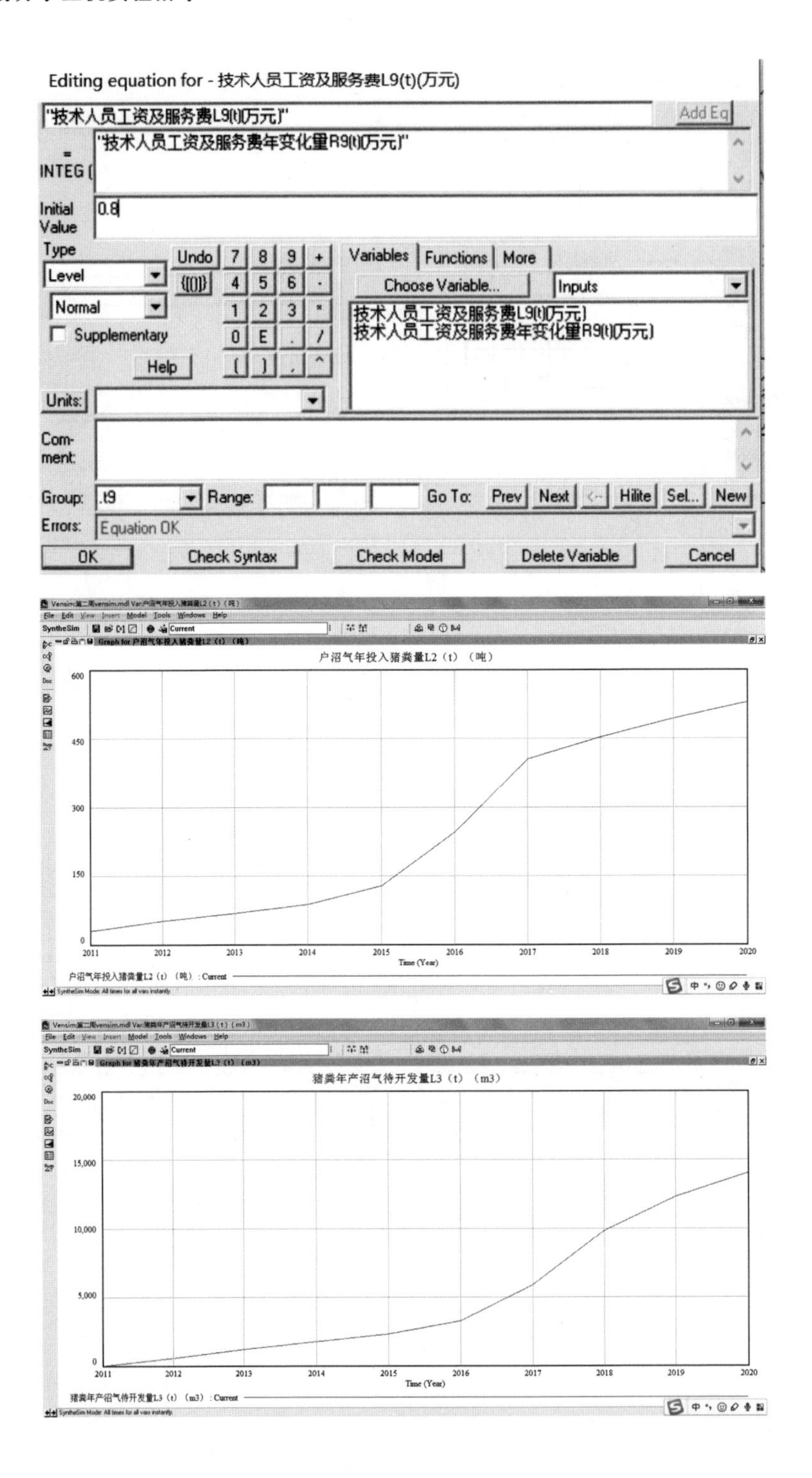
Editing equation for - 技术人员工资及服务费L9(t)(万元)
"技术人员工资及服务费L9(t)(万元)"
Add Eq
=
INTEG (
"技术人员工资及服务费年变化量R9(t)(万元)"
Initial Value
0.8
Type
Level
Normal
Supplementary
Undo
Help
Variables
Functions
More
Choose Variable...
Inputs
技术人员工资及服务费L9(t)(万元)
技术人员工资及服务费年变化量R9(t)(万元)
Units:
Com-ment:
Group:
.t9
Range:
Go To:
Prev
Next
Hilite
Sel...
New
Errors:
Equation OK
OK
Check Syntax
Check Model
Delete Variable
Cancel
户沼气年投入猪粪量L2（t）（吨）
600
450
300
150
0
2011
2012
2013
2014
2015
2016
2017
2018
2019
2020
Time (Year)
户沼气年投入猪粪量L2（t）（吨）: Current
猪粪年产沼气待开发量L3（t）（m3）
20,000
15,000
10,000
5,000
0
2011
2012
2013
2014
2015
2016
2017
2018
2019
2020
Time (Year)
猪粪年产沼气待开发量L3（t）（m3）: Current

上机实验解答 6

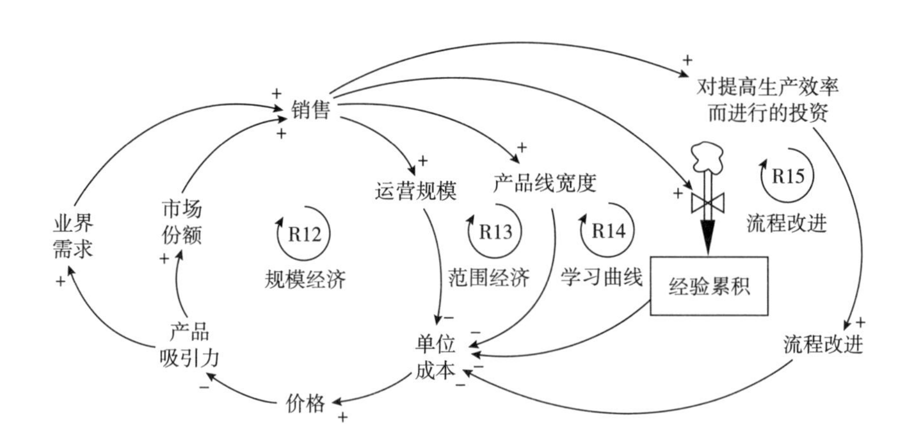

注：图中涉及的每一个效果都包括两个正反馈过程：一个是随着市场份额的增加而导致的销售额的增加；另一个是随着市场总规模的扩大而导致的销售额的增加。

1. 规模经济。分析小型、中型、大型养猪场的规模数量，根据猪周期和历史年每月的平均价格计算利润。

答：根据养猪场年出栏商品肉猪的生产规模，规模化猪场可分为三种基本类型，年出栏 10000 头以上商品肉猪的为大型规模化猪场；年出栏 3000~5000 头商品肉猪的为中型规模化猪场；年出栏 3000 头以下的为小型规模化猪场，现阶段农村适度规模养猪多属此类猪场。

一般而言，猪周期在 3~4 年，循环轨迹一般是：肉价上涨—母猪存栏量大增—生猪供应增加—肉价下跌—大量淘汰母猪—生猪供应减少—肉价上涨。

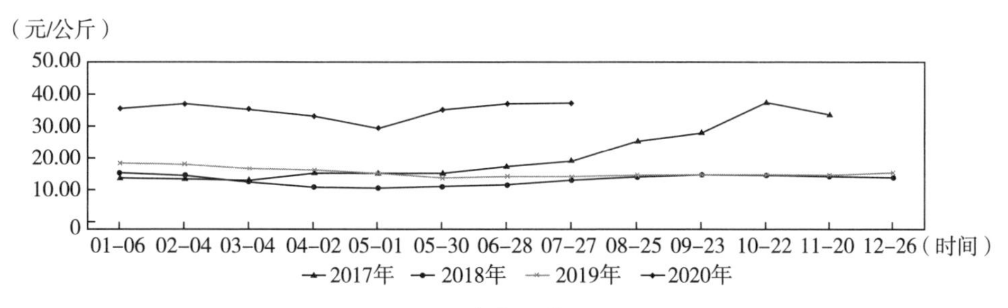

生猪出场价

资料来源：农业农村部，制图；行情宝。

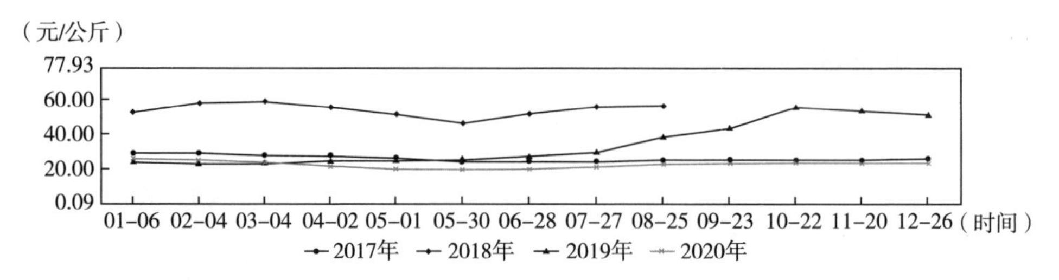

猪肉价

资料来源：农业农村部，制图；行情宝。

时间	品类	指标	地区	单位	数值	
2019E					1404.058595	
2018	大规模生猪	每头(或百只	全国(省份平	元	1530.64279	
2017	大规模生猪	每头(或百只	全国(省份平	元	1668.6393	1.090156
2016	大规模生猪	每头(或百只	全国(省份平	元	1752.705	1.05038
2019E					1506.041313	
2018	小规模生猪	每头(或百只	全国(省份平	元	1641.20215	
2017	小规模生猪	每头(或百只	全国(省份平	元	1788.4931	1.089746
2016	小规模生猪	每头(或百只	全国(省份平	元	1861.2186	1.040663
2019E					1454.049912	
2018	中规模生猪	每头(或百只	全国(省份平	元	1582.84847	
2017	中规模生猪	每头(或百只	全国(省份平	元	1723.0559	1.088579
2016	中规模生猪	每头(或百只	全国(省份平	元	1816.791	1.0544

由此预估全国省份每头猪平均的总成本。

时间	品类	指标	地区	单位	数值
2018	大规模生猪	每头(或百只	全国(省份平	元	43.95721
2017	大规模生猪	每头(或百只	全国(省份平	元	147.3707
2016	大规模生猪	每头(或百只	全国(省份平	元	441.125
2018	小规模生猪	每头(或百只	全国(省份平	元	-45.18215
2017	小规模生猪	每头(或百只	全国(省份平	元	63.4469
2016	小规模生猪	每头(或百只	全国(省份平	元	378.9014
2018	中规模生猪	每头(或百只	全国(省份平	元	31.98153
2017	中规模生猪	每头(或百只	全国(省份平	元	135.5141
2016	中规模生猪	每头(或百只	全国(省份平	元	420.309

计算利润：

大规模：10000×43.96=439600

中规模：3000×31.98=95940

小规模：1000×（-45.18）=-45180

由于国家政策对养猪有补贴，此并非为养猪场最终利润。而且由于2019年、2020年受猪瘟等影响，猪肉价格大幅度上涨。所以2019年、2020年养猪户的利润更大。

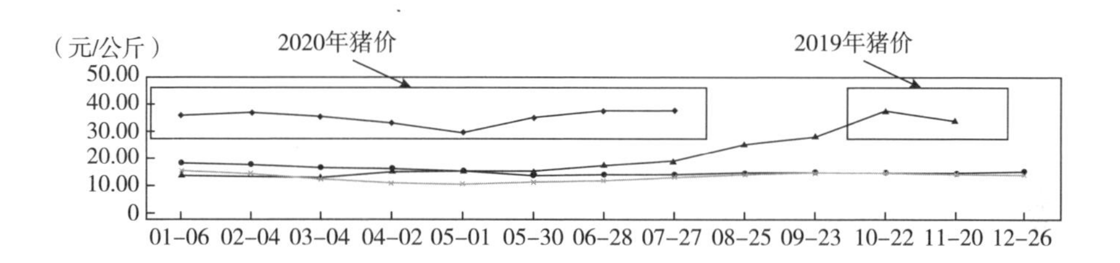

生猪出场价

资料来源：农业农村部，制图；行情宝。

2. 产品线宽度。哪些品种好饲养？

答：①长白猪。它是兰德瑞斯猪在中国的通称，腌肉型猪品种。每胎产仔11~12头，成年公猪体重400~500公斤，母猪300公斤左右。要求有较好的饲养管理条件，遍布于世界各国。

②大白猪。又叫大约克猪。原产于英国，特称为英国大白猪。输入苏联后，经过长期风土驯化和培育，成为苏联大白猪。后者的体躯比前者结实、粗壮，四肢强健有力，适于放牧。

③成华猪。俗话说“家家都有黑毛猪”，这种黑毛猪便是成华猪，全身黑毛、四肢短小、体型膘肥。成华猪是成都猪肉市场的主力品种，也是回锅肉的“最好搭档”。成华猪是成都土生土长的黑毛猪。

④杜洛克。又称万能猪种，原产于美国东部的新泽西州和纽约州等地，用纽约州的杜洛克与新泽西州的泽西红杂交育成，原称杜洛克泽西，后统称杜洛克，分为美系和加系杜洛克；产于中国台湾的杜洛克经过培育自成风格，称台

湾杜洛克或台系杜洛克。

现在国内养殖主要以杜洛克、长白猪、大白猪三元杂交为主。杂交猪具有产仔多、死亡率低等优点，是农村养猪的主要品种。

3. 经验积累。需要哪些经验，如何获得这些经验，需要多长时间（精确到月）？

答：①高度的智能化养殖。在丹麦皇冠集团，高度的机械化和智能化养殖让养猪变得非常轻松。丹麦皇冠的小猪仔们作息时间表完全实行科学统一的管理，定点喂食、定点喝水、定点排便、定点运动、定点听音乐、定点体检、定期检疫。一个1.7万头生猪的养猪场里，只有6名工人，另一个3万头生猪的养猪场里也只有12名工人。大部分繁重的体力活已被机器取代。

②所有猪农都有学位。在皇冠，甚至在整个丹麦，养猪都必须具有专业学位。猪农不但要学习专业养猪知识，还要学习管理和哲学知识。猪农汉森花了8年时间才获得养殖相关学位，他不仅是位农民，更是位有经验的专家。

③人可直接食用猪饲料。丹麦法律规定生猪养殖过程中，饲料不得添加生长激素和瘦肉精。皇冠的饲料更是精品中的极品，它由小麦、大麦和矿物质混合而成，呈黄色小颗粒，口感甜而细软，连人都能直接食用。

④无痛宰杀。在皇冠，猪在被宰杀之前，养殖者会让它们享受音乐、洗澡和两个小时的休息。然后被送入二氧化碳室，40秒内陷入昏迷，之后无痛死亡。皇冠认为这种做法不仅是出于人道主义，更是为了追求高品质的猪肉。因为只有在完全放松的情况下屠宰，猪体内才不会分泌应激激素，肉质才会更好。除了这些规则，皇冠在猪的药品使用和运输等环节还有多达100项的要求。

⑤充分利用猪粪。皇冠通过粪浆热能回收系统，利用热泵产生可持续再生能源，粪浆中的热能被收集用于供暖，使用热泵可以节约75%的供暖成本，并且摆脱了以煤为主的燃料加热方式，减少了环境的污染，节约了大量的能源，不仅不影响粪污清除，而且粪浆冷却后有助于减少氨气和臭味的排放。

以上经验都可以通过外部学习，如向优秀的同行学习，以及通过专业化的学习获得。一些技术上的提升需要依赖于专门的人员研究习得。获得这些经验所需要的时间很短，但是将这些经验投入使用的时间则长短不一。

此处进行估算，假设将这些经验学习并运用到猪身上所需的时间约为 12 个月。

4. 对提高效益而进行的投资。随着规模的扩大，面临哪些问题，如何筹措？

答：①疫病是生猪养殖的第一大风险。生猪规模化养殖导致饲养密度增加，加大了感染病原菌的风险，发生传染病的概率随之增加。

②良种问题。至今，也是没有得到解决的问题。完善的良种确保生猪品质、降低生产成本和疫病风险、提高产品质量和产业素质的高度，必须强化原种猪场、二级扩繁场及猪人工授精中心建设。

③大部分规模养猪采取合作社的形式，没有实现统一购买仔猪饲料、统一防疫、统一销售，而且大部分小规模养殖会员的生产呈分散状态，规模效应尚未生成，由此导致分散生产与集中加工无法形成完善的利益对接机制，使得饲养户无法形成稳定收入的预期。

④养猪规模扩大，需要扩大用地及大规模资金等。实际生产中，申请用地建设手续复杂、条件要求较高，申请难度大。环境方面，随着养猪业进入规模化经营，饲养量的不断增加，使得废水排放量增加，绝大多数养猪场未考虑粪便处理，致使粪便随地堆积，严重地污染了周围的环境。国家现在对养猪环境和排放有严格规定。

投资额通常是通过：①自行循环累积，将所得不断投入扩大规模和规范养猪上；②合作社，多个商户联合在一起，形成生产合作社；③通过银行贷款，获得养殖专用贷款。

年出栏瘦肉猪 1 万头，总投资 1350 万元，其中固定资产 1270 万元，流动资金 80 万元。各项投资情况：猪舍建筑 1 万平方米×300 元/平方米，280 万元；饲料厂（年产 0. 3 万吨），100 万元；办公、生活用房 600 平方米×400 元/平方米，24 万元；兽医室及仪器设备，5 万元；水电、污水处理设施，50 万元；围墙、道路及其他设施，30 万元；种公猪存栏 30 头×1500 元/头，4. 5 万元；母猪存栏 640 头×1500 元/头，96 万元；其他大小猪存栏 0.55 万头×330 元/头，181.5 万元；母猪笼 500 只×220 元/只，11 万元；产仔笼 200 套×1400 元/套，28 万元；保育舍笼架、漏缝地板 50 个×2000 元/只，10 万元；土地 3.3 公顷×120 万元/公顷（平整和出让费），400 万元；运输、交通工具，50 万元；流动资金，80 万元；合计 1350 万元。

关于不同规模的养猪成本之前呈现过，其总成本已经包含固定成本和可变成本，乘以相应的头数即可。本题的统计与表格中数据基本相符。

5. 风险因素。养猪面临的疫病及关键管理技术有哪些，请着重分析非洲猪瘟的影响及关键管理制度。

答：传统养猪主要预防的疫病包括猪瘟（HC）、猪丹毒、猪肺疫、猪副伤寒、猪口蹄疫（FMD）等。对于 HC 和 FMD 必须对猪进行疫苗注射，其他可以通过改变饲养条件和药物预防来发挥作用。

对于种猪的疫病主要有猪乙型Ⅰ脑炎（JE）、猪细小病毒病（PPV）、猪伪狂犬病（PR）、猪繁殖与呼吸综合征［蓝耳病（PRRS）］等。由于这类疫病危害甚烈，主要引起种猪的繁殖障碍，有无治疗可能，只能严格按免疫程序进行接种才有可能控制其危害。

对于疫病的关键管理技术包括制定严格的防疫程序、严把猪进出关、严格消毒和病死猪无害化处理。

严格的防疫程序，猪养殖场应根据当地传染病和寄生虫病发生的频率及规律制定严格的免疫程序。免疫接种的疫苗应从具备生物资质的生产厂家或国家规定的防疫机构处购买，并严格按照疫苗使用说明书运输、储藏、使用。每次免疫后应由专人进行详细的免疫登记并对免疫反应进行观察，以便及时处理接种后产生的免疫副反应。

严把猪进出关，严把检疫关，通过定期的疫情监测，及时清除疫源，把疫病消灭在萌芽状态。引进和出售猪时，必须经过动物防疫监督机构检疫人员的严格检疫，检疫不合格的猪严禁出入。养殖场严禁饲养犬、猫等其他动物，并定期进行灭鼠，夏季及时杀灭蚊蝇，谢绝一切外来参观人员。

严格消毒，养殖场大门入口处要设置宽同大门，长等于进场大型机动车辆车轮一周半长的水泥结构的消毒池，生产区门口设有更衣换鞋、消毒室或淋浴室，圈舍入口处要有高长 1 米的消毒池以供进入人员消毒。外来车辆不得进入生产区。每天坚持打扫场内卫生，保持料槽、水槽干净及地面清洁。应因地制宜选用高效低毒、广谱的消毒药品。坚持定期消毒，定期更换消毒药品，并保持其有效浓度。每批猪调出后圈舍要进行彻底的清扫、冲洗和消毒 5~7 天方可进猪，猪群周转要执行全进全出制。

病死猪无害化处理，当养殖场发生传染病或疑似传染病时应及时向当地防疫机构报告疫情，同时根据疫病种类和流行范围采取封锁、隔离、消毒、紧急防疫、治疗措施、应做到早发现、早确诊、早处理、死畜尸体应深埋或焚烧，以防疾病流行，对一时难以确诊或愈后不良的动物，立即淘汰，以减少用药成本及疾病传播。

非洲猪瘟的影响：2018 年 8 月，非洲猪瘟疫情首次暴发于辽宁沈阳，随后向着全国 30 多个省份快速传播蔓延，造成了严重的发病率，大量生猪被扑杀无害化处理，生猪存栏量显著下降，对我国生猪业造成了严重危害。

随着非洲猪瘟疫情在全国 30 多个省份相继暴发，为了防范非洲猪瘟病毒的进一步传播蔓延，很多省份实施了生猪跨区禁运，加上屠宰场加工分布不均匀，给养殖场的正常经营造成了一定难度，很多养殖场普遍存栏量显著增高，资金回流存在很大压力，养殖空间面临巨大挑战。

关键管理制度：①提高养殖场生物安全水平，在生猪养殖产业发展过程中，应该积极开展社会安全培训，由动物防疫部门和行业协会牵头积极组织有经验的生物安全专家队对重点生猪养殖场的畜牧兽医主管人员进行专业技能培训，要及时发现问题，及时排除隐患。另外加强对养殖环境的有效净化，政府部门应该配合生猪养殖产业，主动清退和收编不符合环保要求散养养殖户和中小规模养殖户，避免这类养殖群体成为非洲猪瘟病毒的传播渠道。在非洲猪瘟防控过程中，还应该设置多样化的安全屏障。养殖场的进出入口应该设置检查站检疫站，避免病死猪无关人员和车辆随意进出养殖区域。养殖场在建造过程中应该做到科学合理的选址，要设置一些防御屏障，避免野生动物进入养殖场。从业人员应该加快构建完善的消毒体系，将各项生物安全措施落到实处。

②提高疫情应对及处置能力，从非洲猪瘟能对我国猪价产生如此之大的影响可以发现，我国对紧急疫情的处理能力较为脆弱。整个养猪市场存在各种各样的问题，猪瘟才使得这个问题爆发出来。一方面应该进一步提高病死生猪无害化处置能力，积极研发和利用符合生物安全环境保护的生猪扑杀无害化处理技术。对发病的养殖场处置之后的圈舍、土壤、植被、水源进行严格监测，确保疫源得到彻底的清除。另一方面应该严格控制从疫点流出的生猪和猪肉制品，并进行严格的无害化处理。

③科学调运生猪。在非洲猪瘟疫情流行大背景下，就需要制定符合实际情况的生猪禁运政策，在始终坚持国家所制定的禁运政策的基础上，结合风险评估和地方的实际情况，科学划定疫区，要确保猪肉销售渠道通畅，保证市场上的猪肉制品正常供给，防范规模化养殖场未被疫情影响，而被禁运政策影响。避免因为禁运政策而造成市场生猪制品供给不足，供给失衡。还应该进一步加强产地检疫和运输检疫，对检疫人员进行严格的培训，确保专业人员能够通过临床观察，并借助快速检测技术及时发现患病猪，及时采取措施处置疫情，控制疫病的扩散蔓延。更为重要的是应该加强对生猪制品的严格检疫检验，对进口屠宰场和市场上的猪肉和相关产品应该按照产地养殖场生产批次进行严格的抽样检测，要确保非洲猪瘟疫情的检测排查工作从养殖领域到生产领域，再到消费者的餐桌，全面覆盖，防止带病的猪肉制品在市场中流通，避免病毒通过猪肉制品传播。

6. 用系统动力学建模，给出仿真方程及仿真图形。

假设销售收入的30%用于规模扩大、20%用于提高生产效率的行为（如改进设备，改进管理制度）、10%用于拓宽产品线。流程改进和投入呈正比关系 $y=x$。经验累积和收入也呈 $y=x$ 正比关系。养猪收入初值设为13500000元，为初始投资额。以大规模养猪场每头成本作为成本的初始值，为1530元，对影响成本的因素乘以-0.00001的系数，并且将其影响因素都相乘。生猪的价格以2018年猪瘟后价格为例，约为1.03倍的成本，产品吸引力为2000年生猪价格。

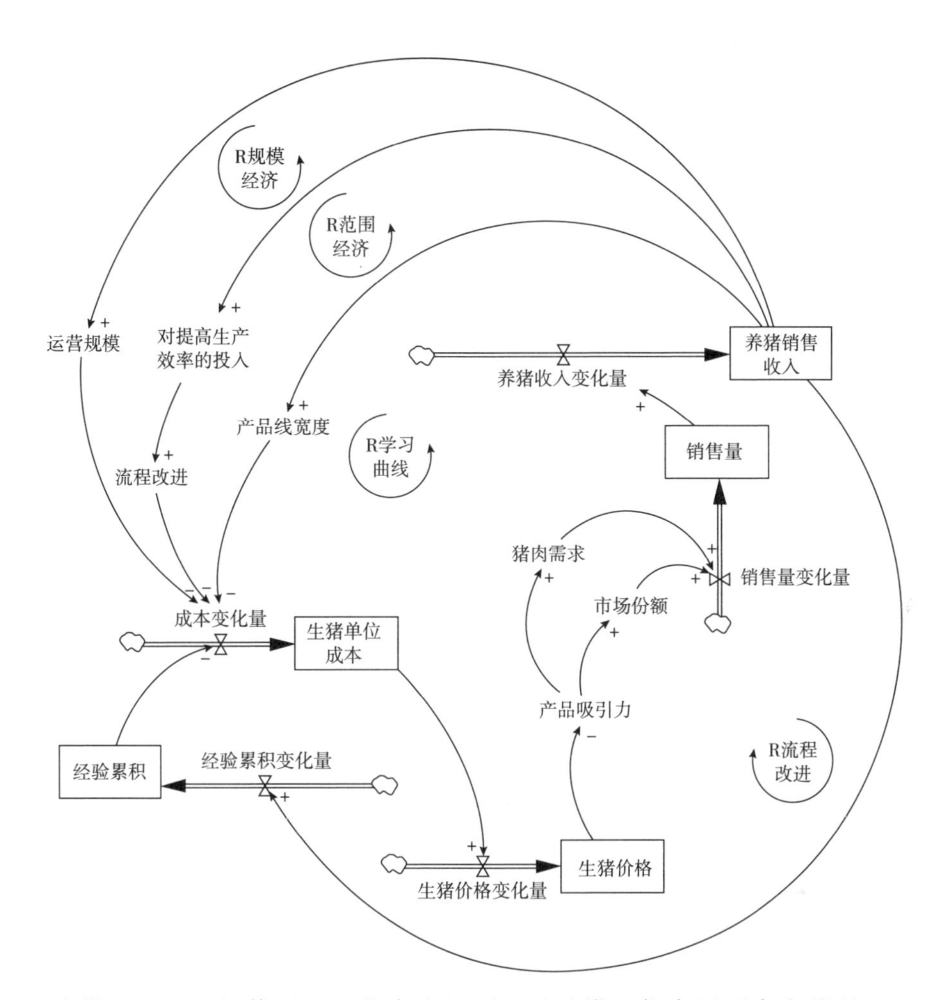

本模型的一些初值设置可能存在问题，导致模型仿真图形存在错误。

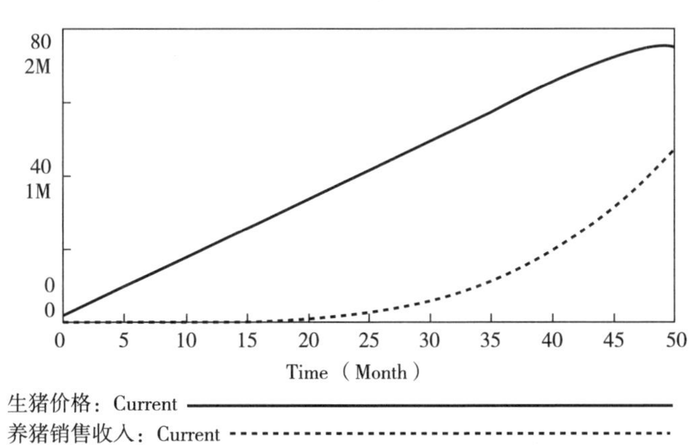

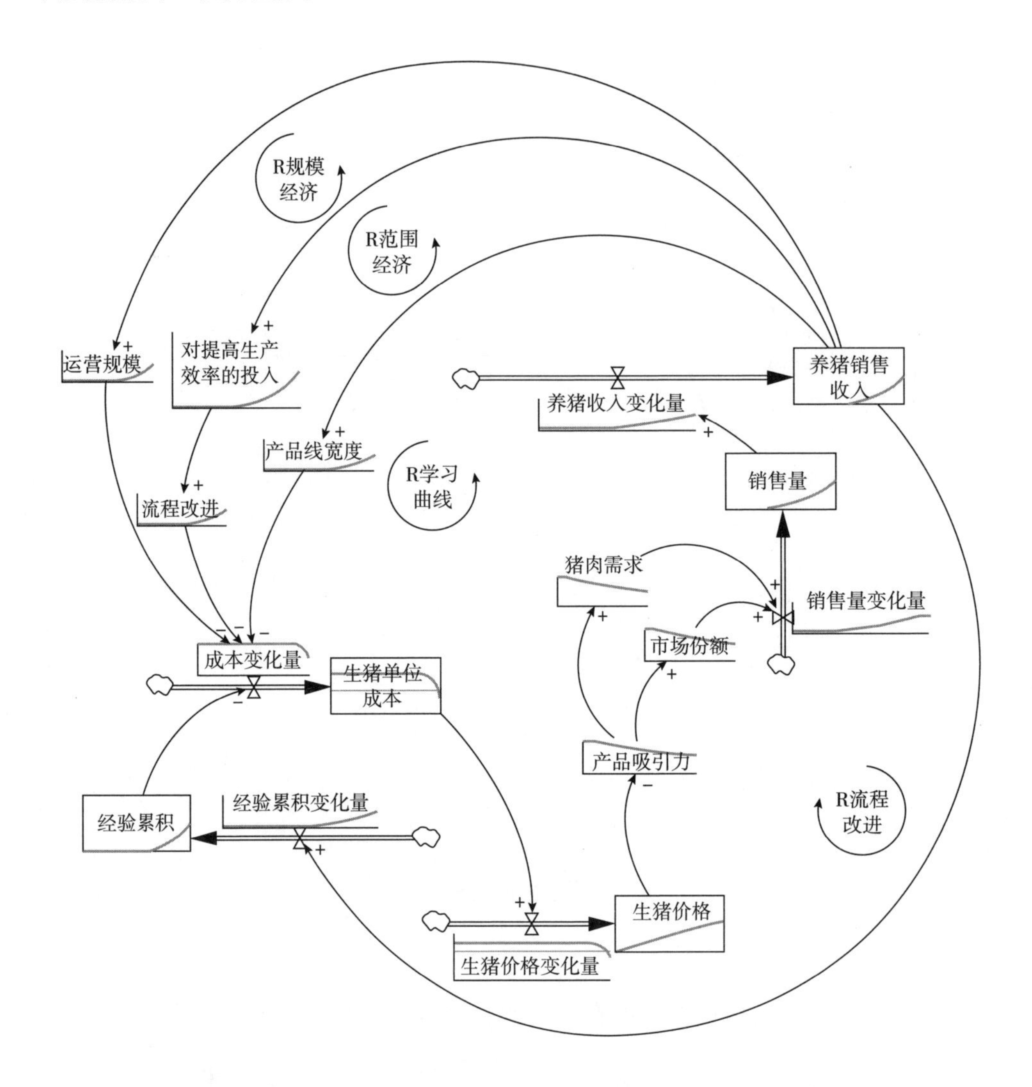

仿真结果可以参看上方模型的小图。因为数值设置问题，本模型有错，需要改进。

上机实验解答 7

1. 疾病传染模型 PPT。

（1）疾病传染的简单模型：SI 模型。

1）假设：不考虑人口的出生、死亡和迁移，模型总人数不变；被感染者无法恢复，且不免疫，不隔离；人群单一同质；生活方式相同。

2）适用：慢性传染。

3）模型。

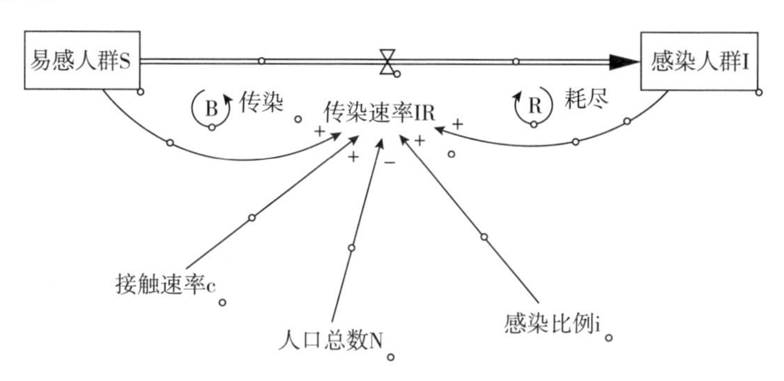

4）仿真方程。

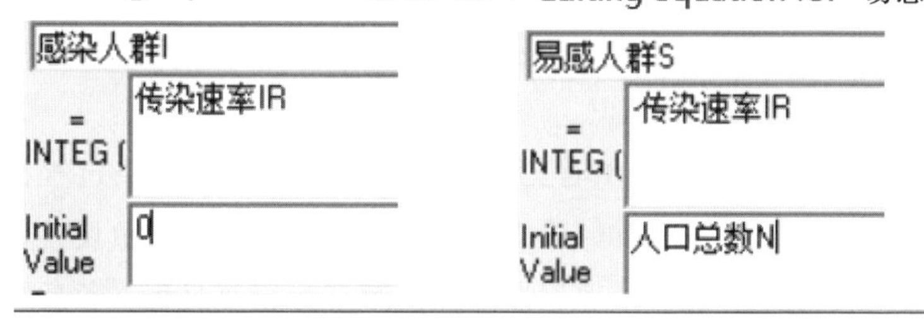

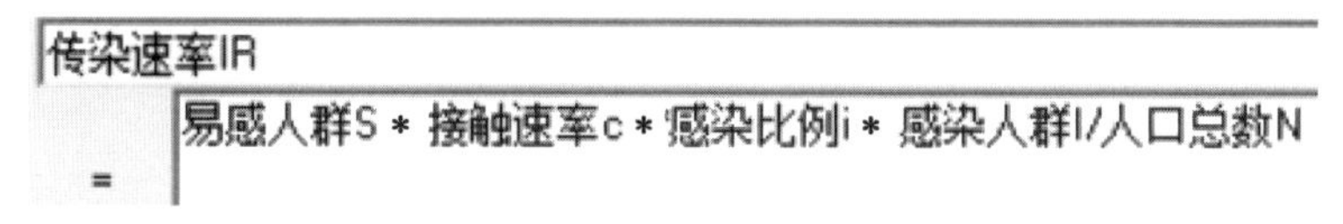

5）符号的解释：如图中所标识。

6）反馈过程描述。该模型有两个反馈过程：第一个为正反馈过程，即传染过程，传染病通过感染人群和易感人群的不断接触而传播；第二个为负反馈过程，即耗尽作用，随着传染病的传播，易感人群速率减小，传染速率不断下降。

7）传染病的基本特征：疾病通过易感人群与感染人群之间的接触而传播。

（2）疾病传染的简单模型：SIR 模型。

1）假设：不考虑人口的出生、死亡和迁移，模型总人数不变；被感染者可以康复且免疫；人群同质；传染病感染之初为急性感染。

2）适用：急性感染。

3）模型。

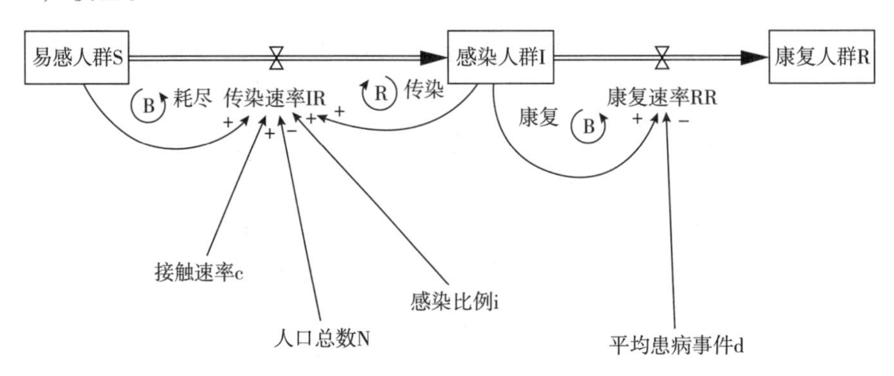

4）仿真方程。

IR = c * i * s * I/N

RR = I/d

5）符号的解释：如图中所标识。

6）反馈过程描述。SIR 模型中有三个反馈结构：第一为感染过程，正反馈，易感者因为与感染者接触增加感染人数；第二为耗尽过程，负反馈，易感者人数减少产生耗尽效果，感染速率逐渐降低；第三为康复过程，负反馈，感染者康复减少感染者人数。

7）传染病的基本特征：疾病通过易感人群与感染人群之间的接触而传播，患者能够康复并产生免疫。

2. S 增长。

（1）Logistic 模型三种表达的公式。

净出生速率 = g(P，C)P = g * (1-P/C)P

$$P(t)=\frac{C}{1+\left[\frac{C}{P(0)}-1\right]\exp(-g^{*}t)}$$

$$P(t)=\frac{C}{1+\exp[-g^{*}t(t-h)]}$$

（2）Gompertz 模型公式。

$$P(t)=C\exp(-k\exp(-g^{*}t))$$

（3）Richards 模型（1959）的净出生率公式。

$$净出生速率=\frac{dp}{dt}=\frac{g^{*}p}{(m-1)}\left[1-\left(\frac{p}{c}\right)^{m-1}\right]$$

$$净出生率=\frac{g^{*}}{(m-1)}\left[1-\left(\frac{p}{c}\right)^{m-1}\right]$$

（4）Weilbull 模型公式。

$P(t)=C\{1-\exp[-(t/b)^{a}]\}$

（5）Rayleigh 分布公式。

$P(t)=C\{1-\exp[-(t/b)^{z}]\}$

3. 创新传播中新观念和新产品的建模。

（1）模型。

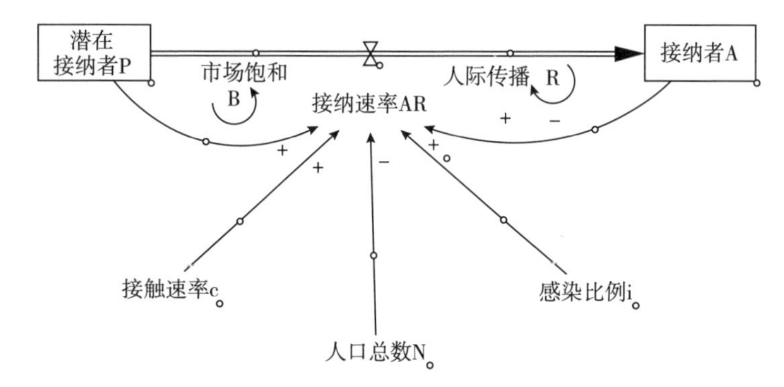

（2）公式。

AR＝A ∗ c ∗ i ∗ P/N

（3）符号解释。

接纳者 A：接受了新观点或购买了新产品的人。

潜在接纳者 P：潜在地接受了新观点或购买了新产品的人。

感染比例 i：潜在接纳者与接纳者接触之后，转化为接纳者的比例。

接触速率 c：单位时间内潜在接纳者与接纳者之间接触的次数。

人口总数 N：即人口总数。

接纳速率 AR：单位时间内潜在接纳者被转化为接纳者的人数。

4. 创新传播的 Logistic 模型。

（1）公司：数字设备公司，销售 VAX11/750 小型计算机及外设。

（2）客户：大公司、研究组织和大学之中的研发实验室、产品开发部门和科研机构。

（3）用途：数据处理。

（4）图形。

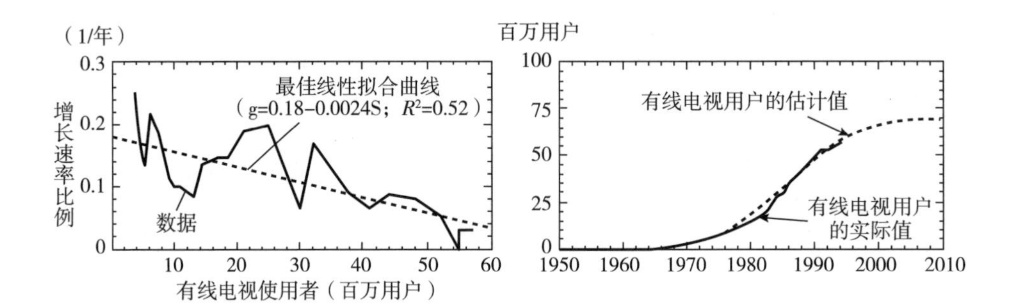

（5）实际曲线。

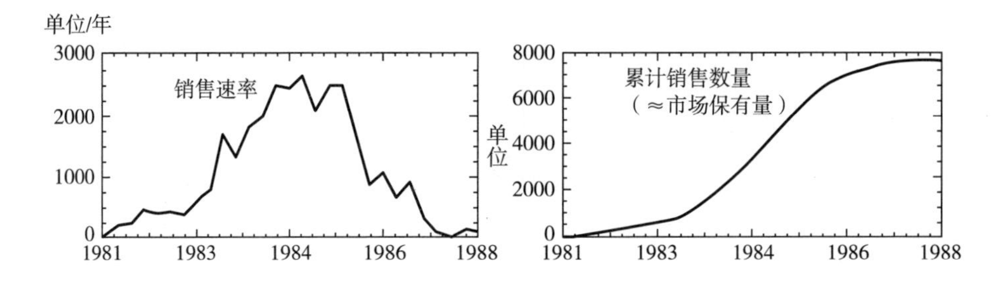

注：左图为销售速率（每季度做一次统计，接年度发布）；右图为累计销传量（粗略等于市场保有量）。

（6）模型得出过程。

1）将创新传播之中的符号代入 Logistic 方程中并进行数学处理，得

$$\ln\left[\frac{A}{N-A}\right]=\ln\left[\frac{A_0}{N-A_0}\right]+g_0t$$

2）将 N−A=P（总人数−接纳者=潜在接纳者）代入得

$$\ln\left(\frac{A}{P}\right)=\ln\left(\frac{A_0}{P_0}\right)+g_0t$$

3）对参数进行线性回归处理，代入作图。

（7）符号表示。

P 为潜在接纳者。

P_0 为潜在接纳者的初始值。

A 为接纳者，在数量上等于计算机保有量。

A_0 为计算机保有量的初始值。

N 为最终接纳者的数量，也即最终计算机保有量。

g 为计算机保有量增长比例。

g_0 为计算机保有量增长比例的初始值。

（8）不足：①这一过程是追溯的，建模使用了所有的时间数据，而在实际的商业活动中往往需要对产品可能的销售趋势进行预测，当拥有足够时间数据使得我们能够看出来产品销售趋势时，就没有预测的意义了。②不能够解决初始值确定问题。

5. Bass 模型。

（1）优点：解决了创新传播的初始值确定问题，能够很好地分析新产品的初始增长。

（2）适用：市场分析、新技术管理等多个领域。

（3）建模目的：用于对新产品销售进行预测。

（4）模型。

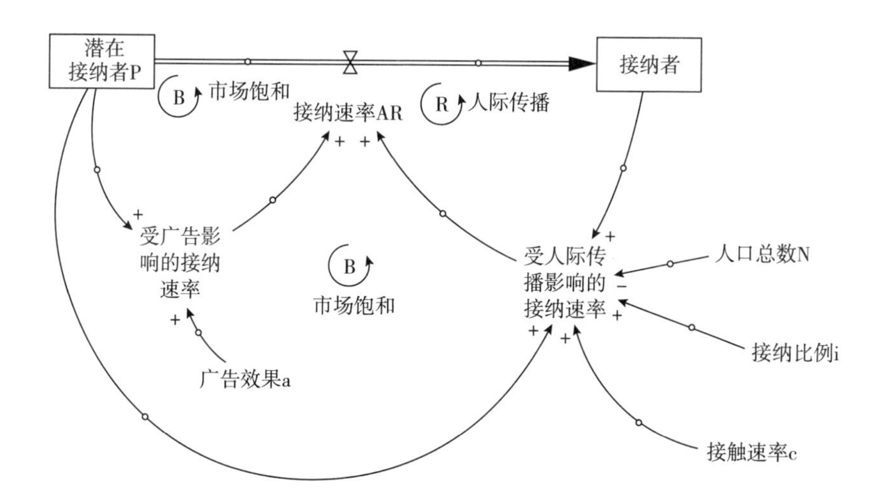

（5）两个反馈过程：人际传播与广告影响。

（6）符号解释。

AR=受广告影响的接纳速率+受人际传播影响的接纳速率

受广告影响的接纳速率$=aP$（a 表示广告的效果，它是单位时间内受广告影响的接纳者占接纳者总数的百分比）

受人际传播影响的接纳速率$=ciPA/N$

（7）仿真方程：$AR=aP+ciPA/N$。

（8）三种条件下的相变图（横轴为接纳者，纵轴为接纳者增加比例）（趋势均为先下降后上升，不过是具体的变化情况不同）。

1）广告宣传效果为零。

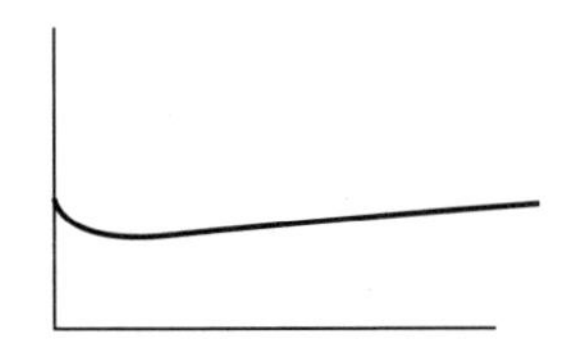

2）人际传播效果为零。

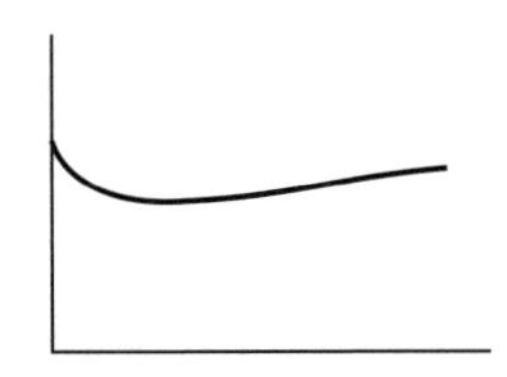

3）人际传播与广告宣传同时起作用。

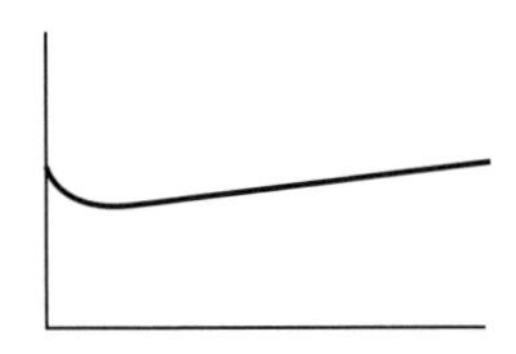

（9）模型的行为特征：解决 Logistic 创新传播模型的初始值确定问题，假设

初始的接纳者来自系统外的广告而不是接纳者自身。

它是对 Logistic 模型在初始阶段的修正，使得预测值不再偏低，与实际值相吻合。

（10）评论 Bass 模型：对反馈结构进行的微小修正，使得模型更好地契合了头两年的销售数据，以及销售数据的最大值。从理论上解决了 Logistic 的初始值确定问题。

（11）拓展：可以应用于众多新产品的推广和增长现象的分析中。

6. 重置消费模型。

（1）产品淘汰和重置模型。

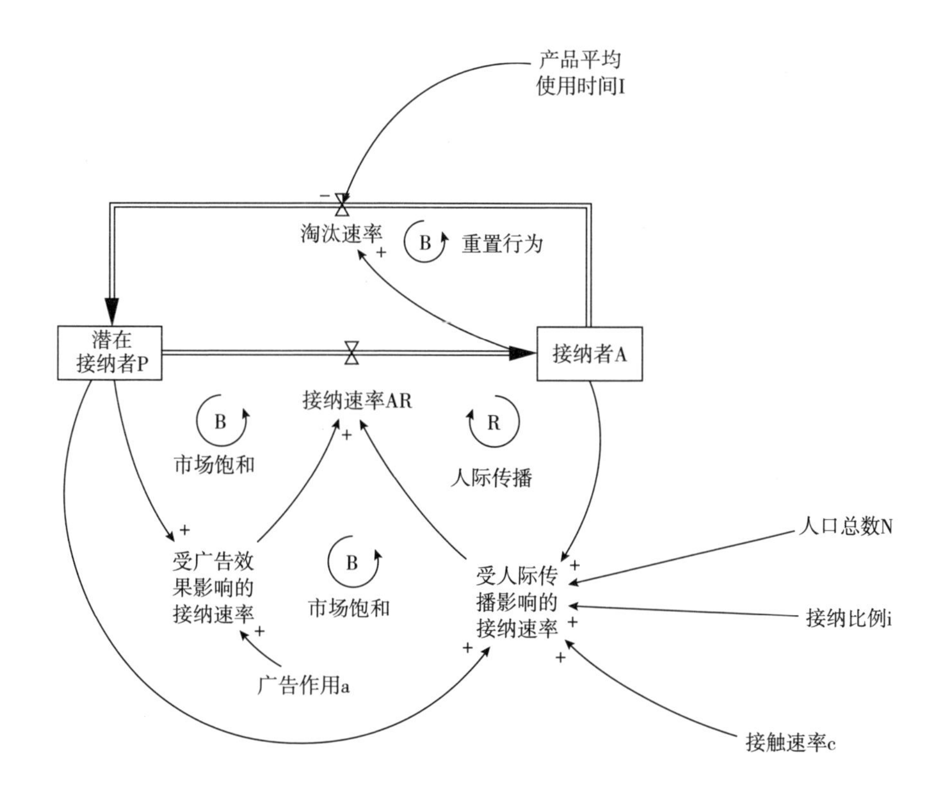

（2）仿真曲线。

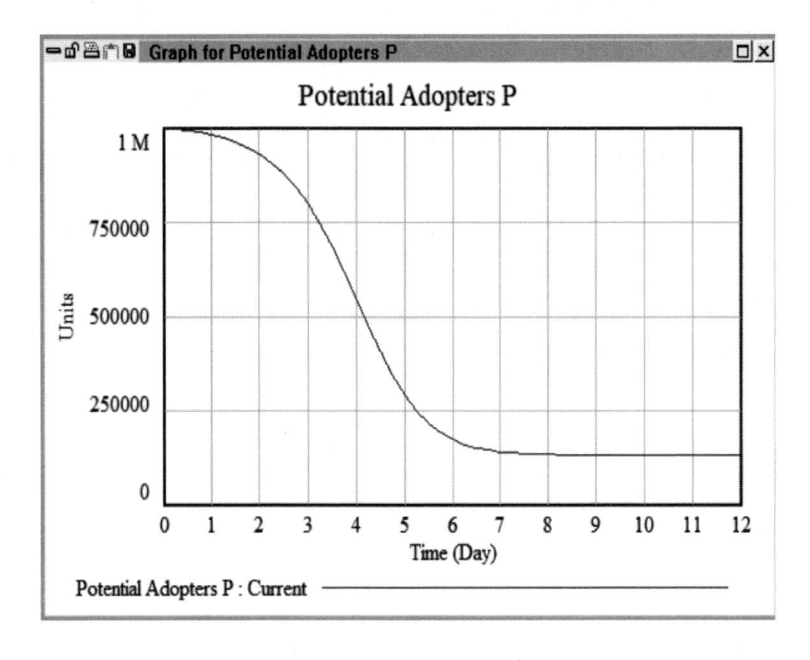
Graph for Potential Adopters P
Potential Adopters P
1 M
750000
500000
250000
0
Units
0 1 2 3 4 5 6 7 8 9 10 11 12
Time (Day)
Potential Adopters P : Current

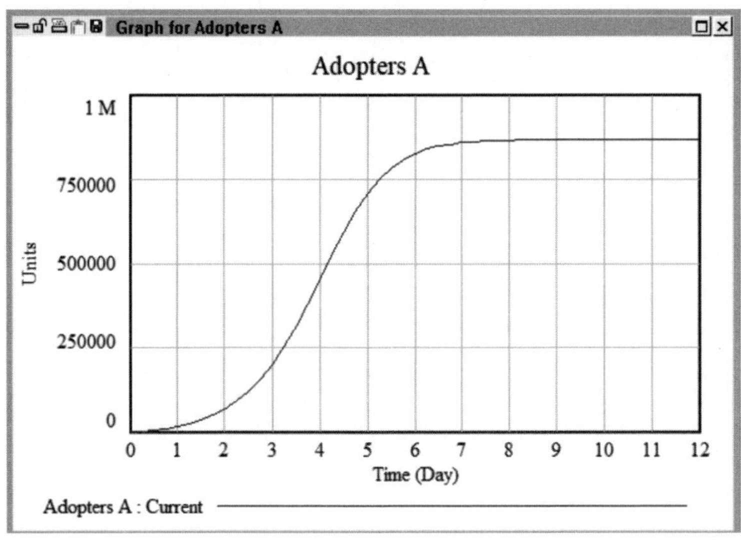
Graph for Adopters A
Adopters A
1 M
750000
500000
250000
0
Units
0 1 2 3 4 5 6 7 8 9 10 11 12
Time (Day)
Adopters A : Current

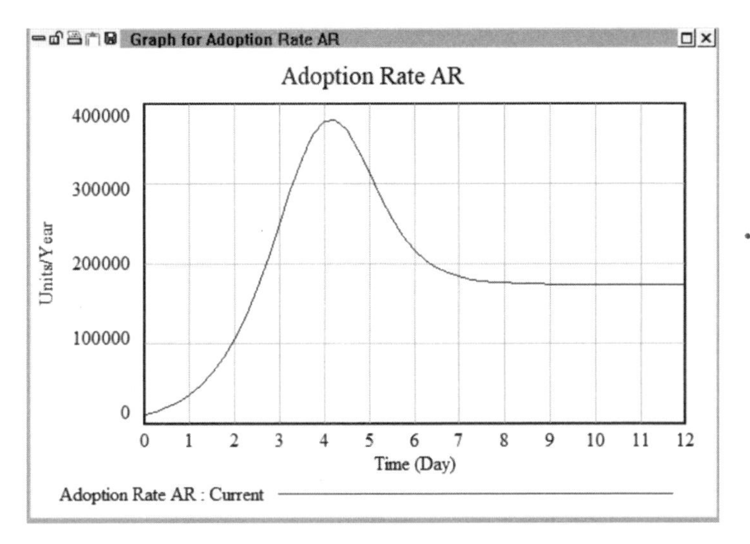

（3）重置购买行为建模。与 Bass 中的产品淘汰和重置建模相比，重置购买行为的建模将潜在接纳者与从接纳者之中重置的潜在接纳者区分开来，将他们的购买决策分为初始购买决策与重置购买决策。其余部分的建模与修正之后的 Bass 模型一致。

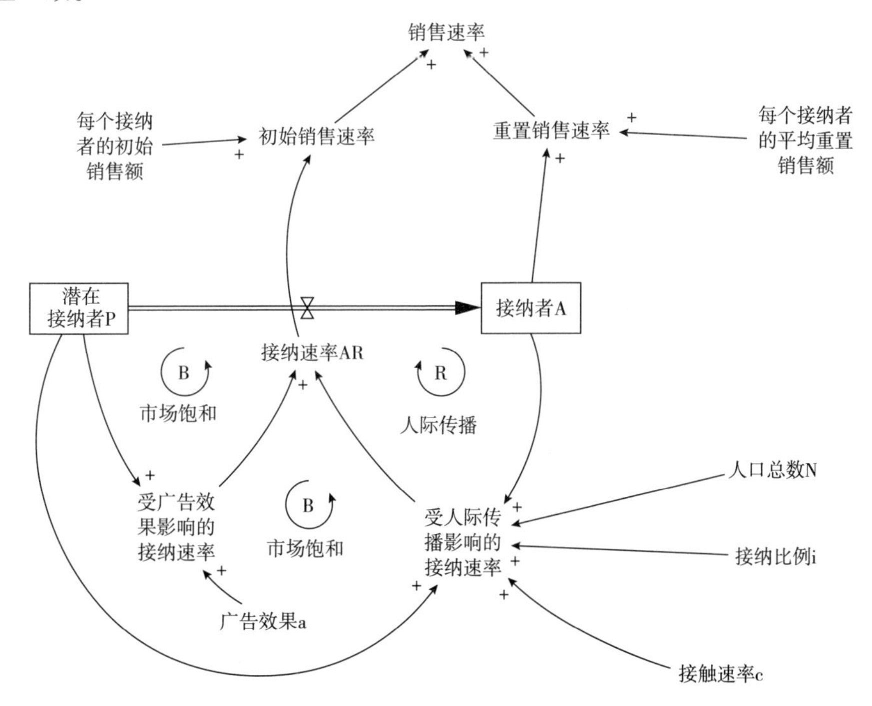

（4）重置消费模型的行为特征。

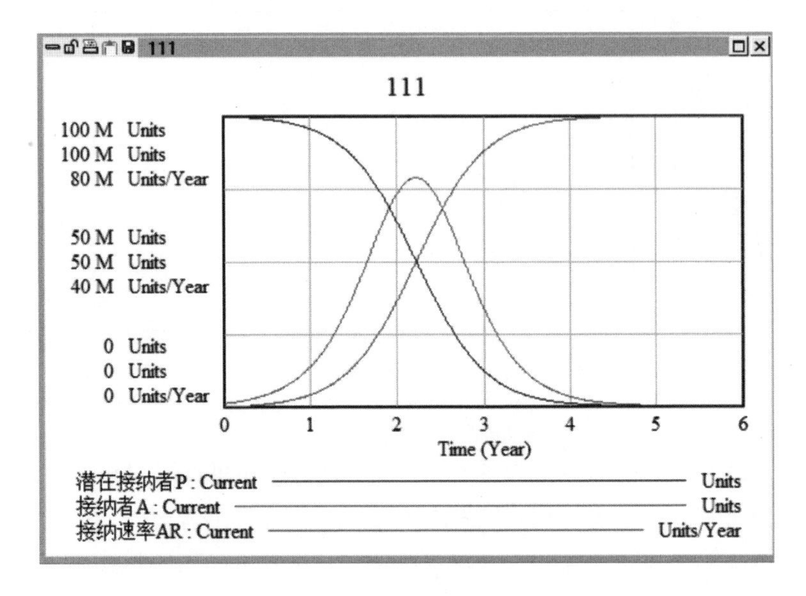

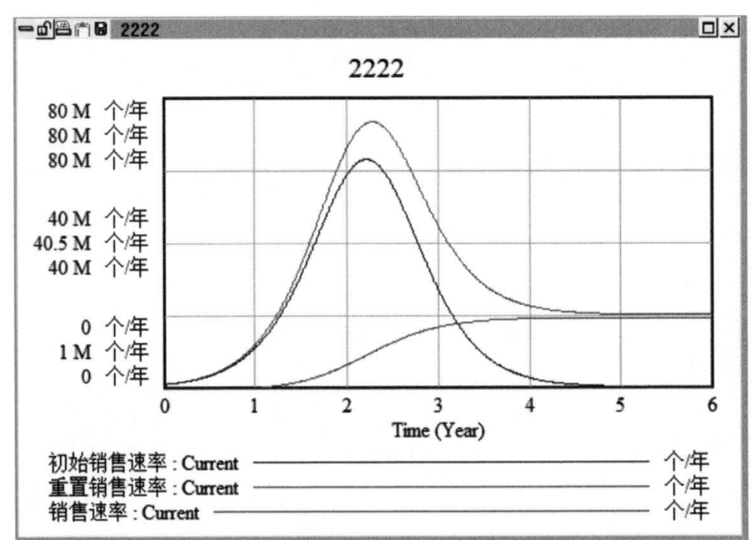

上机实验解答 8

1. 第 1 种人口模型。

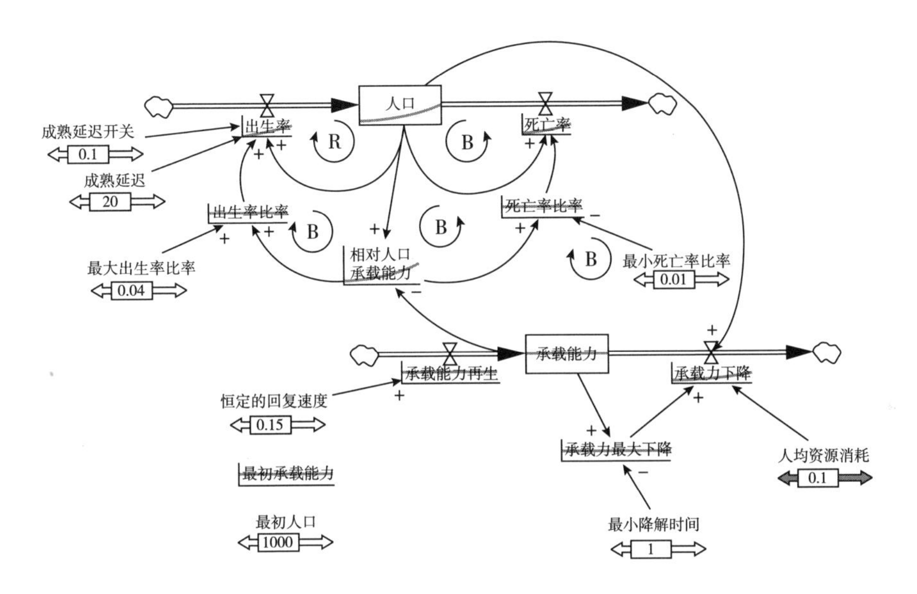

仿真方程：

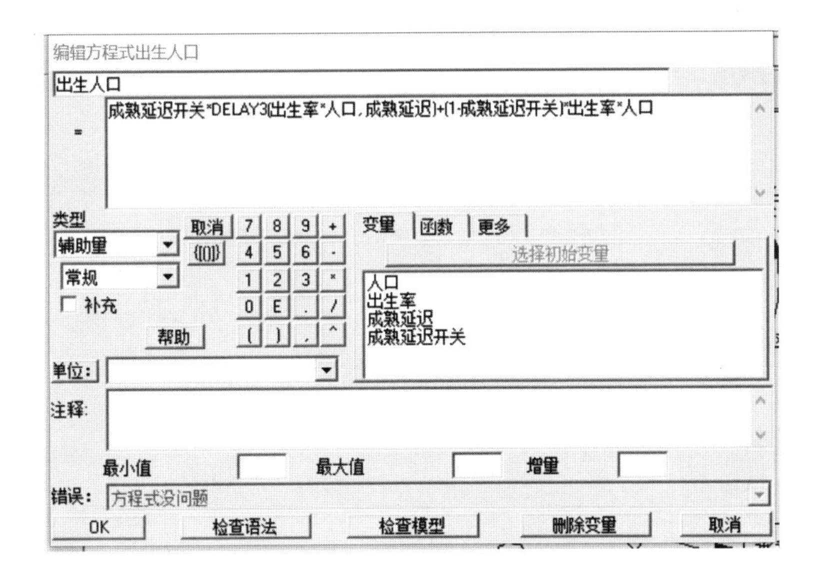

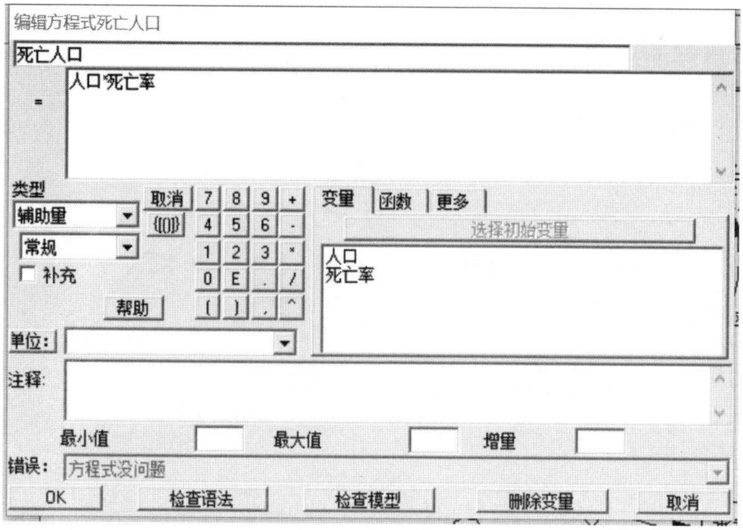

编辑方程式死亡人口

死亡人口

= 人口*死亡率

类型 辅助量 常规 补充 帮助

取消 7 8 9 + {()} 4 5 6 - 1 2 3 * 0 E . / ( ) , ^

变量 函数 更多

选择初始变量

人口
死亡率

单位:

注释:

最小值 最大值 增量

错误: 方程式没问题

OK 检查语法 检查模型 删除变量 取消

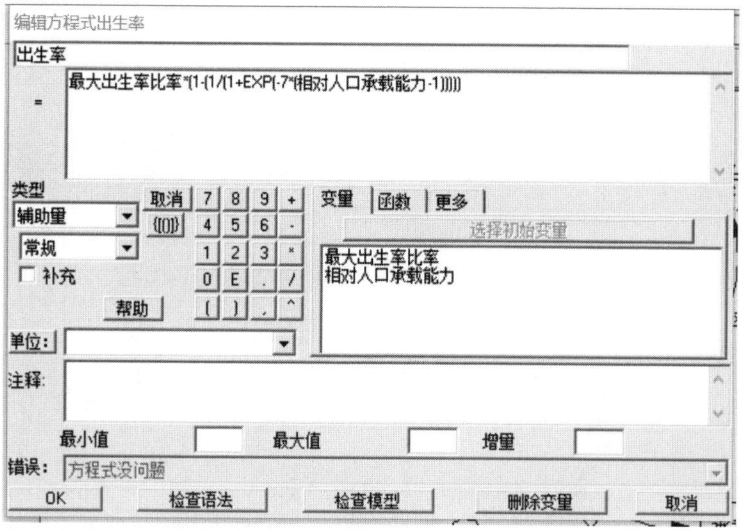

编辑方程式出生率

出生率

= 最大出生率比率*(1-(1/(1+EXP(-7*(相对人口承载能力-1)))))

类型 辅助量 常规 补充 帮助

取消 7 8 9 + {()} 4 5 6 - 1 2 3 * 0 E . / ( ) , ^

变量 函数 更多

选择初始变量

最大出生率比率
相对人口承载能力

单位:

注释:

最小值 最大值 增量

错误: 方程式没问题

OK 检查语法 检查模型 删除变量 取消

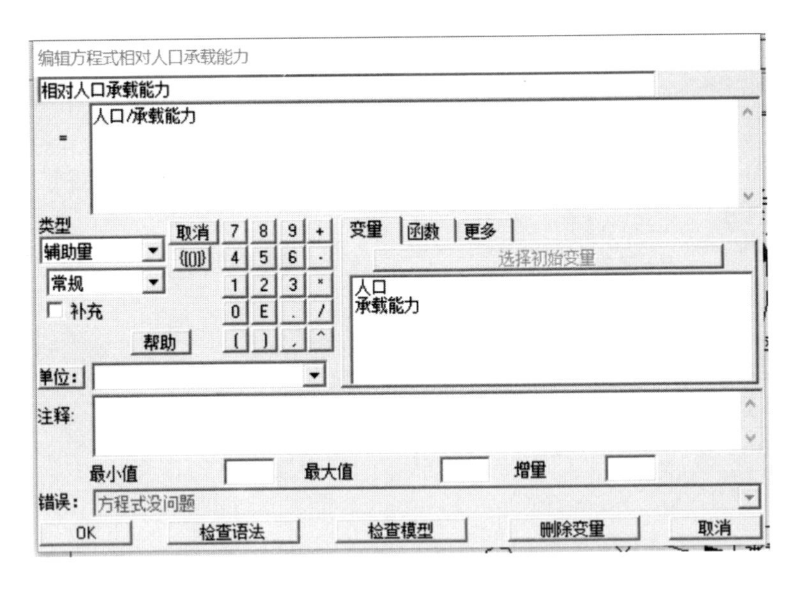
编辑方程式相对人口承载能力
相对人口承载能力
=
人口/承载能力
类型
辅助量
常规
补充
帮助
取消
7 8 9 +
4 5 6 -
1 2 3 *
0 E . /
( ) , ^
变量
函数
更多
选择初始变量
人口
承载能力
单位:
注释:
最小值
最大值
增量
错误:
方程式没问题
OK
检查语法
检查模型
删除变量
取消

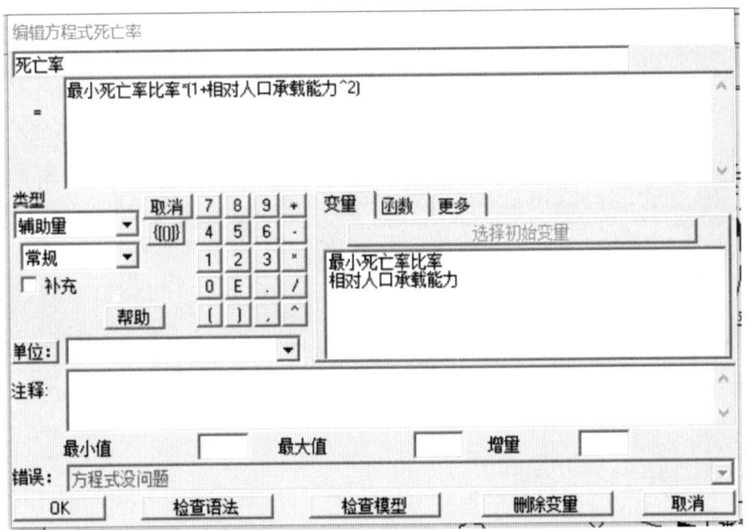
编辑方程式死亡率
死亡率
=
最小死亡率比率*(1+相对人口承载能力^2)
类型
辅助量
常规
补充
帮助
取消
7 8 9 +
4 5 6 -
1 2 3 *
0 E . /
( ) , ^
变量
函数
更多
选择初始变量
最小死亡率比率
相对人口承载能力
单位:
注释:
最小值
最大值
增量
错误:
方程式没问题
OK
检查语法
检查模型
删除变量
取消

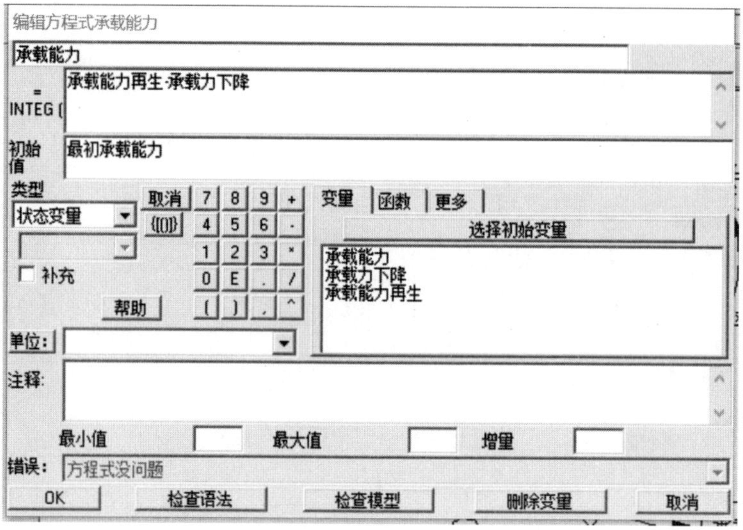

编辑方程式承载能力

承载能力

= INTEG (承载能力再生-承载力下降

初始值 最初承载能力

类型 状态变量

取消 7 8 9 + {[()]} 4 5 6 - 1 2 3 * 0 E . / ( ) , ^

变量 函数 更多

选择初始变量

承载能力
承载力下降
承载能力再生

补充 帮助

单位:

注释:

最小值 最大值 增量

错误: 方程式没问题

OK 检查语法 检查模型 删除变量 取消

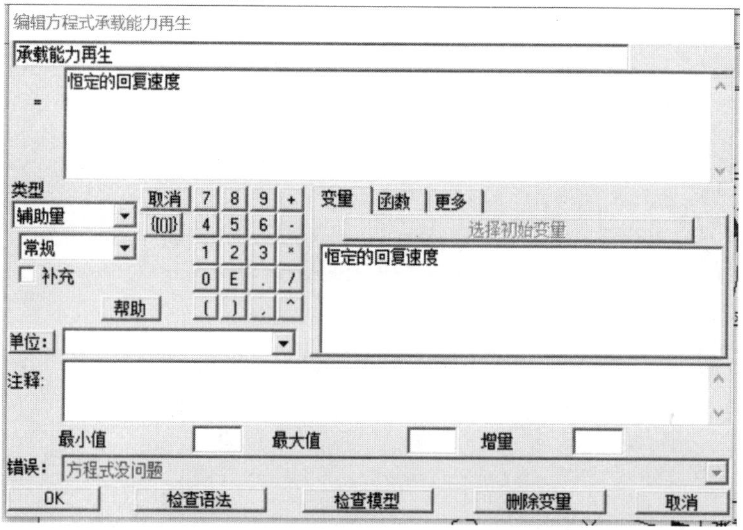

编辑方程式承载能力再生

承载能力再生

= 恒定的回复速度

类型 辅助量 常规

取消 7 8 9 + {[()]} 4 5 6 - 1 2 3 * 0 E . / ( ) , ^

变量 函数 更多

选择初始变量

恒定的回复速度

补充 帮助

单位:

注释:

最小值 最大值 增量

错误: 方程式没问题

OK 检查语法 检查模型 删除变量 取消

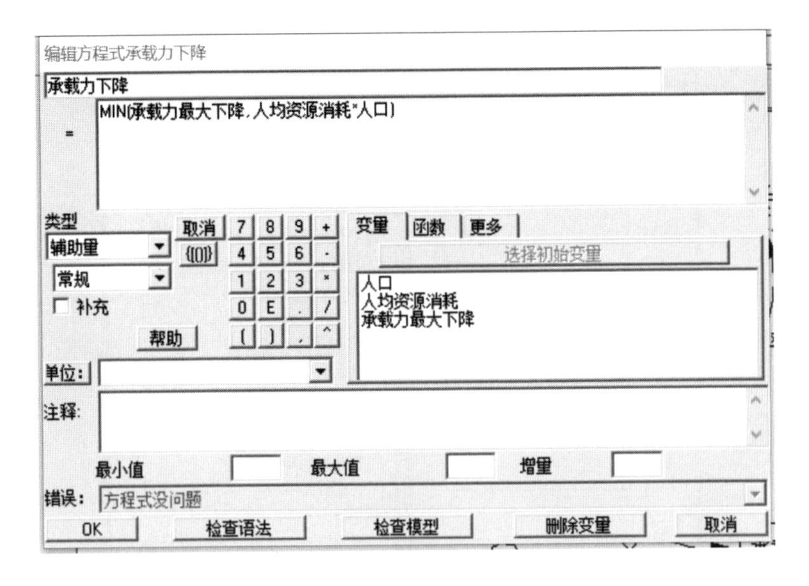

编辑方程式承载力最大下降
承载力最大下降
=
承载能力/最小降解时间
类型
辅助量
常规
补充
取消
帮助
变量
函数
更多
选择初始变量
承载能力
最小降解时间
单位:
注释:
最小值
最大值
增量
错误:
方程式没问题
OK
检查语法
检查模型
删除变量
取消

仿真结果：

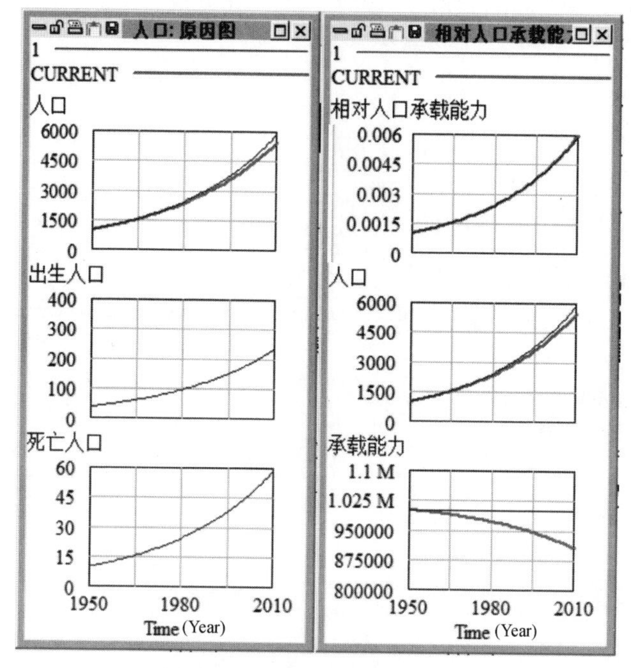
人口: 原因图
1
CURRENT
人口
6000
4500
3000
1500
0
出生人口
400
300
200
100
0
死亡人口
60
45
30
15
0
1950
1980
2010
Time (Year)
相对人口承载能力
1
CURRENT
相对人口承载能力
0.006
0.0045
0.003
0.0015
0
人口
6000
4500
3000
1500
0
承载能力
1.1 M
1.025 M
950000
875000
800000
1950
1980
2010
Time (Year)

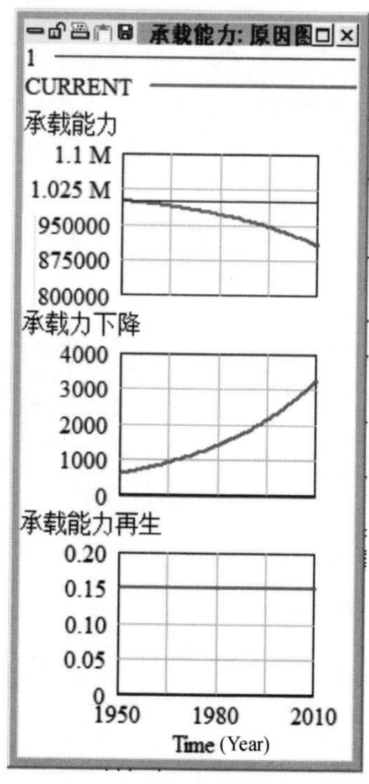
承载能力: 原因图
1
CURRENT
承载能力
1.1 M
1.025 M
950000
875000
800000
承载力下降
4000
3000
2000
1000
0
承载能力再生
0.20
0.15
0.10
0.05
0
1950
1980
2010
Time (Year)

深色曲线为当成熟延迟开关为0.1、恒定恢复速度为0.15、人均资源消耗为0.1时的仿真结果。浅色曲线为当成熟延迟开关为0、恒定回复速度为0、人均资源消耗为0时的仿真结果。

该模型满足仿真定理条件，可以仿真，变化趋势与预期实际变化趋势大致相同，仿真结果通过行为检验。

2. 第2种人口模型。

（1）建模。

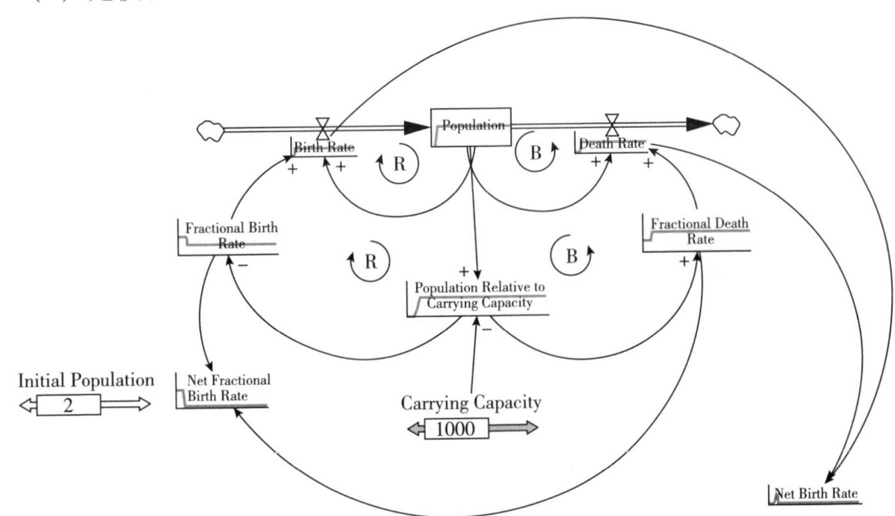

（2）仿真方程。

编辑方程式Population
Population
= INTEG (Birth Rate-Death Rate
初始值 Initial Population
类型 状态变量
补充
帮助
取消 7 8 9 + 4 5 6 - 1 2 3 * 0 E . / ( ) , ^
变量 函数 更多
选择初始变量
Population
Birth Rate
Death Rate
单位：
注释：
最小值 最大值 增量
错误：方程式没问题
OK 检查语法 检查模型 删除变量 取消

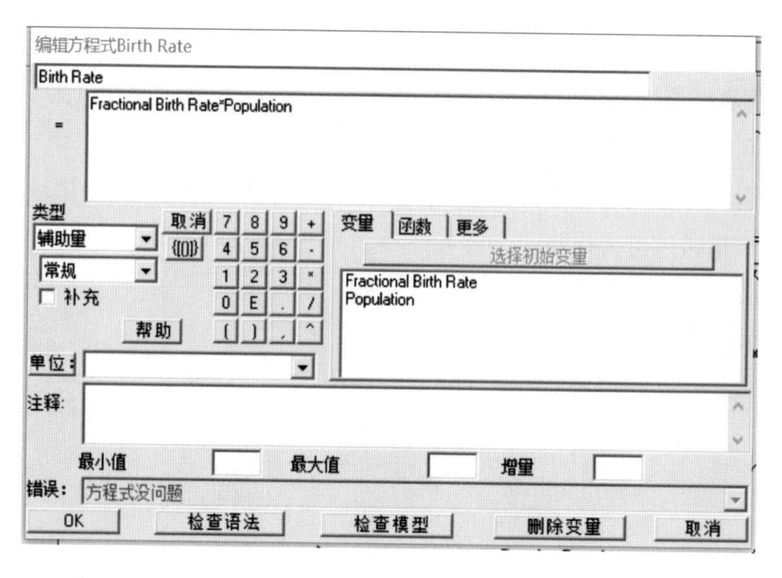
编辑方程式Birth Rate
Birth Rate
=
Fractional Birth Rate*Population
类型
辅助量
常规
补充
取消
帮助
7 8 9 +
4 5 6 -
1 2 3 *
0 E . /
( ) , ^
变量
函数
更多
选择初始变量
Fractional Birth Rate
Population
单位：
注释：
最小值
最大值
增量
错误：
方程式没问题
OK
检查语法
检查模型
删除变量
取消

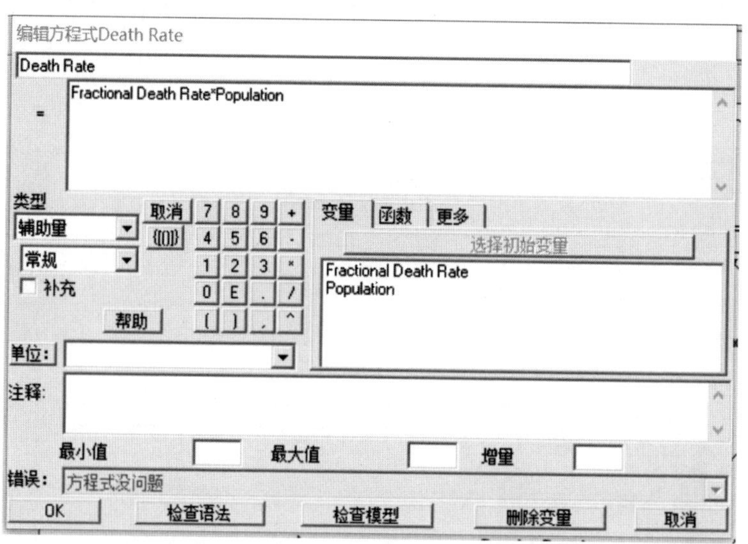
编辑方程式Death Rate
Death Rate
=
Fractional Death Rate*Population
类型
辅助量
常规
补充
取消
帮助
7 8 9 +
4 5 6 -
1 2 3 *
0 E . /
( ) , ^
变量
函数
更多
选择初始变量
Fractional Death Rate
Population
单位：
注释：
最小值
最大值
增量
错误：
方程式没问题
OK
检查语法
检查模型
删除变量
取消

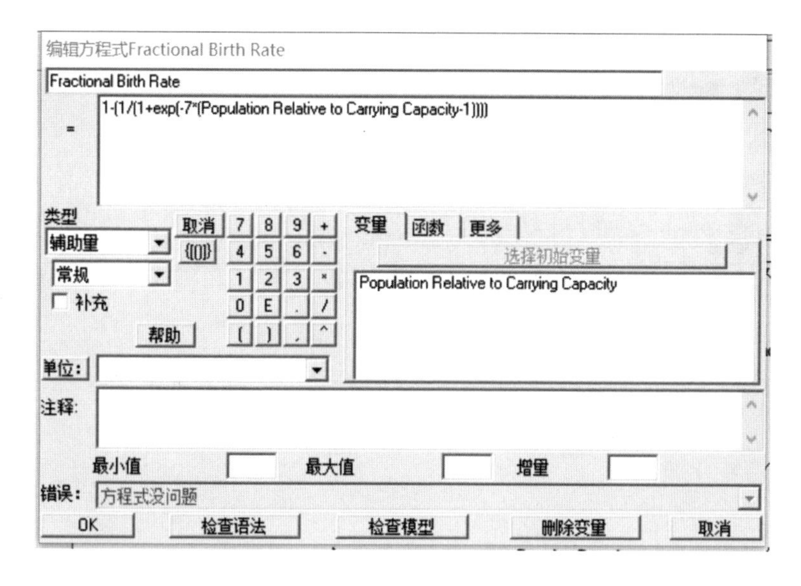
编辑方程式Fractional Birth Rate
Fractional Birth Rate
=
1-(1/(1+exp(-7*(Population Relative to Carrying Capacity-1))))
类型
辅助量
常规
补充
取消
帮助
变量
函数
更多
选择初始变量
Population Relative to Carrying Capacity
单位:
注释:
最小值
最大值
增量
错误:
方程式没问题
OK
检查语法
检查模型
删除变量
取消

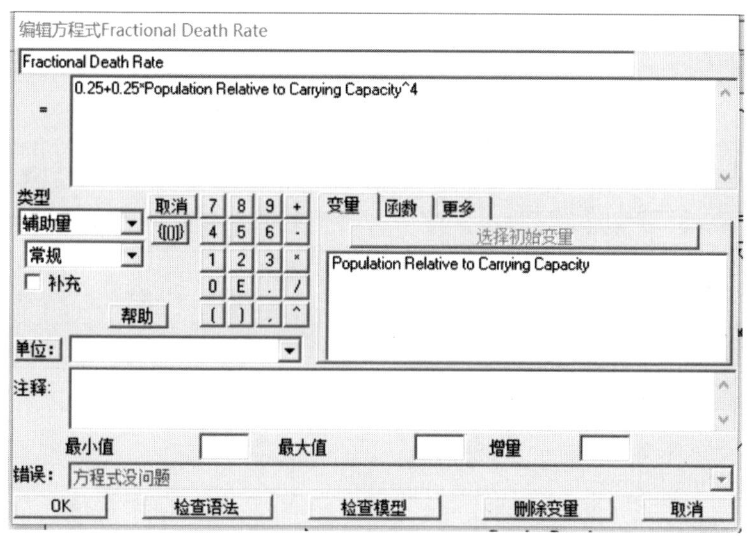
编辑方程式Fractional Death Rate
Fractional Death Rate
=
0.25+0.25*Population Relative to Carrying Capacity^4
类型
辅助量
常规
补充
取消
帮助
变量
函数
更多
选择初始变量
Population Relative to Carrying Capacity
单位:
注释:
最小值
最大值
增量
错误:
方程式没问题
OK
检查语法
检查模型
删除变量
取消

编辑方程式Population Relative to Carrying Capacity
Population Relative to Carrying Capacity
= Population/Carrying Capacity
类型 辅助量 常规 补充 帮助
取消 {[()]} 7 8 9 + 4 5 6 - 1 2 3 * 0 E . / ( ) , ^
变量 函数 更多
选择初始变量
Carrying Capacity
Population
单位:
注释:
最小值 最大值 增量
错误: 方程式没问题
OK 检查语法 检查模型 删除变量 取消

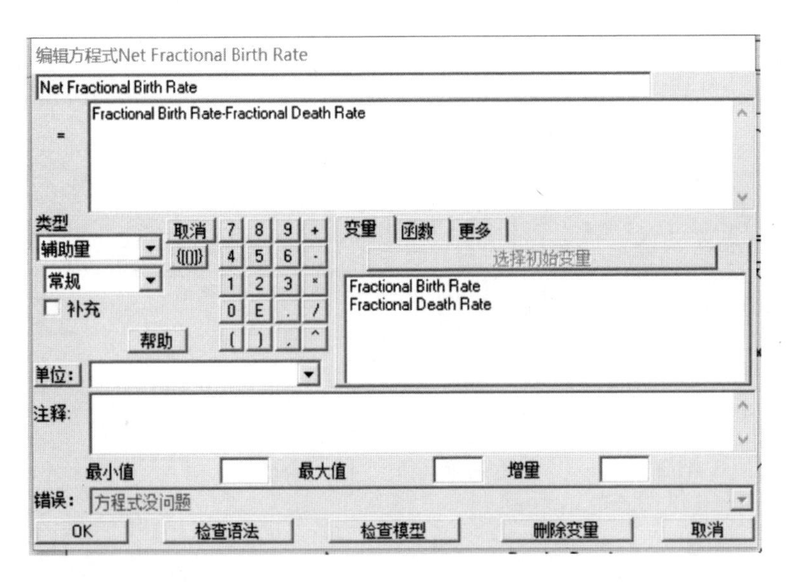

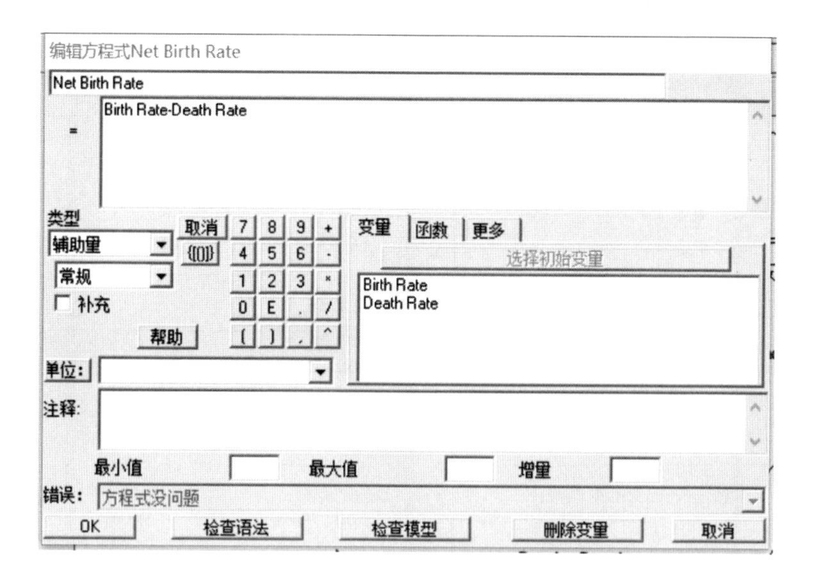
编辑方程式Net Birth Rate
Net Birth Rate
=
Birth Rate-Death Rate
类型
辅助量
常规
补充
取消
帮助
变量
函数
更多
选择初始变量
Birth Rate
Death Rate
单位:
注释:
最小值
最大值
增量
错误:
方程式没问题
OK
检查语法
检查模型
删除变量
取消

（3）仿真结果。

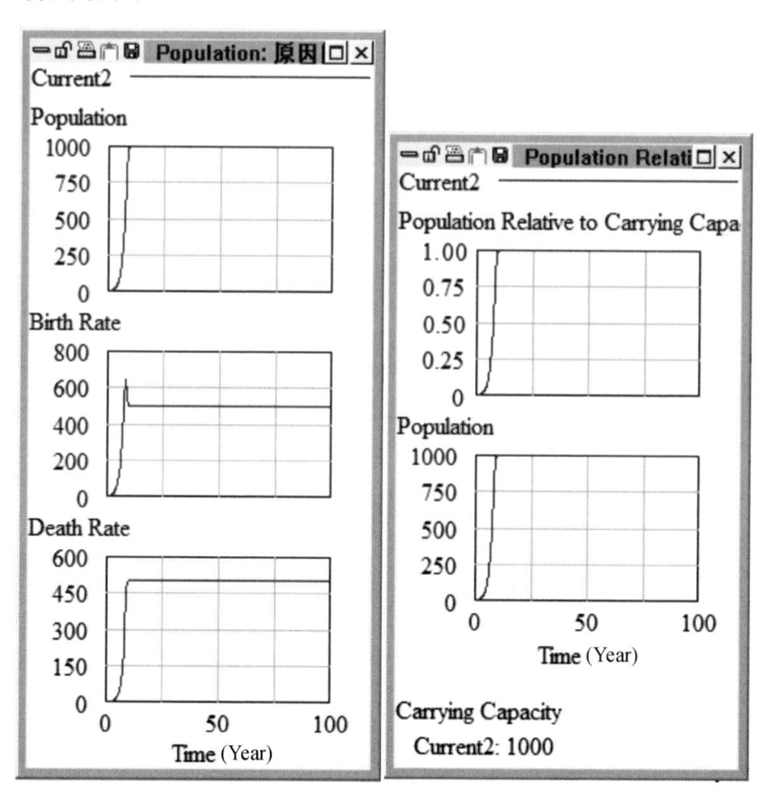
Population: 原因
Current2
Population
1000
750
500
250
0
Birth Rate
800
600
400
200
0
Death Rate
600
450
300
150
0
0
50
100
Time (Year)
Population Relati
Current2
Population Relative to Carrying Capa
1.00
0.75
0.50
0.25
0
Population
1000
750
500
250
0
0
50
100
Time (Year)
Carrying Capacity
Current2: 1000

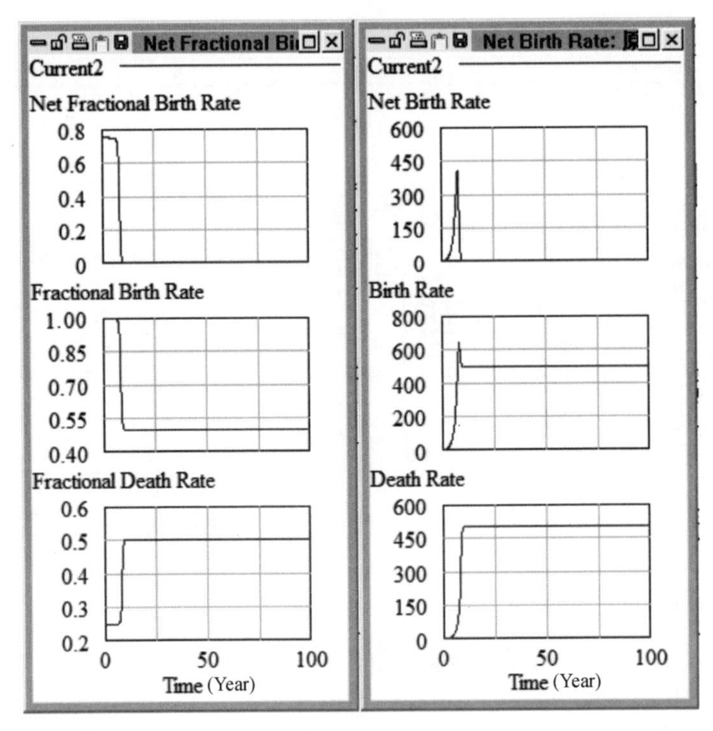

该模型满足仿真定理条件，曲线呈 S 形变化趋势，与预期相符，模型通过行为检验。

3. 第 3 种人口模型。

（1）建模。

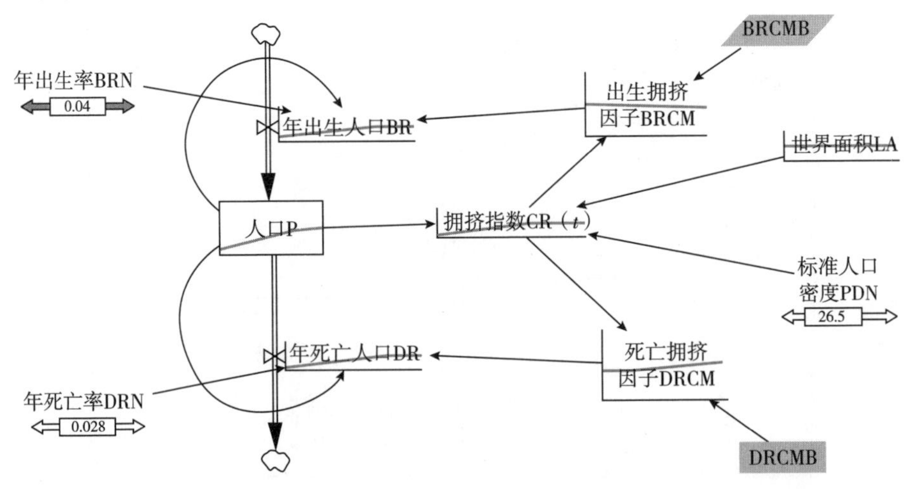

（2）仿真方程。

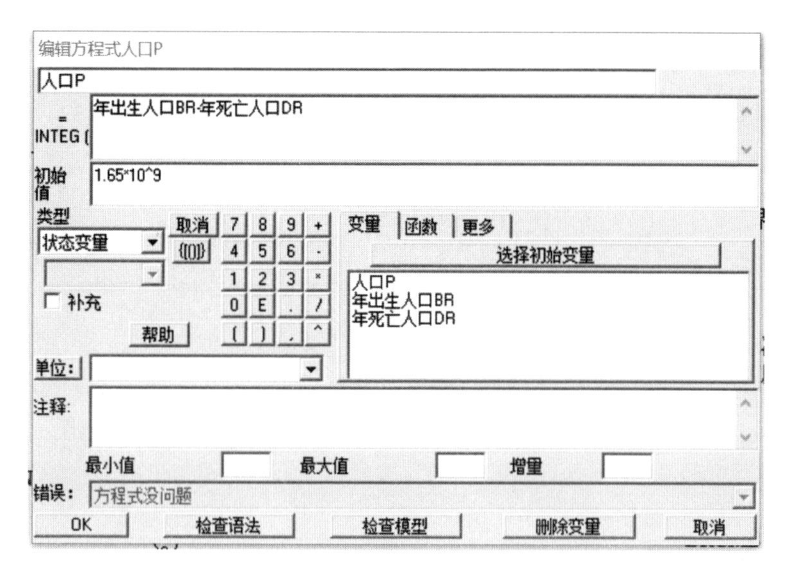

编辑方程式年死亡人口DR

年死亡人口DR

= 年死亡率DRN*人口P*死亡拥挤因子DRCM

类型：辅助量 常规

补充 帮助 取消

变量 函数 更多

选择初始变量

人口P
年死亡率DRN
死亡拥挤因子DRCM

单位:

注释:

最小值 最大值 增量

错误: 方程式没问题

OK 检查语法 检查模型 删除变量 取消

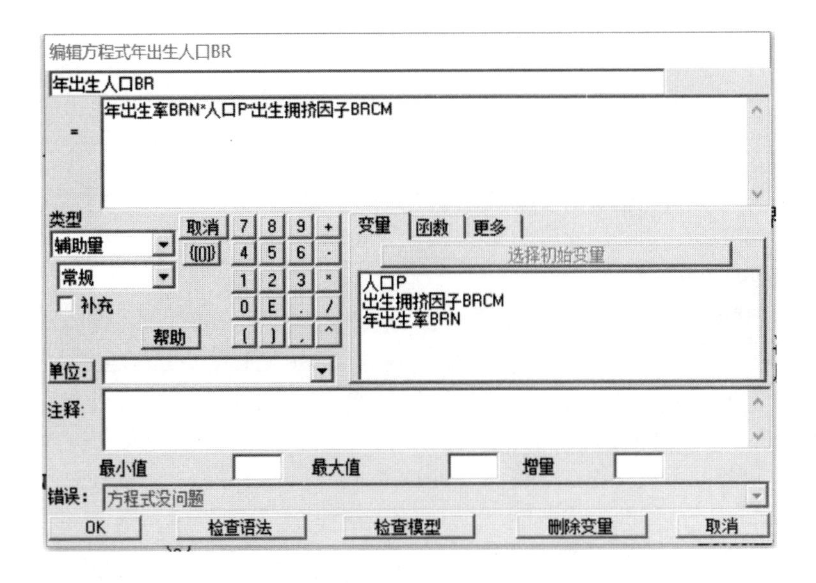
编辑方程式年出生人口BR
年出生人口BR
=
年出生率BRN*人口P*出生拥挤因子BRCM
类型
辅助量
常规
补充
帮助
取消
7 8 9 +
4 5 6 -
1 2 3 *
0 E . /
( ) , ^
变量
函数
更多
选择初始变量
人口P
出生拥挤因子BRCM
年出生率BRN
单位:
注释:
最小值
最大值
增量
错误:
方程式没问题
OK
检查语法
检查模型
删除变量
取消

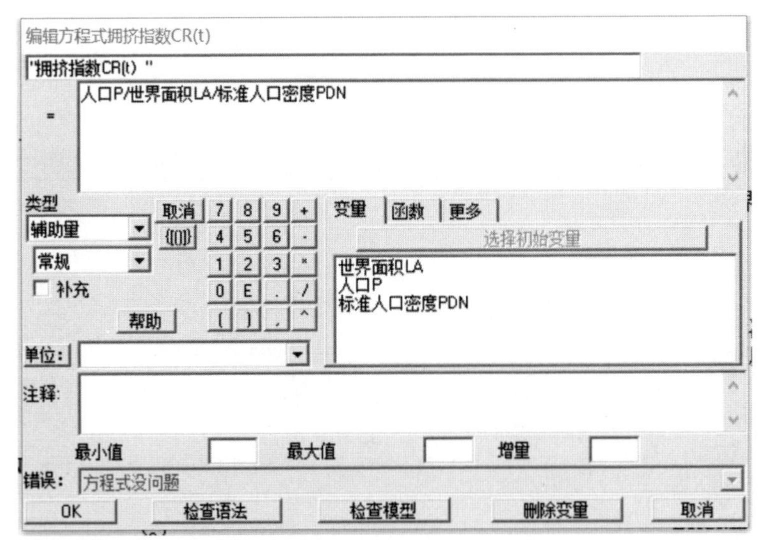
编辑方程式拥挤指数CR(t)
"拥挤指数CR(t) "
=
人口P/世界面积LA/标准人口密度PDN
类型
辅助量
常规
补充
帮助
取消
7 8 9 +
4 5 6 -
1 2 3 *
0 E . /
( ) , ^
变量
函数
更多
选择初始变量
世界面积LA
人口P
标准人口密度PDN
单位:
注释:
最小值
最大值
增量
错误:
方程式没问题
OK
检查语法
检查模型
删除变量
取消

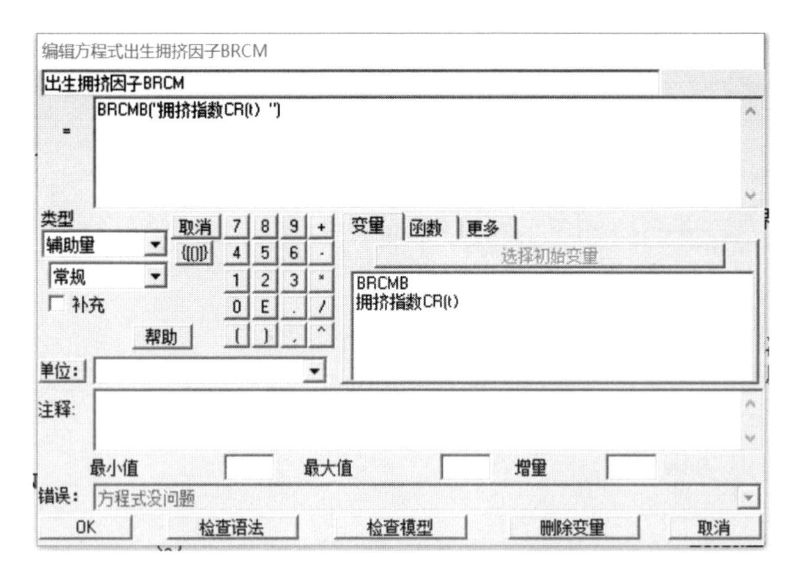
编辑方程式出生拥挤因子BRCM
出生拥挤因子BRCM
=
BRCMB('拥挤指数CR(t) ')
类型
辅助量
常规
补充
取消
帮助
变量
函数
更多
选择初始变量
BRCMB
拥挤指数CR(t)
单位:
注释:
最小值
最大值
增量
错误:
方程式没问题
OK
检查语法
检查模型
删除变量
取消

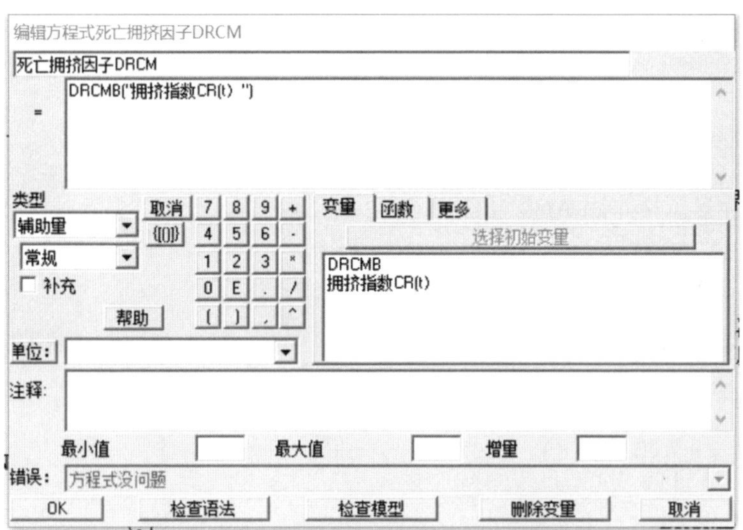
编辑方程式死亡拥挤因子DRCM
死亡拥挤因子DRCM
=
DRCMB('拥挤指数CR(t) ')
类型
辅助量
常规
补充
取消
帮助
变量
函数
更多
选择初始变量
DRCMB
拥挤指数CR(t)
单位:
注释:
最小值
最大值
增量
错误:
方程式没问题
OK
检查语法
检查模型
删除变量
取消

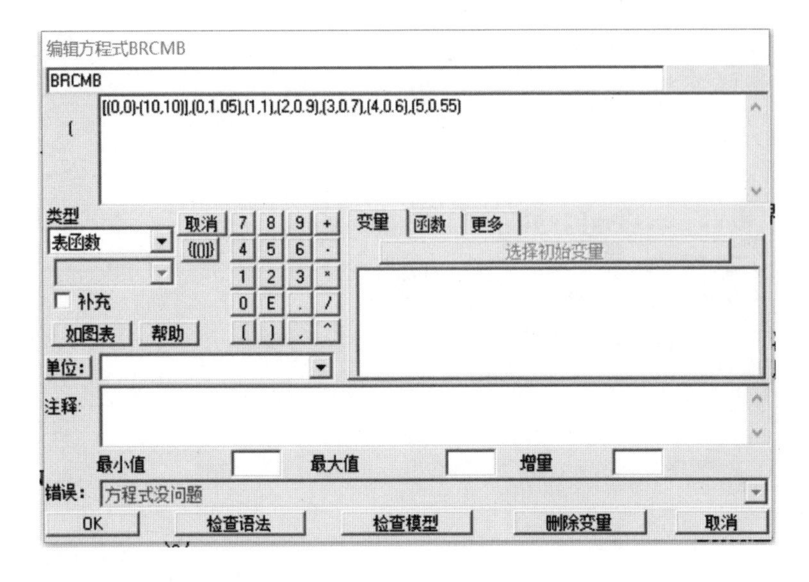
编辑方程式BRCMB
BRCMB
[(0,0)-(10,10)],(0,1.05),(1,1),(2,0.9),(3,0.7),(4,0.6),(5,0.55)
类型
表函数
补充
如图表
帮助
取消
变量
函数
更多
选择初始变量
单位:
注释:
最小值
最大值
增量
错误:
方程式没问题
OK
检查语法
检查模型
删除变量
取消

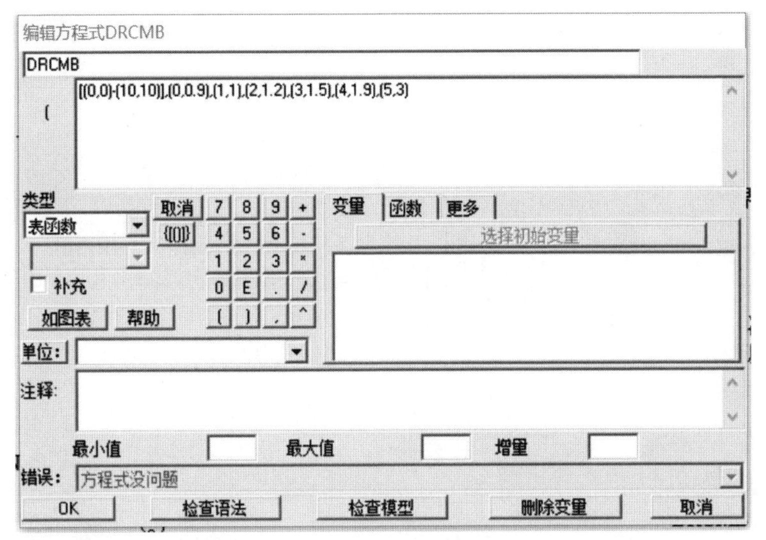
编辑方程式DRCMB
DRCMB
[(0,0)-(10,10)],(0,0.9),(1,1),(2,1.2),(3,1.5),(4,1.9),(5,3)
类型
表函数
补充
如图表
帮助
取消
变量
函数
更多
选择初始变量
单位:
注释:
最小值
最大值
增量
错误:
方程式没问题
OK
检查语法
检查模型
删除变量
取消

（3）仿真结果。

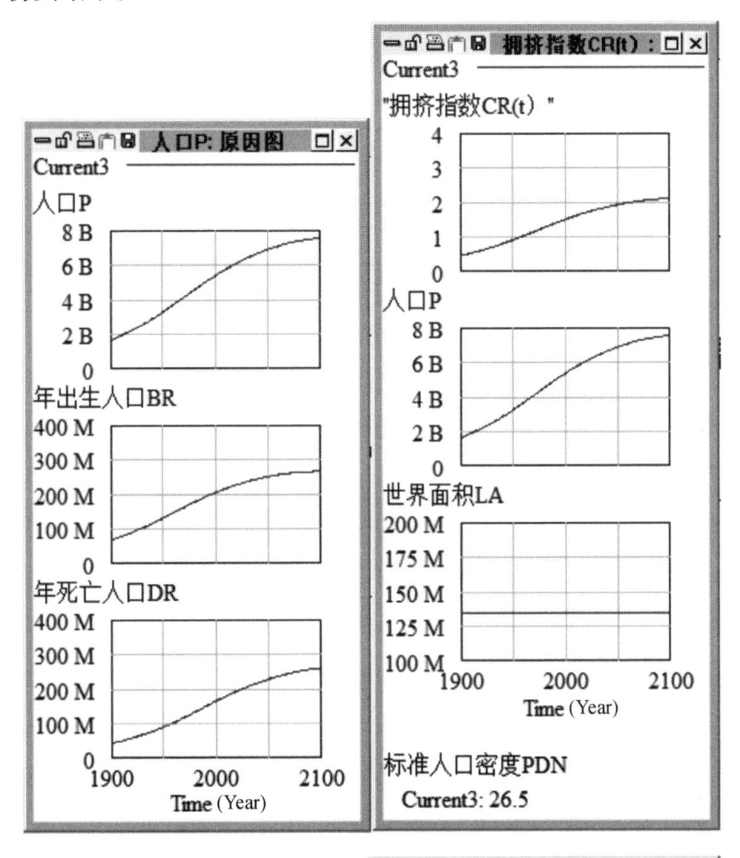
人口P: 原因图
Current3
人口P
8 B
6 B
4 B
2 B
0
年出生人口BR
400 M
300 M
200 M
100 M
0
年死亡人口DR
400 M
300 M
200 M
100 M
0
1900
2000
2100
Time (Year)
拥挤指数CR(t) :
Current3
"拥挤指数CR(t) "
4
3
2
1
0
人口P
8 B
6 B
4 B
2 B
0
世界面积LA
200 M
175 M
150 M
125 M
100 M
1900
2000
2100
Time (Year)
标准人口密度PDN
Current3: 26.5

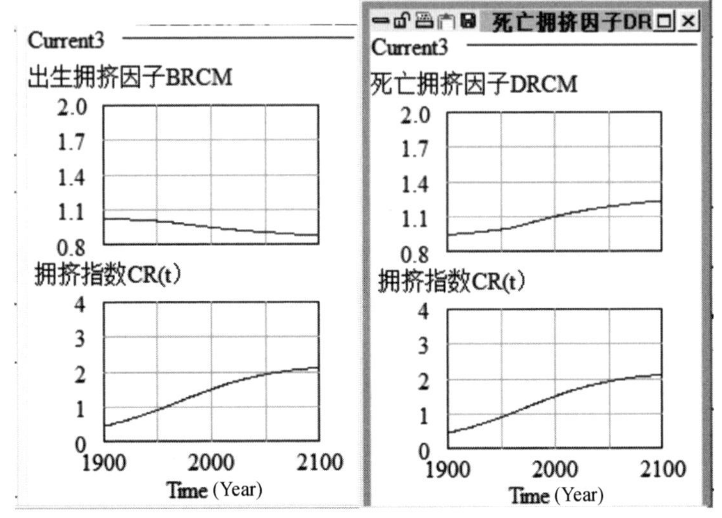
Current3
出生拥挤因子BRCM
2.0
1.7
1.4
1.1
0.8
拥挤指数CR(t)
4
3
2
1
0
1900
2000
2100
Time (Year)
死亡拥挤因子DR
Current3
死亡拥挤因子DRCM
2.0
1.7
1.4
1.1
0.8
拥挤指数CR(t)
4
3
2
1
0
1900
2000
2100
Time (Year)

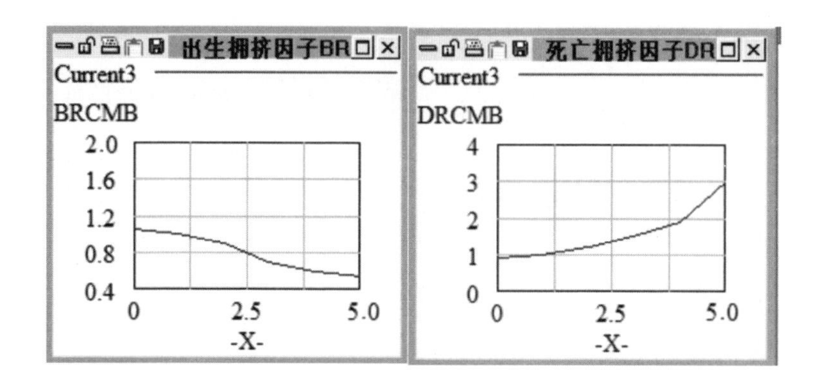

模型符合仿真原理，可以仿真，且拥挤指数仿真值与表函数中预期的拥挤指数值相符，所有仿真曲线的变化趋势符合实际变化趋势，故仿真结果通过行为检验。

4. 三种建模方法区别。

第 1 种建模方法从人口与环境的关系进行研究，考虑到环境的变化对人口的影响，同时添加了人口的成熟延迟。第 2 种方法偏重以出生率和死亡率影响人口，环境承载力此时作为人为设定的定值，没有像第一种方法中考虑到环境的破坏与再生。第 3 种方法从世界面积和人口的容纳关系进行研究，以在一定的世界面积限制下，人口的变化导致的拥挤程度对人口的影响作为变化量，引入了表函数，需要大量历史统计数据作为基础。

上机实验解答 9

1. 模型背景。

以四川雅安地震灾害为例，在所构建的应急物资供应系统下对供应策略的选择进行仿真研究。此次地震灾害的灾区人口密度为 101 人/平方公里，以对雅安地震进行需求预测得到的不同种类的物资需求量作为系统的输入值进行仿真分析。已知 3 个重灾区芦山县、宝兴县和天全县的总人口数为 385157 人，第一天的受伤人数为 5500 人。为了系统地描述方便，选择一个灾区对其受灾人口设置为 1 万人，受伤人数为 0. 5 万人，在应急救援活动中，不仅有静态的物资需求如瓶装水、帐篷等生活必需品发放给每位受灾群众，还有动态的物资需求如医药品供应给伤员。配送中心的 AT 指收到受灾地订单到配送中心仓库发货这段时间里，周转库存所需的时间，初步设为 1 周期；在没有自然灾害的影响的情况下，由配送中心发货到受灾地的正常运输时间为 0.5 天。系统内相关变量的输入参数的设定如下。

2. 模型建立。

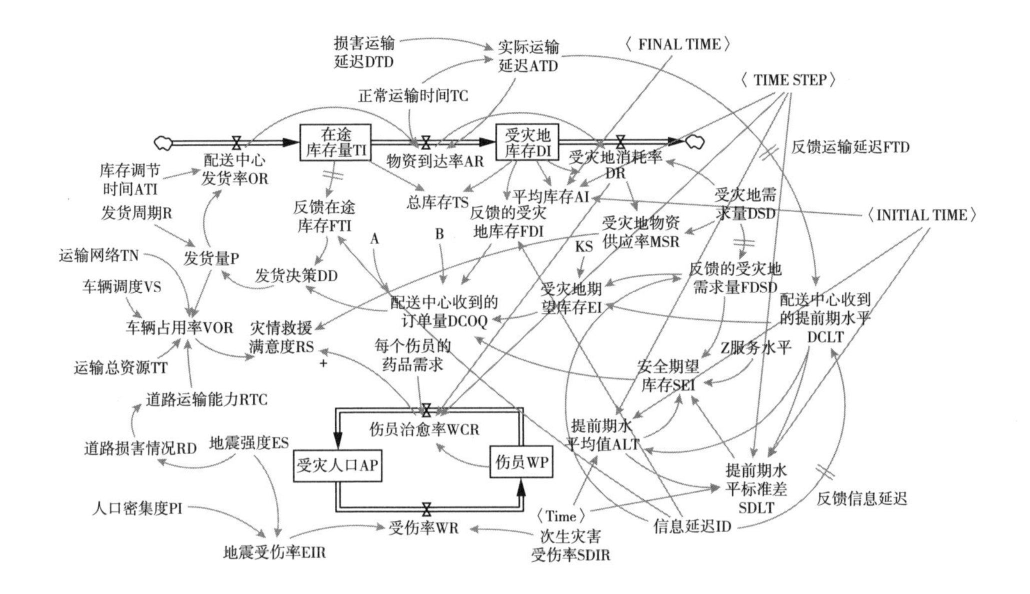

3. 仿真方程。

项目	变量	仿真方程
模块发货决策板块	配送中心发货率 OR	=MAX(P, 0)/ATI
	物资到达率 AR	=DELAY MATERIAL(OR, ATD+TC, 0, 0)
	受灾地消耗率 DR	=MIN(DSD, DI+AR)
	受灾地需求量 DSD	=2+RANDOM UNIFORM(50, 200, 1)（万瓶）
	反馈的受灾地需求量 FDSD	=DELAY INFORMATION(DSD, ID, 0)
	物资供应率 MSR	=DR/DSD
	发货决策 DD	=DCOQ-FTI
	发货量 P	=DD * PULSE TRAIN(1, 0, 2, 60)
	接收的订单量 DCOQ	=(EI-FDI) * A+(SEI-FDI) * B

项目	变量	仿真方程
库存模块	在途库存 TI	=INTEG(OR-AR, 0)
	受灾地库存 DI	=INTEG(AR-DR, 1)(万瓶)
	反馈在途库存 FTI	=DELAY INFORMATION(TI, ID, 0)
	反馈受灾地库存 FDI	=DELAY INFORMATION(DI, ID, 0)
	平均库存 AI	=DI/((FINAL TIME-INITIAL TIME)/TIME STEP)(万瓶)
	总库存 TS	=TI+DI
	期望库存 EI	=DCLT * FDSD * KS
	安全期望库存 SEI	=(ALT+Z * SDLT) * FDSD
	库存调节时间 ATI	1
延迟模块	损害运输延迟 DTD	=EXP(-RAMP(1/10, 0, 60)) * RANDOM UNIFORM(0, 8, 3)
	实际运输延迟 ATD	=TC+DTD
	信息延迟 ID	=EXP(-RAMP(1/5, 0, 60)) * RANDOM UNIFORM(0, 6, 2)
救援满意度板块	提前期水平均值 ALT	= (DCLT * TIME STEP) / (Time-INITIAL TIME+TIME STEP)
	提前期水平标准差 SDLT	=SQRT (ZIDZ (((DCLT-ALT) ^2 * TIME STEP), (Time-INITIAL TIME+TIME STEP-1)))
	车辆占用率 VOR	=(P/TT) * (1-(0.3 * VS+0.3 * TN+0.3 * RTC))
	伤员治愈率 WCR	=MIN (WP/TIME STEP, DR/每个伤员的药品需求)
	伤员 WP	=INTER (WR-WCR, 0)
	受伤率 WR	=EIR+SDIR
	地震受伤率 EIR	=EXP (-22.73+10.6 * LN (ES) +0.34 * LN (PI))
	次生灾害受伤率 SDIR	=RANDOM UNIFORM (50, 200, 1)
	受灾人口 AP	=INTER (WCR-WR, 0)
	灾情救援满意度 RS	=WCR+VOR+MSR
	道路损害情况 RD	=0.1 * ES
	道路运输能力 RTC	=1-RD
	车辆调度 VS	1
	运输网络 TN	1

4. 仿真结果。

（1）首先给变量赋初值。

变量	初始值	变量	初始值
人口密集度	101	库存调节时间	1
地震强度	7 级	受灾地需求量	2 万
车辆调度	1	服务水平 Z	1.65
运输网络	1	决策补偿系数 KS	1
运输总资源	5500 件	期望库存系数 A	1
每个伤员的药品需求	3 瓶	安全期望库存系数 B	0
正常运输时间	0.5 天	在途初始库存	1 万

（2）模型设置。

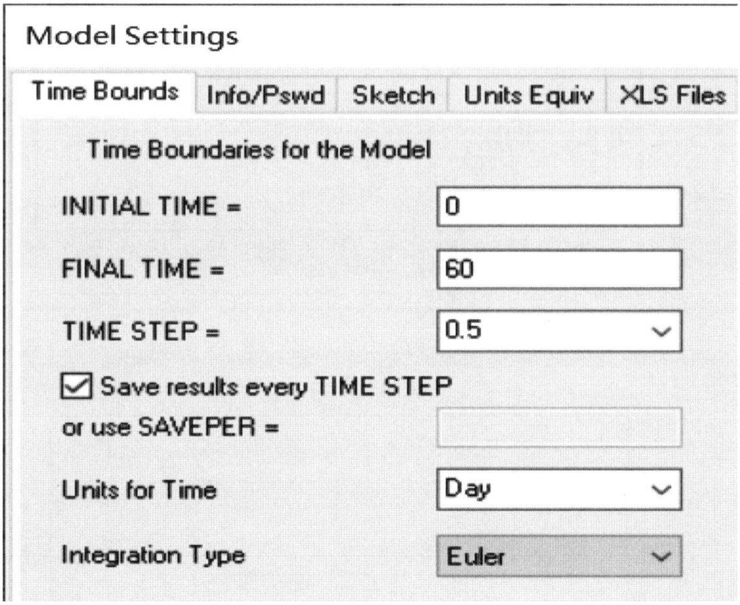

Model Settings

Time Bounds | Info/Pswd | Sketch | Units Equiv | XLS Files

Time Boundaries for the Model

INITIAL TIME = 0

FINAL TIME = 60

TIME STEP = 0.5

☑ Save results every TIME STEP

or use SAVEPER =

Units for Time: Day

Integration Type: Euler

（3）仿真图形。

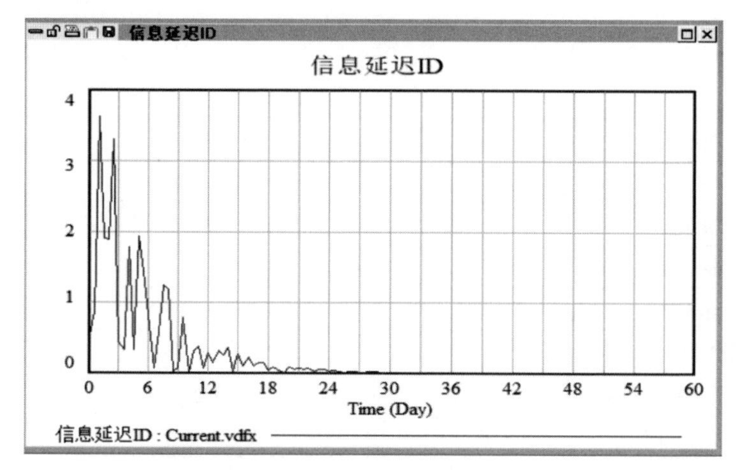
信息延迟ID
信息延迟ID
4
3
2
1
0
0 6 12 18 24 30 36 42 48 54 60
Time (Day)
信息延迟ID : Current.vdfx

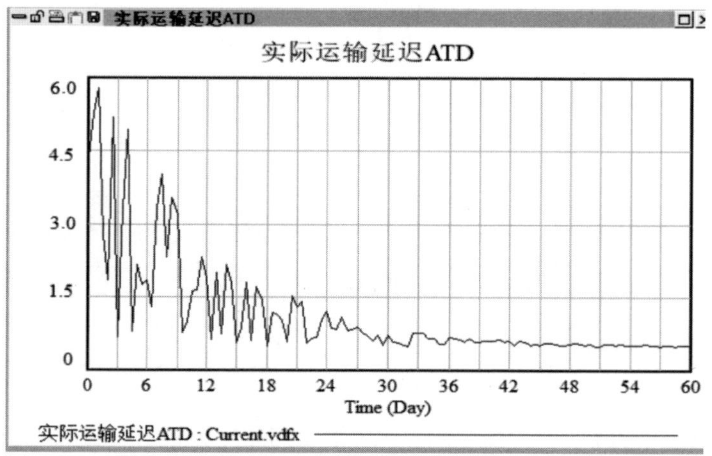
实际运输延迟ATD
实际运输延迟ATD
6.0
4.5
3.0
1.5
0
0 6 12 18 24 30 36 42 48 54 60
Time (Day)
实际运输延迟ATD : Current.vdfx

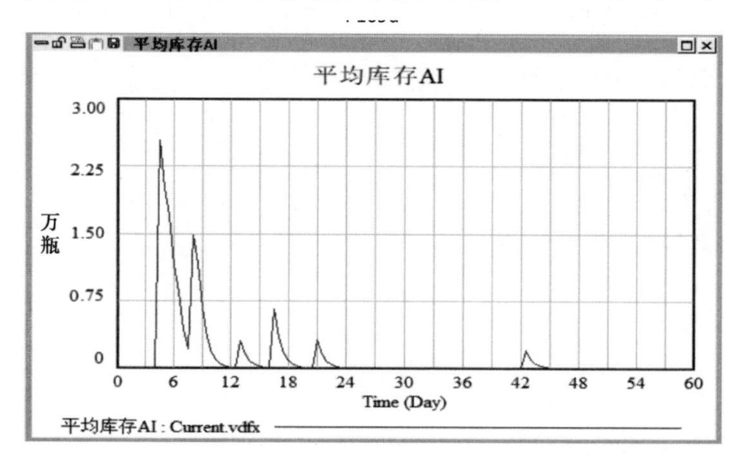
平均库存AI
平均库存AI
3.00
2.25
万瓶
1.50
0.75
0
0 6 12 18 24 30 36 42 48 54 60
Time (Day)
平均库存AI : Current.vdfx

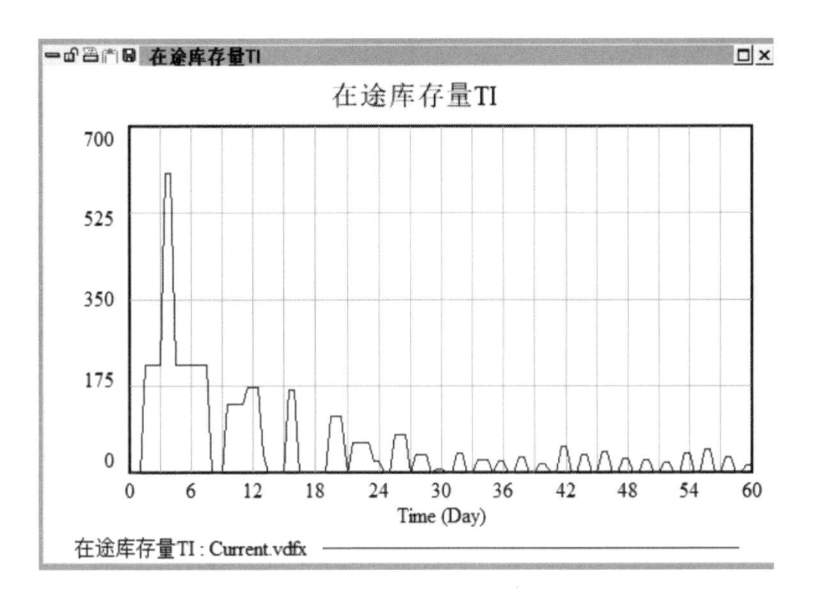
在途库存量TI
在途库存量TI
700
525
350
175
0
0 6 12 18 24 30 36 42 48 54 60
Time (Day)
在途库存量TI : Current.vdfx

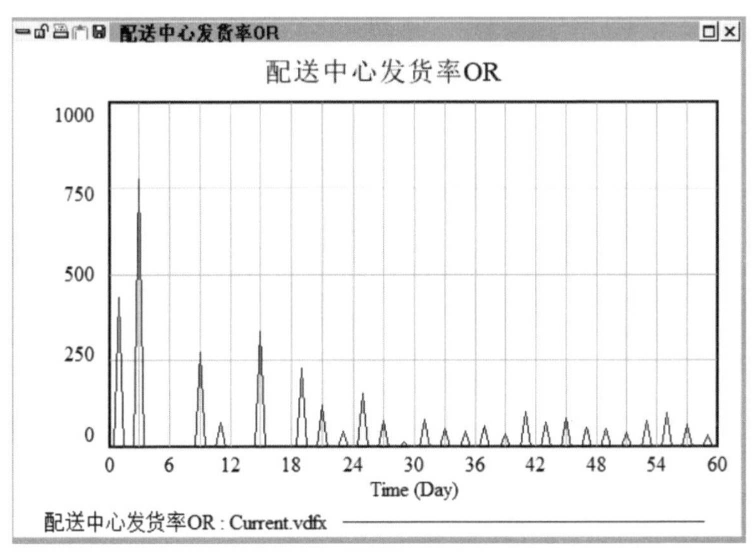
配送中心发货率OR
配送中心发货率OR
1000
750
500
250
0
0 6 12 18 24 30 36 42 48 54 60
Time (Day)
配送中心发货率OR : Current.vdfx

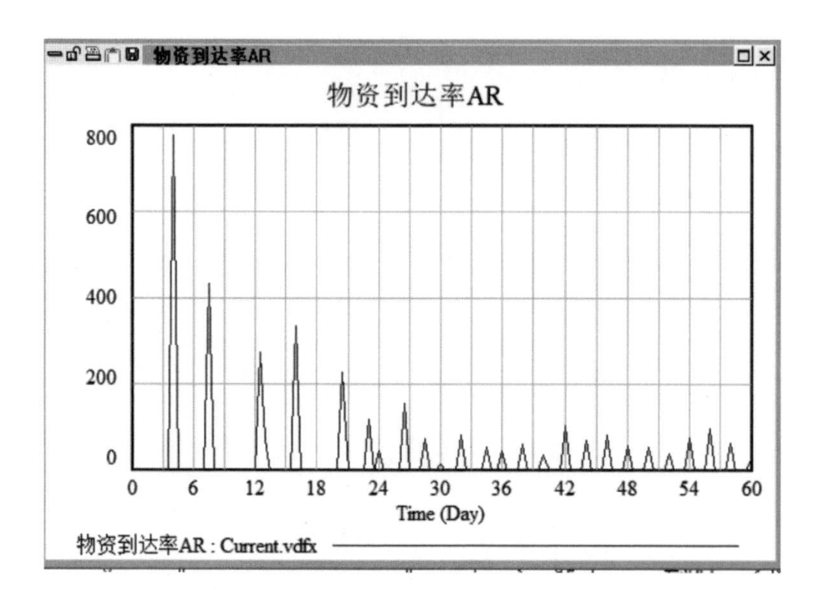
物资到达率AR
物资到达率AR
800
600
400
200
0
0
6
12
18
24
30
36
42
48
54
60
Time (Day)
物资到达率AR : Current.vdfx

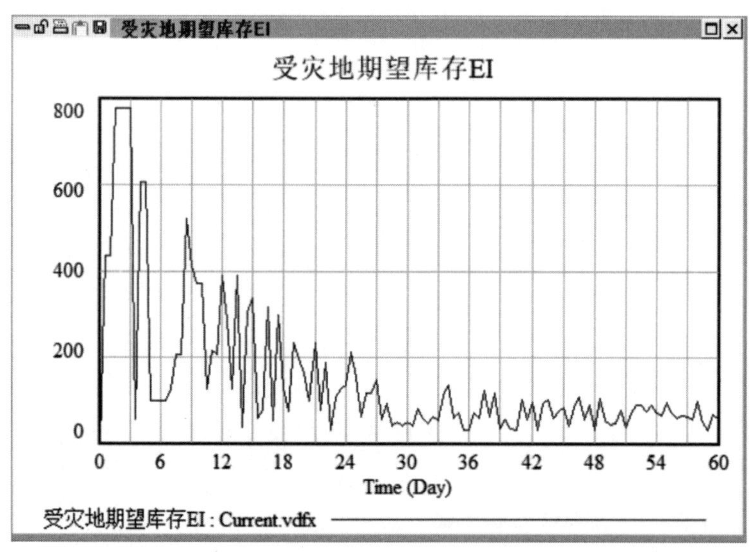
受灾地期望库存EI
受灾地期望库存EI
800
600
400
200
0
0
6
12
18
24
30
36
42
48
54
60
Time (Day)
受灾地期望库存EI : Current.vdfx

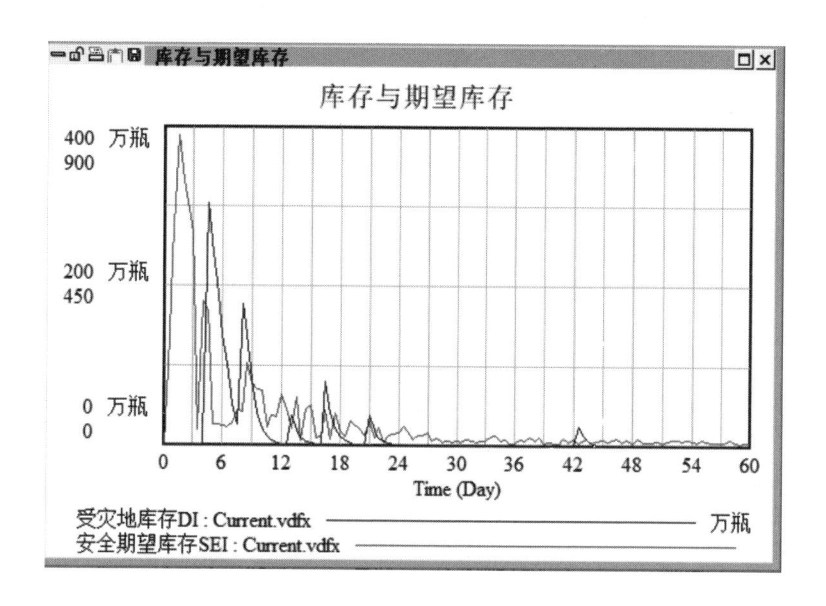

（4）仿真结果分析。

在地震灾害发生初期，灾区的通信信号和道路运输损害严重，运输延迟和信息延迟处于较高水平，以发货周期 R=2 的供应策略能弥补动荡的灾区环境和动态需求的不确定性，因为该方案以上游订单为主，下游提出的订单作为辅助对上游订单做出微调，使救灾物资能够平稳安全地到达灾区，提高救援效率。另外，安全期望库存所采用的提前期水平平均值 ALT 平滑了提前期的波动，提前期水平标准差 SDLT 则使安全缓冲库存地数量增加了，不仅能使灾区获得持续性的救灾物资，而且每次供货的数量充足，这在一定程度上也弥补了持续供应方案的不足，为灾区送去了更多的物资供应。

上机实验解答 10

1. 近 5 年的生猪出栏价格和猪肉价格查到（精确到月或天）。

时间	品类	指标	指标类型	地区	周期	单位	数值
202010	猪肉	市场价	市场价	全国	月	元/公斤	44.42
202009	猪肉	市场价	市场价	全国	月	元/公斤	47.38
202008	猪肉	市场价	市场价	全国	月	元/公斤	48.39
202007	猪肉	市场价	市场价	全国	月	元/公斤	48.05
202006	猪肉	市场价	市场价	全国	月	元/公斤	42.11
202005	猪肉	市场价	市场价	全国	月	元/公斤	40.46
202004	猪肉	市场价	市场价	全国	月	元/公斤	44.86
202003	猪肉	市场价	市场价	全国	月	元/公斤	47.79
202002	猪肉	市场价	市场价	全国	月	元/公斤	49.68
202001	猪肉	市场价	市场价	全国	月	元/公斤	46.83
201912	猪肉	市场价	市场价	全国	月	元/公斤	43.45
201911	猪肉	市场价	市场价	全国	月	元/公斤	47.09
201910	猪肉	市场价	市场价	全国	月	元/公斤	44.61
201909	猪肉	市场价	市场价	全国	月	元/公斤	35.92
201908	猪肉	市场价	市场价	全国	月	元/公斤	28.49
201907	猪肉	市场价	市场价	全国	月	元/公斤	23.65
201906	猪肉	市场价	市场价	全国	月	元/公斤	21.59
201905	猪肉	市场价	市场价	全国	月	元/公斤	20.63
201904	猪肉	市场价	市场价	全国	月	元/公斤	20.31
201903	猪肉	市场价	市场价	全国	月	元/公斤	19.48
201902	猪肉	市场价	市场价	全国	月	元/公斤	18.33
201901	猪肉	市场价	市场价	全国	月	元/公斤	18.59
201812	猪肉	市场价	市场价	全国	月	元/公斤	19.49
201811	猪肉	市场价	市场价	全国	月	元/公斤	19.30
201810	猪肉	市场价	市场价	全国	月	元/公斤	19.81

日期	生猪出场价格
2015/12/9	16.85
2015/12/16	16.87
2015/12/23	16.79
2015/12/30	16.8
2016/1/6	17.13
2016/1/13	17.73
2016/1/20	17.78
2016/1/27	18.03
2016/2/17	18.46

A	B
	生猪出栏价（单位：元/公斤）
2016年3月	19.23
2016年4月	20.14
2016年5月	20.74
2016年6月	20.03
2016年7月	18.77
2016年8月	18.68
2016年9月	18.12
2016年10月	16.65
2016年11月	16.72
2016年12月	17.11
2017年1月	17.77
2017年2月	17.15

2. 建立 Sd 模型。

答：一般来说，体型小而成熟早的本地猪以 8~12 月龄，体重约 75 公斤为宜，饲料条件较好的地区，同时又是杂交改良猪种，则以 8~10 月龄、体重 100 公斤左右为宜。大型晚熟猪种，则以 12~18 个月龄、体重 100~120 公斤为宜。而猪成熟时间为 5~6 个月，平均配种时间为 7 个月。成熟到宰杀有 2 个月延迟，幼年到成熟有 5 个月延迟，怀孕的平均时间是 114 天，约为 3.8 个月。从挑选种猪到等待种猪第三次发情所需时间约为 21×3=63（天），约为 2.1 个月。用 2018 年猪出栏数除以 12 个月得到市场对猪的购买速率。2018 年出栏数 69.3824 千万头，除以 12 为 5.782 千万头/月。关于初始值设定，以 2015 年的猪存栏和能繁殖母猪设为初值，将其单位化为千万，约为 4.693 千万头。生猪存栏数为 45.11250。

年份	品类	指标名称	地区	周期	单位	数值
2018	畜禽本期出栏数_猪	数量	全国	年	万头	69382.40
2017	畜禽本期出栏数_猪	数量	全国	年	万头	70202.10
2016	畜禽本期出栏数_猪	数量	全国	年	万头	70073.90
2015	畜禽本期出栏数_猪	数量	全国	年	万头	70825.00
2014	畜禽本期出栏数_猪	数量	全国	年	万头	73510.40

年份	品类	指标名称	地区	周期	单位	数值
2016	能繁母猪存栏	数量	全国	年	万头	4456.18
2015	能繁母猪存栏	数量	全国	年	万头	4693.05

年份	品类	指标名称	地区	周期	单位	数值
2018	畜禽存栏数（年末数）_猪	数量	全国	年	万头	42817.10
2017	畜禽存栏数（年末数）_猪	数量	全国	年	万头	44158.90
2016	畜禽存栏数（年末数）_猪	数量	全国	年	万头	44209.20
2015	畜禽存栏数（年末数）_猪	数量	全国	年	万头	45112.50
2014	畜禽存栏数（年末数）_猪	数量	全国	年	万头	46582.70

年份	品类	指标名称	地区	周期	单位	数值
2018	猪肉	产量	全国	年	万吨	5403.70
2017	猪肉	产量	全国	年	万吨	5451.80
2016	猪肉	产量	全国	年	万吨	5425.50
2015	猪肉	产量	全国	年	万吨	5486.50
2014	猪肉	产量	全国	年	万吨	5671.40

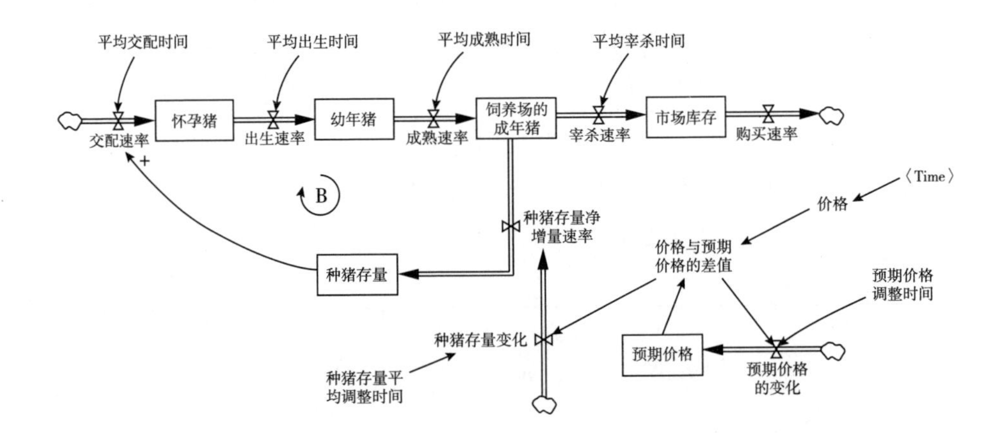

关于初值和仿真方程设置如下。

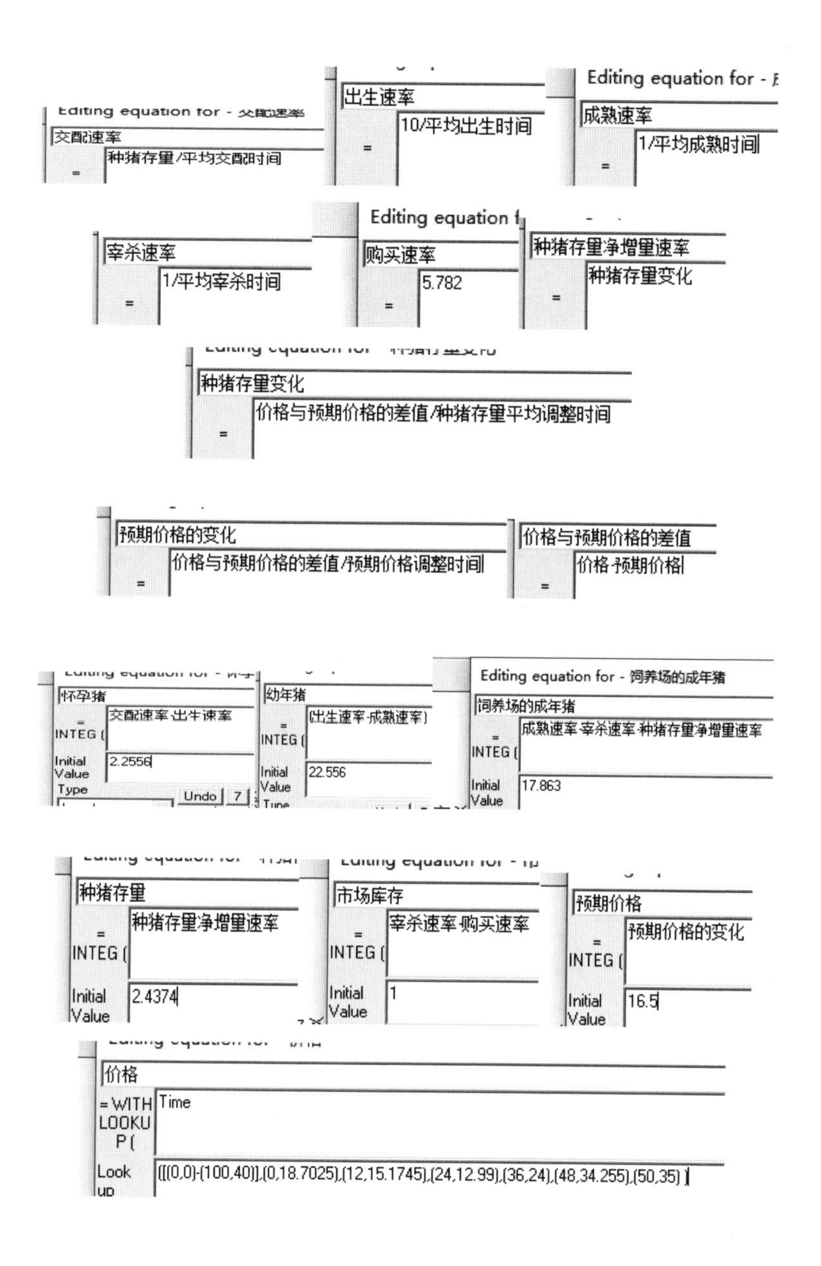

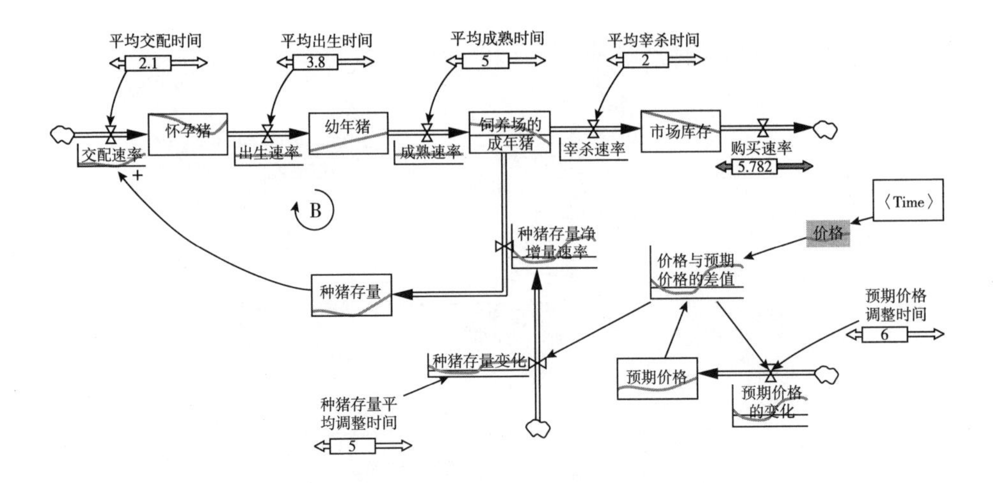

3. 将非洲猪瘟作为外生变量，分析下面的专业判断是否正确：猪市场有周期性，一般为 2~3 年或 3~5 年。

答：世界动物卫生组织数据显示，2018 年，中国暴发非洲猪瘟疫情，随后蔓延至韩国、菲律宾和越南。在一年的时间里，中国生猪死亡数量多达 1 亿头，约为总存栏的 23.4%。

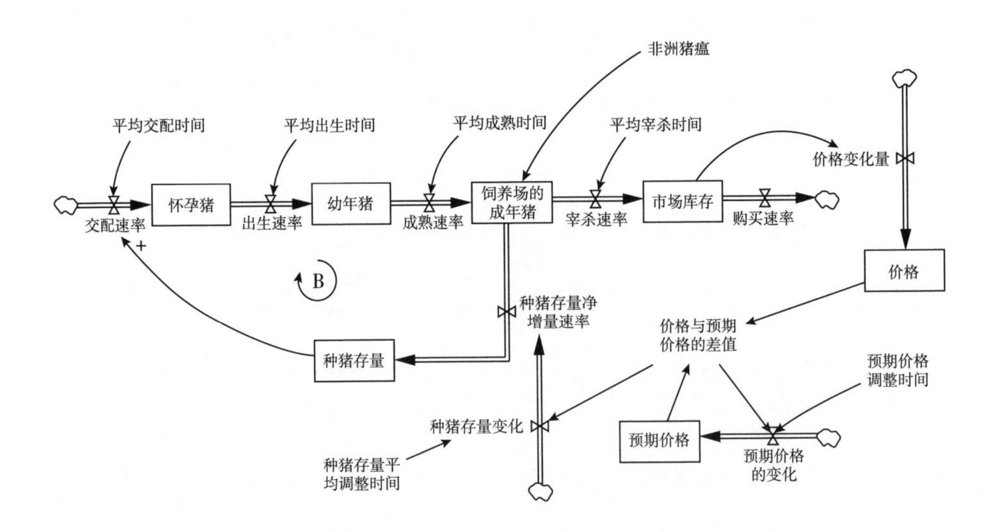

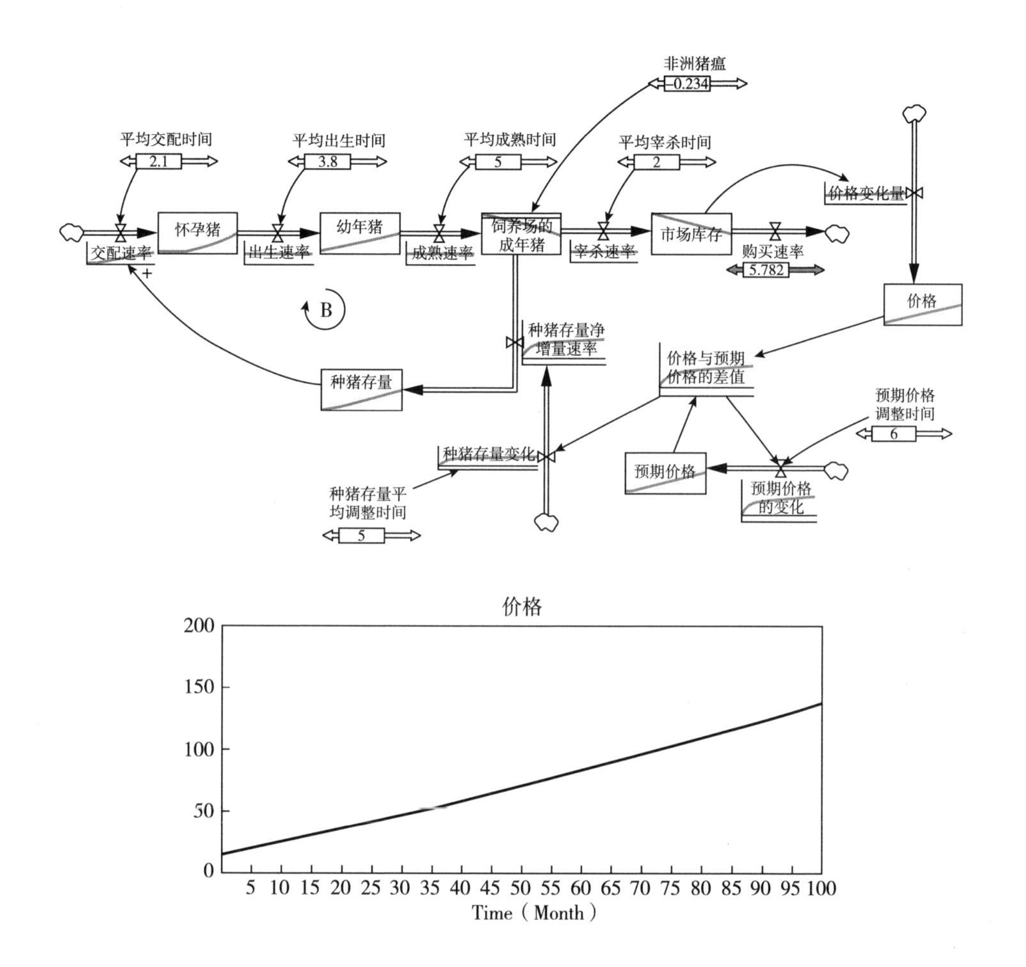

非洲猪瘟导致价格上涨（此处只考虑 2018 年非洲猪瘟之后的趋势）。

由于 2016/2017 季价格较低，所以导致养猪户离场，养猪周期为 11 个月，加上调整时间 11 个月，需要 22 个月左右的反应时间。所以，2016 年价格低导致 2018 年市场的供给量减少，猪肉价格上涨。而非洲猪瘟则是在供给原本减少的基础上变得更少，甚至限制了猪的流通，当时我国仅由几个省份负责提供全国市场的大部分猪肉，导致个别省份价格上升严重。原本供给就少，价格就升，猪瘟一来价格更加上涨。

上机实验解答 11

1.1

Question	Answer
Causal Loop Diagram-CLD	GDP + RGDP
Stock and Flow diagram-SFD	GDP RGDP
	RGDP = GDP * g Hypothesis g = 5% per year (RMB) Account is yuan RGPD = 5% * GDP
Graph	Graph 5.0 2.5 0 2.5 5.0 7.5 10.0 GDP

1.2

I think 5 instataneous relationships are appropriate as follows

(a) Effect of interest rate on interest payment.

Yes, Interest payment = Loan * Interest Rate.

hypothesis Loan = 100000per year (RMB) Account is yuan

Interest payment = 5% * Interest Rate

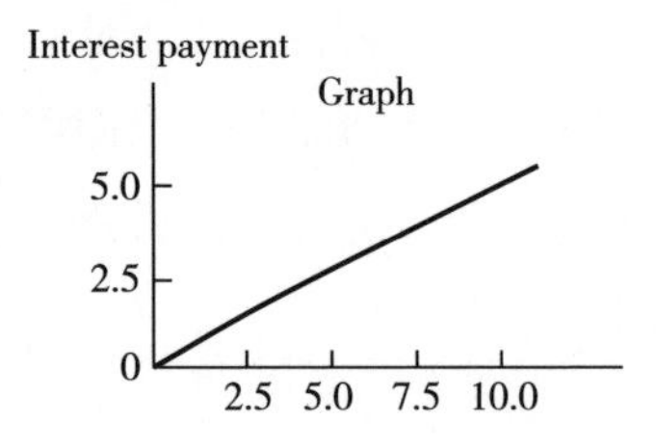

(b) Effect of per unit production costs on profits.

Yes, Profits = (per unit sales price-per unit production costs) * Sales Quantities hypothesis Price is 600 kr, Sales Quantities are 300.

Profits = (600−per unit production costs) * 300

(c) Effect of amount of water in a funnel on the outflow from the funnel.

(d) Effect of hours studied per day on learning per day.

The graphs I draw should be linear.

Mathematical formula Y = a * X+b a, b are constant; X is cause, Y is effect.

(a) Effect of interest rate on interest payment.

Interest payment = Account * Interest Rate

Learning per day = hours studied per day * knowledge per hour

Hypothesis knowledge per hour is 20 pages.

(b) learning per day = 20 * hours studied per day.

2. 1

(a)

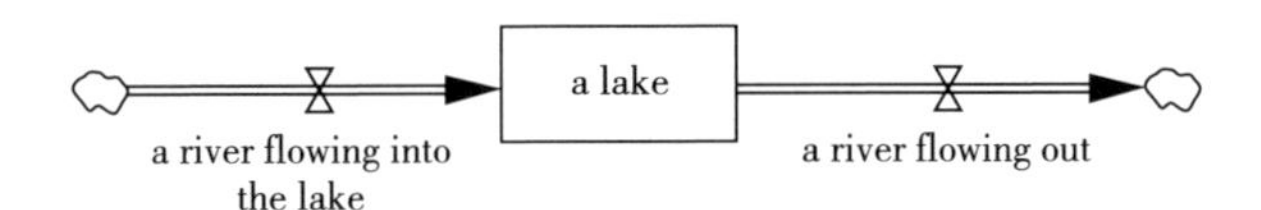

(b)

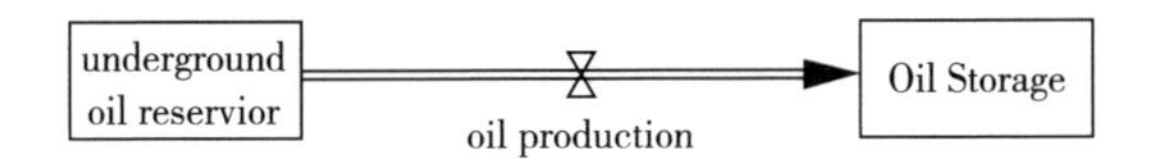

(c)

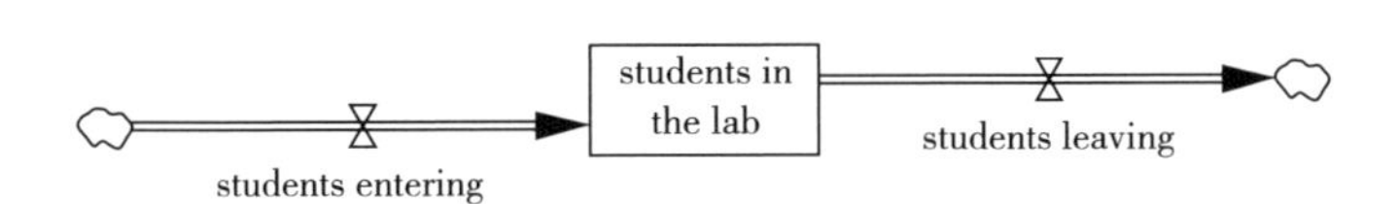

(d)

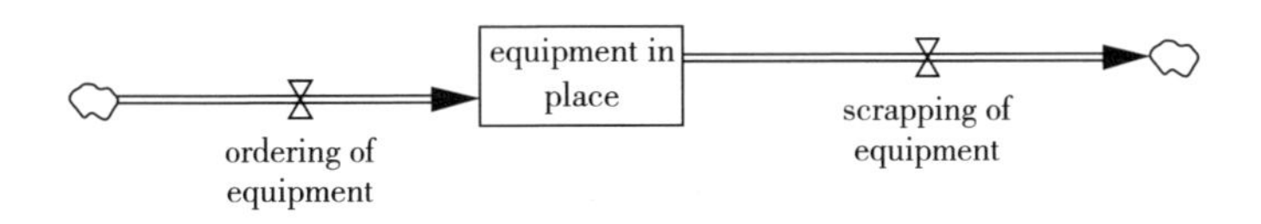

(e)

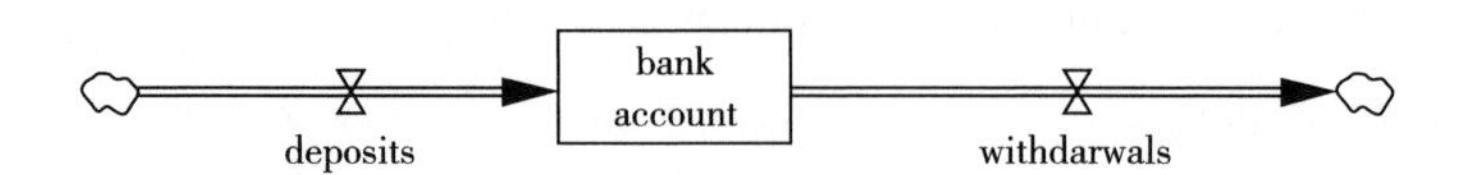

(f) Because two units are person, so they are not stock and flow.

(g) Because they're not related, so they are not stock and flow.

(h)

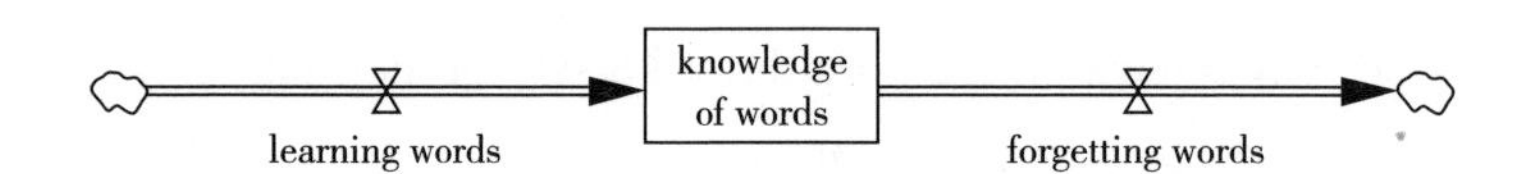

(i)

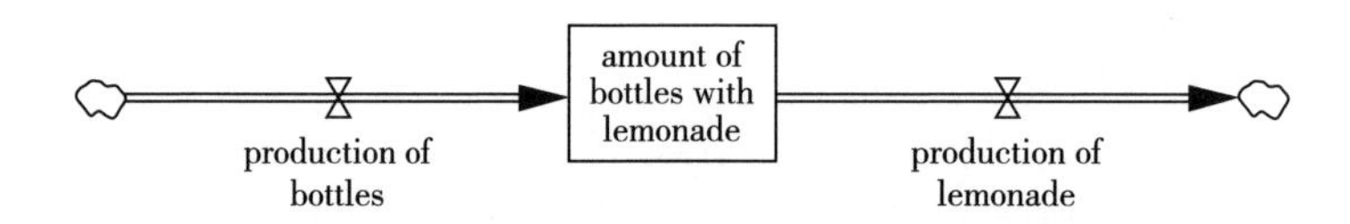

(j)

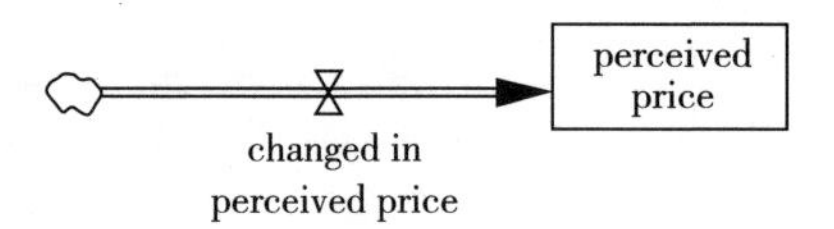

2.2

(a)

a) A lake (stock): Cubic meters.

A river flowing into the lake and a river flowing out (flow): Cubic meters per week.

b) Underground oil reservoir (stock), oil storage (stock): Ten thousand tons (10 kt), oil production (flow): 10 kt per year.

c) Students in the lab (stock): Students.

Students entering (flow), students leaving (flow): Students per hour.

d) Equipment in place (stock): Sets of equipment.

Ordering of equipment (flow), scraping ofequipment (flow): Sets of equipment per year.

e) Bank account (stock): RMB.

Withdrawals (flow), deposits (flow): RMB per month or per day.

f) Mature fish (stock), young fish (stock): Fish.

Maturation (flow), natural death (flow), recruitment (flow): Fish per year.

g) Customers using a product (stock): Customers.

Customers no longer using a product (flow): Customers per year.

h) Knowledge of words (stock): Words.

Learning words (flow), forgetting words (flow): Words per week.

i) Amount of bottles with lemonade (stock): Bottles.

Production ofbottles (flow), production oflemonade (flow): Bottles per day.

j) Perceived price (stock): RMB.

Changed in perceived price (flow): RMB per week.

(b) The actual object, international standard and make sense.

(c) There is one and only Timestep (dt).

2. 3

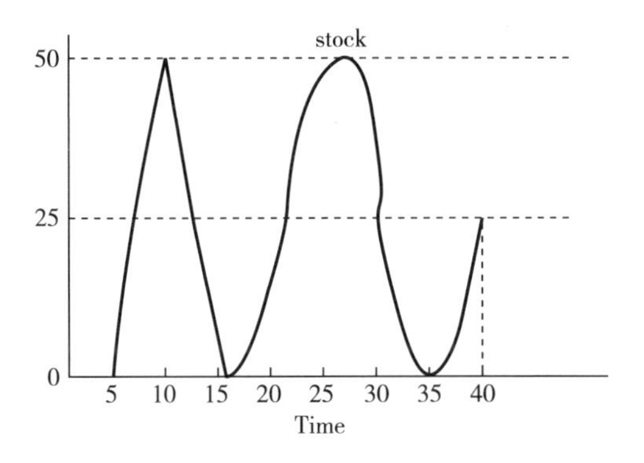

2.4

(a) $Stock_t = Stock_{t-1}$ +Timestep * ($Inflow_{t-1} - Outflow_{t-1}$).

$Stock_{t-1} = Stock_{t-2}$ +Timestep * ($Inflow_{t-3} - Outflow_{t-3}$)

$\vdots$

$Stock_t$ =Timestep * ($Inflow_{t-1} - Outflow_{t-1}$) +Timestep * ($Inflow_{t-3} - Outflow_{t-3}$) +…+ Timestep * ($Inflow_0 - Outflow_0$)

Stocks add up flow over time.

(b) When $Inflow_{t-1} = Outflow_{t-1}$.

$Stock_t = Stock_{t-1}$ +Timestep * ($Inflow_{t-1} - Outflow_{t-1}$)

$Stock_t = Stock_{t-1}$

Stocks create delays

In question 2.3 when t=0-5, $Inflow_{t-1} = Outflow_{t-1}$

$Stock_5 = Stock_4 = \cdots = Stock_0$

Stocks create 5 delays

(c) $$Stock_{(t)} = \int_{t0}^{t} (inflow - outflow)\,dt + Stock_{(t0)}.$$

When t=5-10

$$Stock_{(t)} = \int_{5}^{t} (10 - 0)\,dt + 0 = 10t$$

When t=15-20

$$Stock_{(t)} = \int_{15}^{t} ((10 + 2t) - 0)\,dt + 0 = (t-15)^2$$

Stock in time t is actually caculated by add net flow (t-1->t) to stock (t-1). so the previous steps. Therefore, every amount ever added to stock will affect the current status, which makes stocks serve as a "memory" or "summary".

(d) Stocks created the behaviour endogenously. Effect=f (cause) requires exdogenous cause.

2.5

No, the government hasn't. Because there is accumulation.

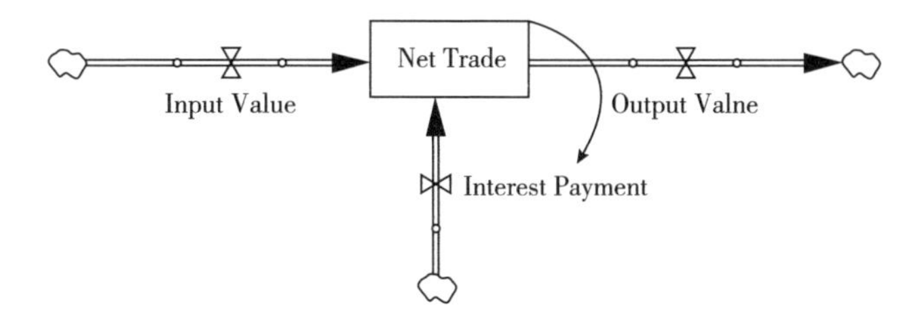

$NetTrade_t = NetTrade_{t-1} + \text{Timestep} * (Inflow_{t-1} - Outflow_{t-1})$

When output$value_{t0-1}$ = input $value_{t0-1}$

$NetTrade_t = nettrade_{t0-1}$

Interest Payment = Interest Rate * Net Trade, If Interest Rate = 5% per year

Interest Payment = 0.05 * Net Trade

There is trade deficit for debt accumulation.

2.6

(a) 07. 01. 2001.

(b) 03. 01. 2001.

(c) 13. 01. 2001.

(d) 31. 12. 2000, 01. 01. 2001, 06. 01. 2001.

(e) 08. 01. 2001.

(f) 31. 12. 2000.

2.7

(a) N would look for a cause that is increasing steadily.

(b) S would explain a steady increase as an accumulation.

(c) N would have a difficulty finding a plausible cause.

(d) S would have a difficulty finding a plausible net flow.

(e) People explain things based experienced view.

(f) Behavior that is not generated by external (out of the system) cause but is an effect of the structure of the system itself.

2.8

P-The problem is an inventory that is out, in the future, customers will be dissatisfied because of delivery delays. (Problem)

H-production capacity is insufficient (Cause)

A-production capacity up to 120% of sales

Structure test should be conducted in communication/accompanied, therefore people from the operator, to know what the reality is. Analyze

P-buy the machine and work by turns (Policy)

Production must be raised above sales for a while to rebulid inventories

I-Higher production may require use of overtime and investment. Therefore one should also consider if raising prices to curtail sales is a more profitable policy.

Money and time, salary increase in spending (Implement) .

上机实验解答 12

2.9

(a)

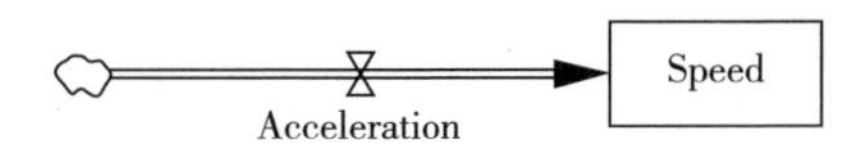

Speed = Acceleration * *Time* = 10 m/s^2 * *Time* (s) = 10 * *Time* (m/s)

Question 1 Time = 4s Speed = ? Answer 1 Speed = 40 (m/s)

Question 2 Time = 8s Speed = ? Answer 2 Speed = 80 (m/s)

Question 3 Time = 12s Speed = ? Answer 3 Speed = 120 (m/s)

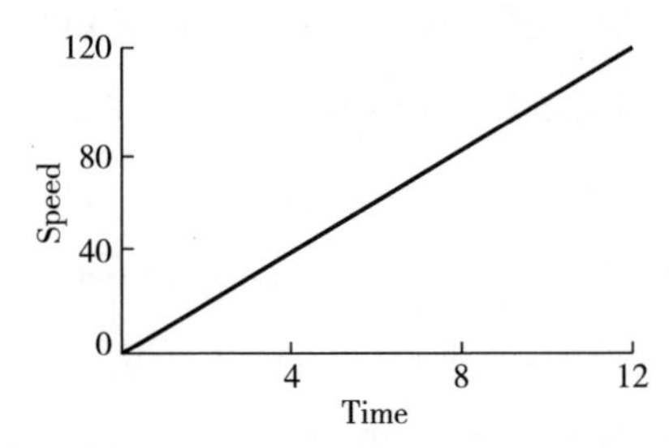

(b)

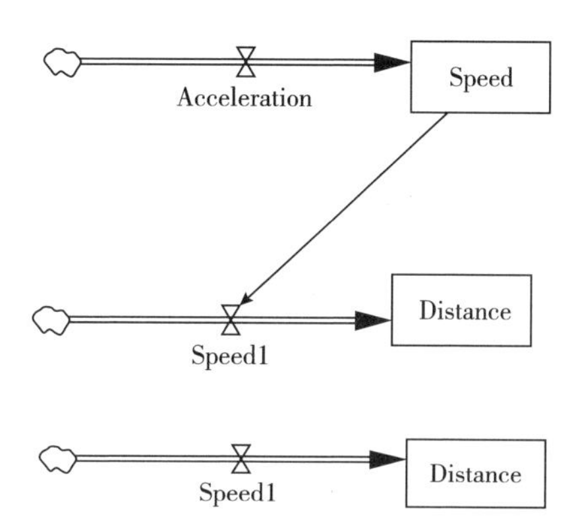

Distance = 1/2 * Speed * Time

Question 1 Time = 4s Speed = 40 (m/s) Distance = ?

Answer 1 Distance = 80 (m)

Question 2 Time = 8s Speed = 80 (m/s) Distance = ?

Answer 1 Distance = 320 (m)

Question 3 Time = 12s Speed = 120 (m/s) Distance = ?

Answer 1 Distance = 720 (m)

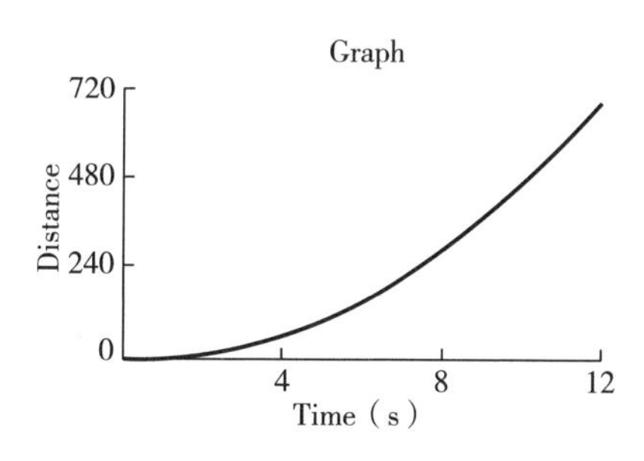

(c) 1Yes, I do.

Explain *Distance* = f ($Time^2$)

Because *Distance* = 1/2 * *Speed* * *Time* *Speed* = Acceleration * *Time*

So *Distance* = 1/2 * Acceleration * $Time^2$ = f ($Time^2$)

(d) Yes, they did. The stock is accumulated variable and effect. The flow is change of stock and cause.

2.10

(a) Vensim.

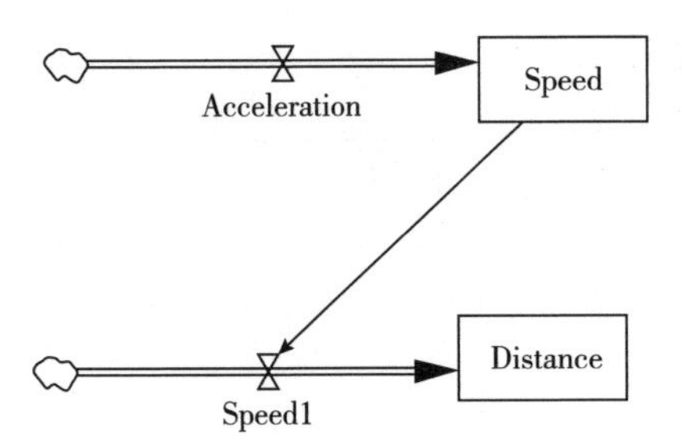

Equations Acceleration = 10m/s^2 Change of distance = Speed (m/s) .

Time units is seconds. DT = 1. Speed unit is m/s.

Distance unit is meter.

(b)

a)

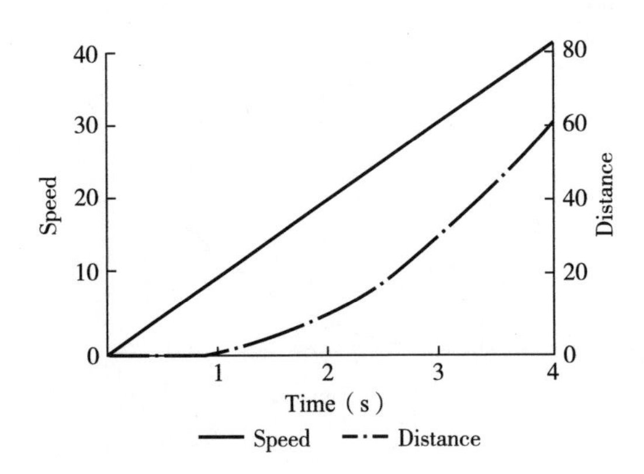

b) Time from 0 is easily understandable.

c) DT = 1.

Table		
	Speed	Distance
0	0.0	0.0
1	10.0	0.0
2	20.0	10.0
3	30.0	30.0
Final	40.0	60.0

d) DT=0. 001 Euler's method.

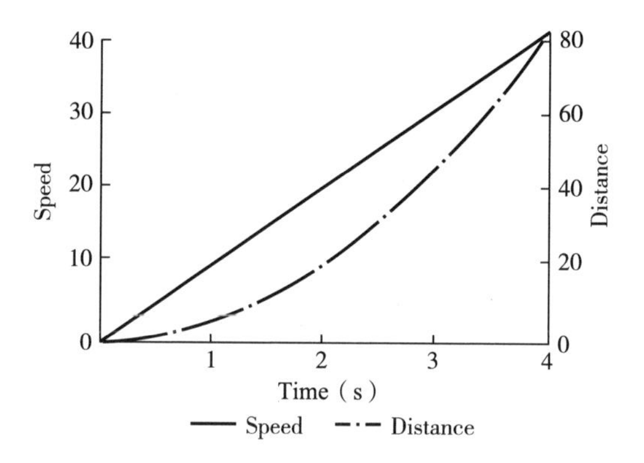

Table		
	Speed	Distance
0	0.0	0.0
1	10.0	5.0
2	20.0	20.0
3	30.0	45.0
Final	40.0	80.0

(c) DT=1.

Stock (t) = Stock (t-1) +Timestep * (Inflow (t-1) -Outflow (t-1))

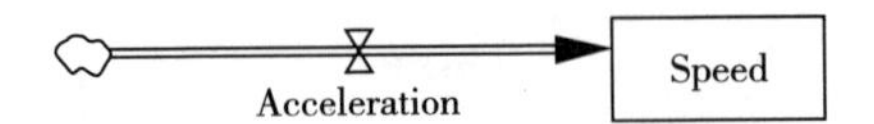

When Timestep = DT = 1, t = 0, Acceleration (0) = Inflow (0-1) = 10 Speed (0) = 0

t = 1, Speed (1) = Speed (0) + 1 * Acceleration (0) = 0 + 1 * 10 = 10

t = 2, Speed (2) = Speed (1) + 1 * Acceleration (1) = 10 + 1 * 10 = 20

t = 3, Speed (3) = Speed (2) + 1 * Acceleration (2) = 20 + 1 * 10 = 30

t = 4, Speed (4) = Speed (3) + 1 * Acceleration (3) = 30 + 1 * 10 = 40

So speed is correct.

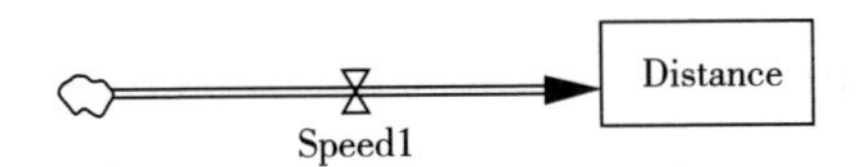

When Timestep = DT = 1, t = 0, Acceleration (0) = Inflow (0-1) = 10 Speed (0) = 0

t = 1, Distance (1) = Distance (0) + 1 * Speed (0) = 0 + 1 * 0 = 0

t = 2, Distance (2) = Distance (1) + 1 * Speed (1) = 0 + 1 * 10 = 10

t = 3, Distance (3) = Distance (2) + 1 * Speed (2) = 10 + 1 * 20 = 30

t = 4, Distance (4) = Distance (3) + 1 * Speed (3) = 30 + 1 * 30 = 60

So distance is too low.

(d) DT = 1 RK4 method.

Table

	Speed	Distance
0	0.0	0.0
1	10.0	5.0
2	20.0	20.0
3	30.0	45.0
Final	40.0	80.0

(e) DT = 1, 1/2, 1/4, 1/8.

Distance

0　0

1　5

2　20

3　45

4　80

5　125

6　180

7　245

Final　320

Conlusion: When using RK4 method, it seems that DT doesnot matter.

2.11

P′: How to control emission ' s development for stabling the level of CO_2 by 2030.

H: Structure (Accumulating cause and effect).

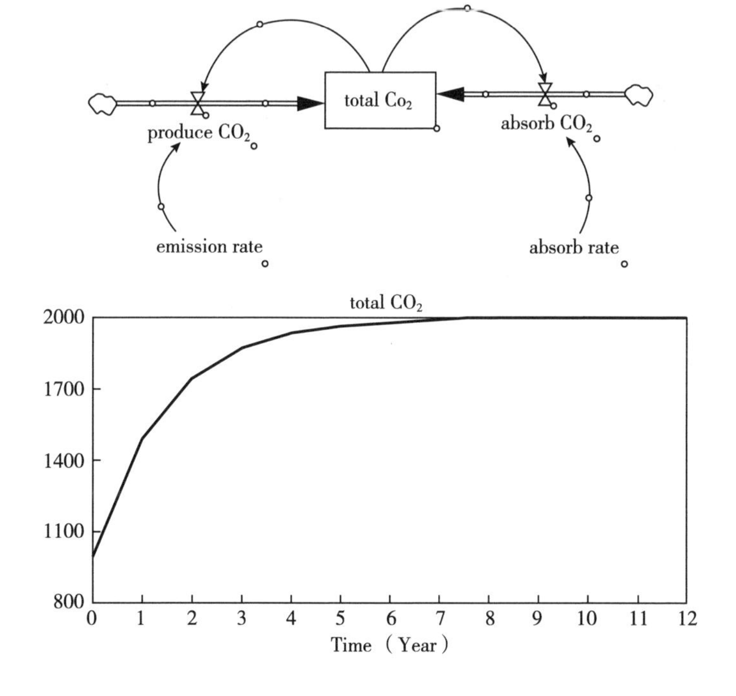

同学 A 答案摘要：

A：光合作用吸收 CO_2 量小于人工生产排放 CO_2 量，CO_2 的总量增加。当光合作用吸收 CO_2 量等于人工生产排放 CO_2 量，系统维持稳定，CO_2 总量维持稳定水平。

P：减少 CO_2 排放量，增加植物（树木、水生植物等）以消耗更多的 CO_2，从而使得 CO_2 的净排放量为负值。

I：由于城市化发展，生活中及工业生产中产生的 CO_2 无法短时间内降低，从而无法从根本上降低排放量。

评价：图表中的时间设置成真实年份可以更直观地看到每年的变化；CO_2 的吸收量应该是流出量。

同学 B 答案摘要：

P′：工业化的发展产生了过量的 CO_2，形成了温室效应，使地球气温越来越高、海平面上升等，因此控制 CO_2 的排放十分重要。

H：假设我国开始调整产业结构，发展低能耗的产业；提倡大众乘坐公共交通工具，减少 CO_2 的排放量。

A：在人们的努力下，CO_2 的排放量不断减少，直至 CO_2 的排放量小于其吸收量。

I：假设中所提出的政策建议可能无法及时被采用，采用后的效果可能也没有如此明显。

评价：没有建立一个模型进行仿真模拟，无法确定哪年能达到二氧化碳排放量最小。

3.1

-Population growth.

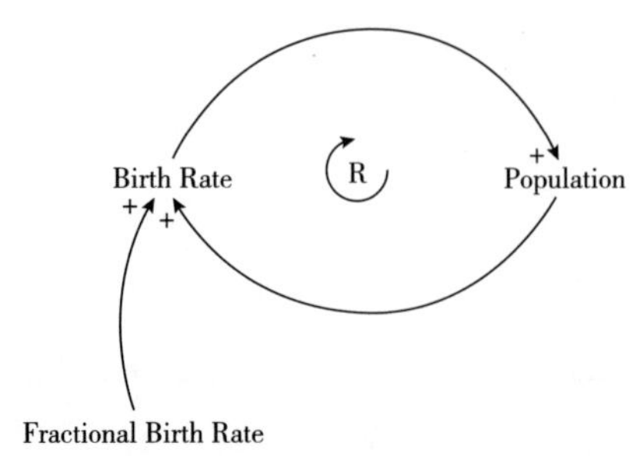

-A quarrel（include level of anger and provoking comments）.

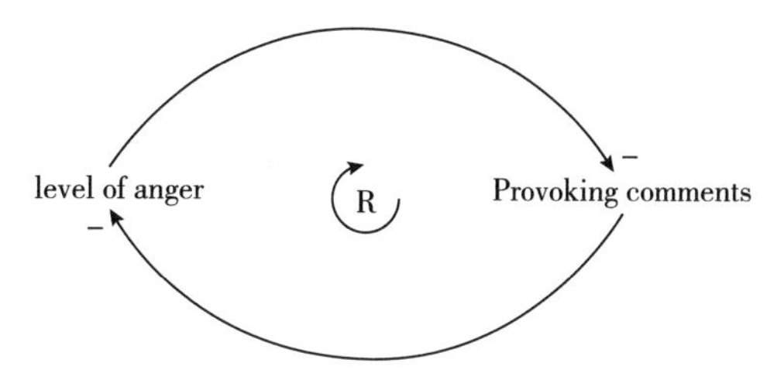

-A war between two countried（provocaion and retaliation）.

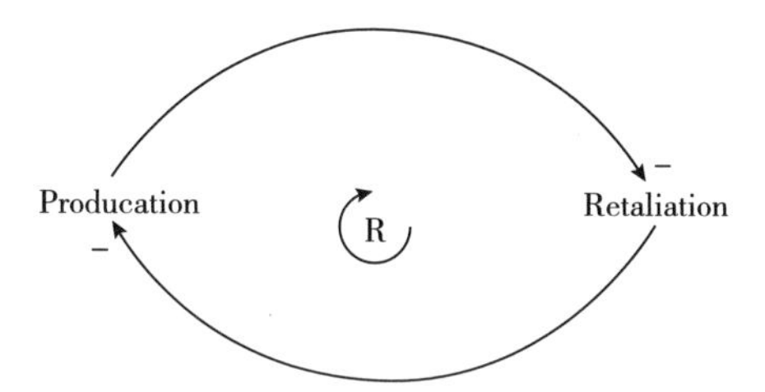

-Economic growth（include machines, labour, technology, production）.

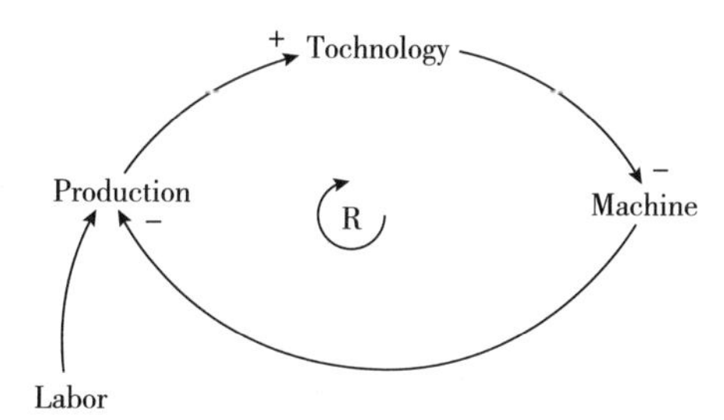

-Spread of an epidemic（consider those infected and the infection rate）.

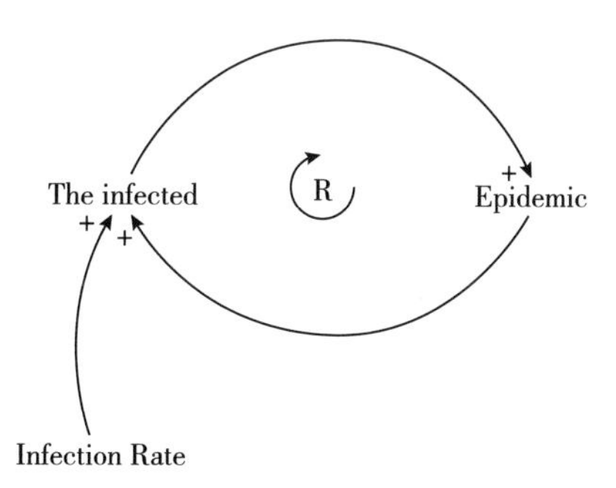

-Market growth for a new product (assume that production costs and produc prices vary (decrease) with accumulated production.

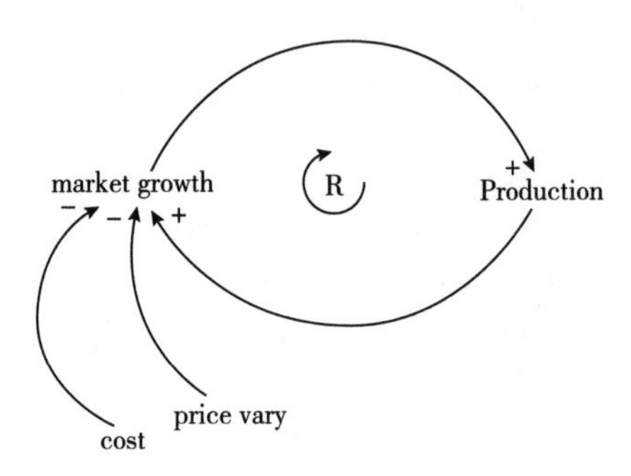

3.2

(a) bacteria(t) = bacteria(t−1) +bacteria(t−1) * net growth rate(3.5%).

Note: t unit per minute

(b)

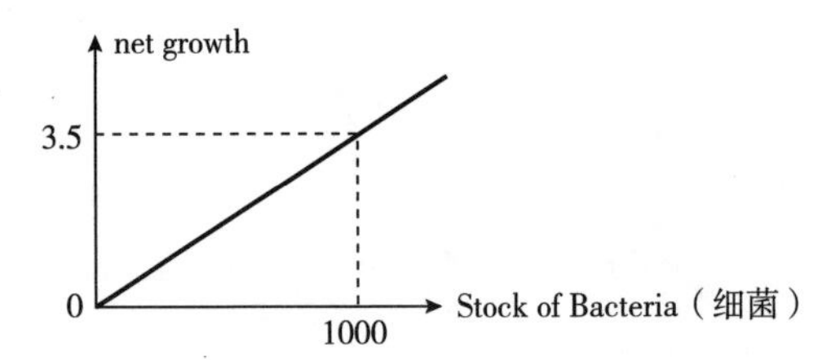

(c) 70/3.5 = 20 minutes.

(d) Stock and flow diagram.

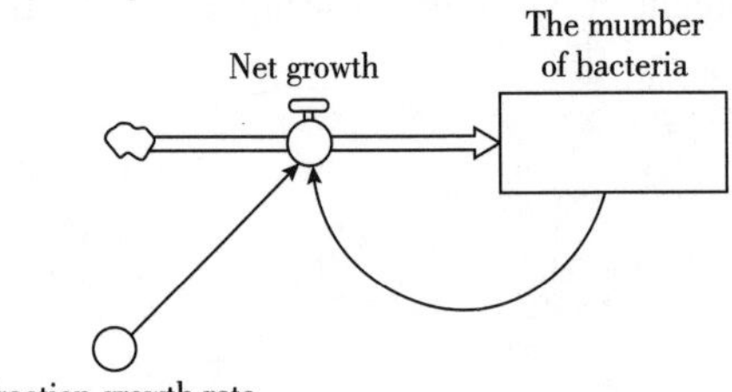

Use Runge−Kutta 4 (RK4) and set DT = 4

1 hour bacteria The number of bacteria = 8166

6 hour bacteria The number of bacteria = 296547918.436

24 hour bacteria The number of bacteria = $7.73356107578 * 10^{24}$

Use Runge-Kutta 4 (RK4) and 8164

DT The number of bacteria

1/64 8161

1/16 8147

1/8 8128

1/4 8091

1/2 8019

1 7878

2 7612

3 7366

4 7137

(e) Use the Euler integration method and.

DT The number of bacteria

1/256 8164.99781156

1/128 8163.82609236

1/64 8161.48379865

1/32 8156.80378489

1/16 8147.46201151

1/8 8128.85115826

1/4 8091.91766577

1/2 8019.18343134

1 7878.09090076

2 7612.25504266

3 7366.23484193

4 7137.93797839

5 6925.55207623

(f) No, I say development after a 24 hours simulation is not realistic.

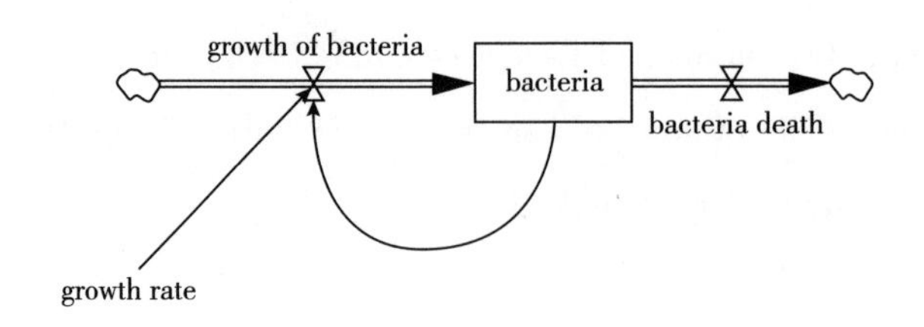

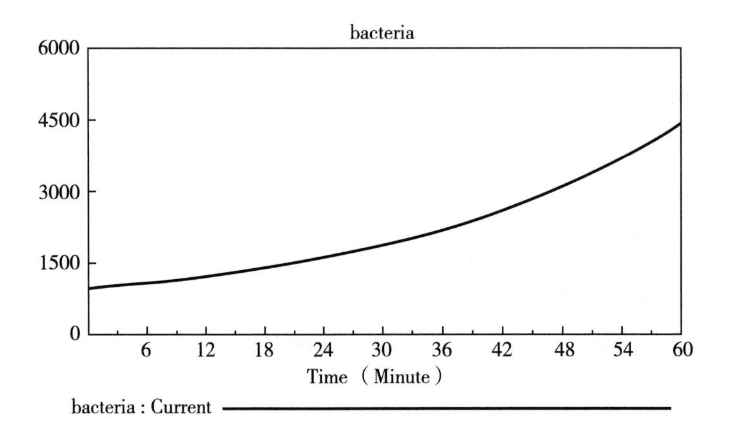

The model need care about the death rate.

(g)

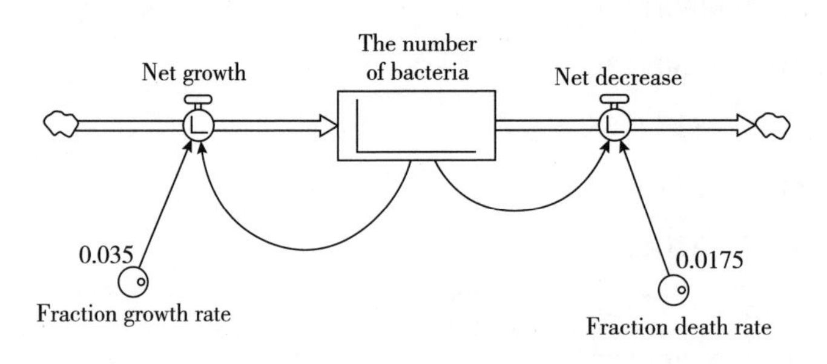

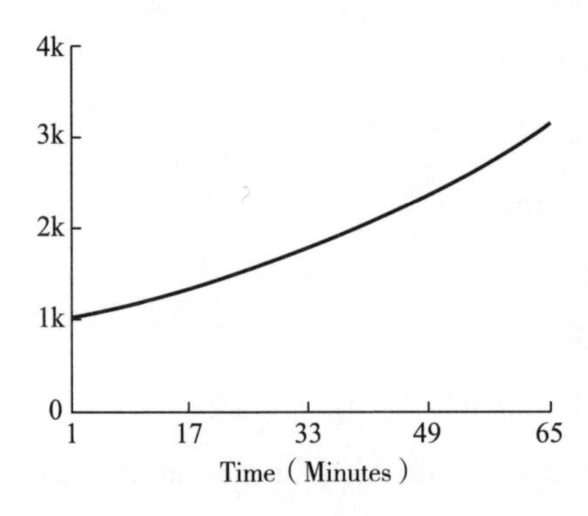

Simulationg in Stella, RK4 DT = 1/4

Initial Bacteria	1000	500
After 10 minutes	1209	595
20 mins	1506	709
30 mins	1928	845
40 mins	2527	1006
50 mins	3377	1199
60 mins	4583	1428

(h)

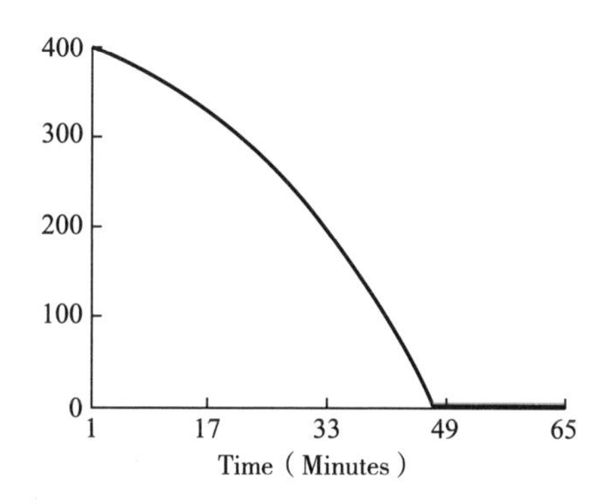

Simulationg in Stella, RK4 DT = 1/4

Initial Bacteria	400
After 10 minutes	358
20 mins	298
30 mins	214
40 mins	94
50 mins	2
60 mins	2

(i) Unstable equilibrium point is 500 (17.5/0.035)

3.3

(a) 假设 the desire stock = 100, the initial stock = 50, time = 5, 模型如下：

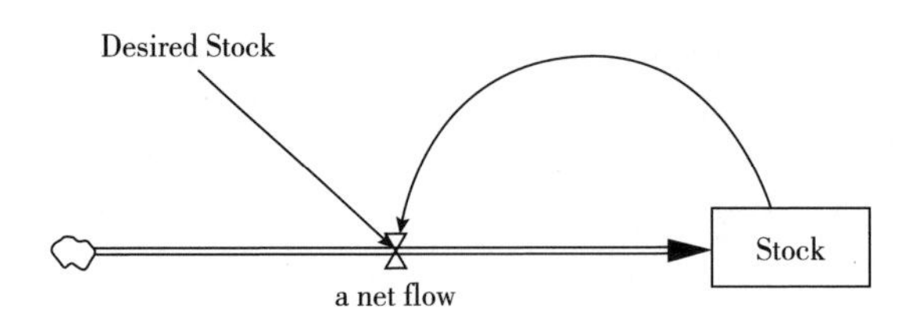

a net flow= （Desired Stock-Stock） /5

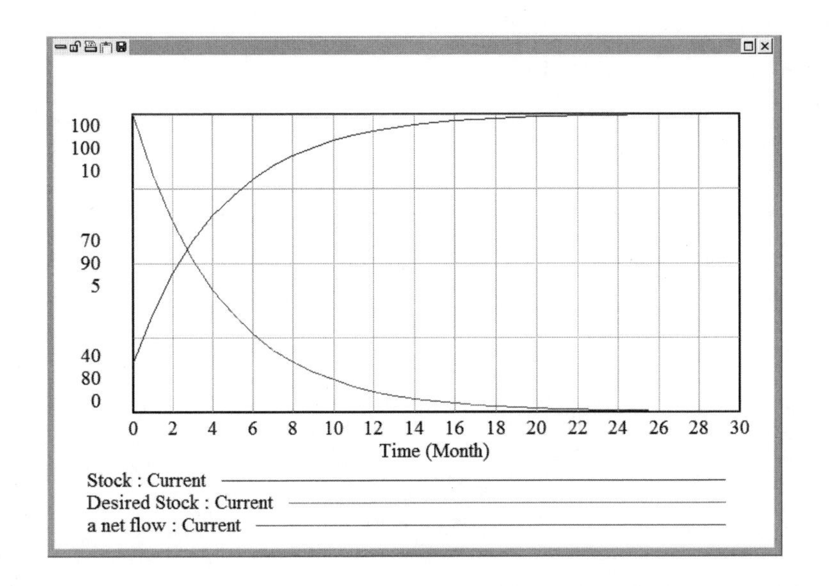

（b）由图可以看出，当 the desire stock=actual stock 时，X 也就是 the net inflow 为 0，即随着时间的推移，X 的值逐渐降低，当 the desire stock=actual stock 时，X=0。

（c）假设 the desire stock=10，the initial stock=0，随着时间的推移，stock 的量随之增加，但其增加的速度降低，最终趋近于 0。

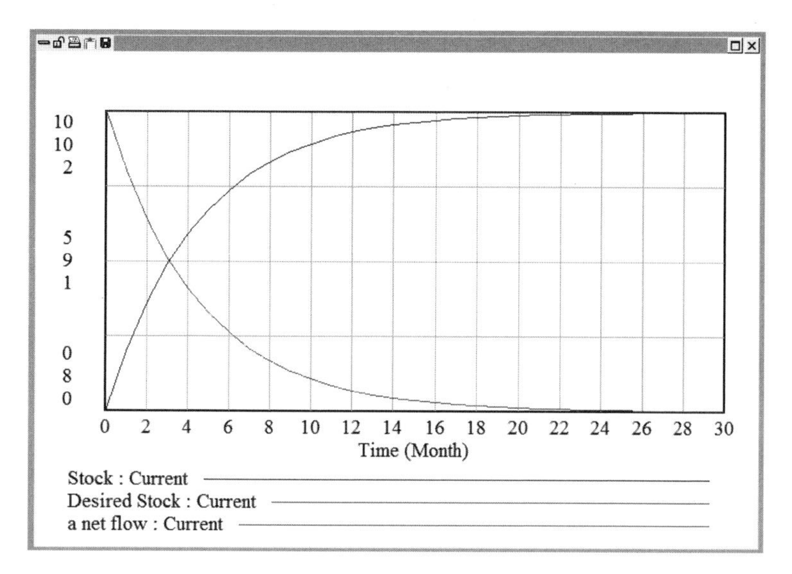

(d) Develops as shown in the figure. Assume that the initial stock value is zero. Use the method with tangents and the 63% rule. Show the stock development and *show the tangents* that you use! Be as accurate as you can, but do not use a calculator or stella.

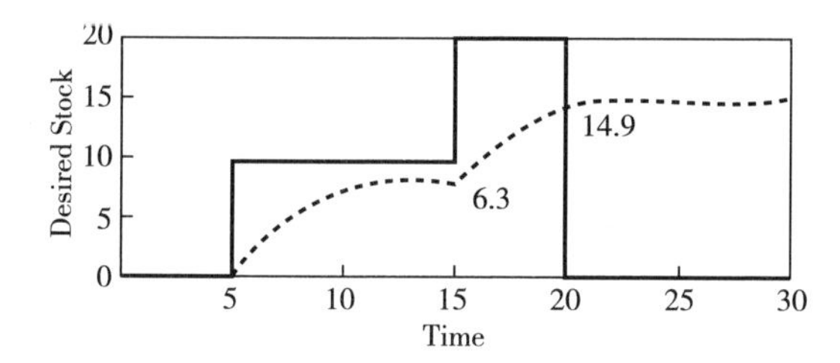

(e) Look at the development of Stock in part d) and make a rough sketch of how the net.

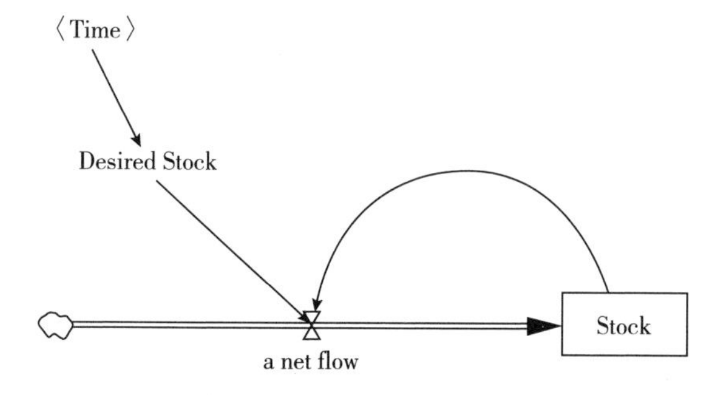

a net flow= （Desired Stock-Stock）/5

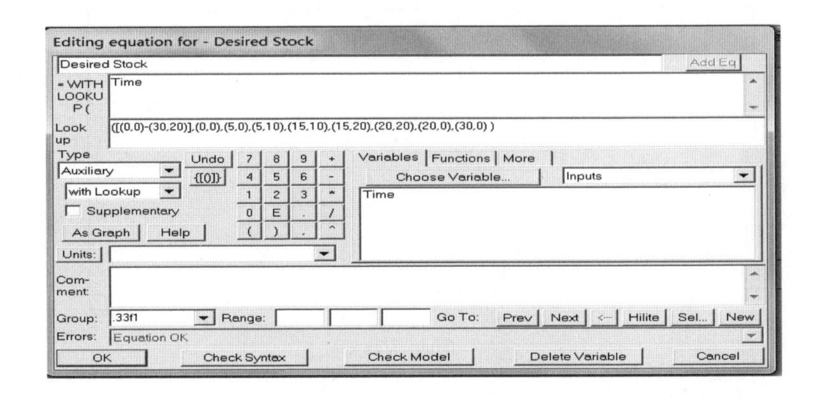

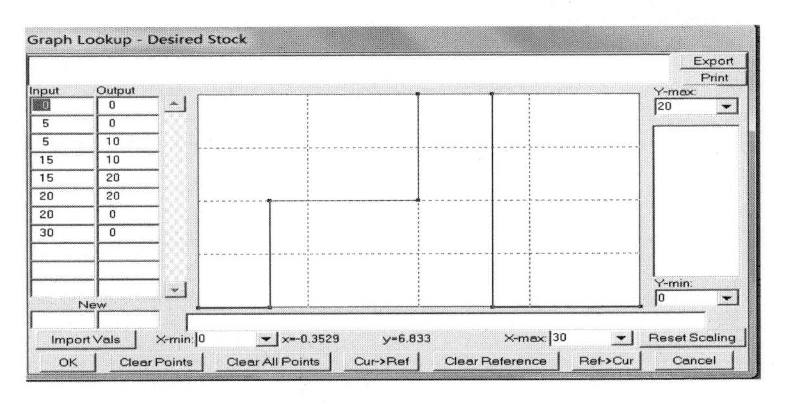

Input	Output
0	0
5	0
5	10
15	10
15	20
20	20
20	0
30	0

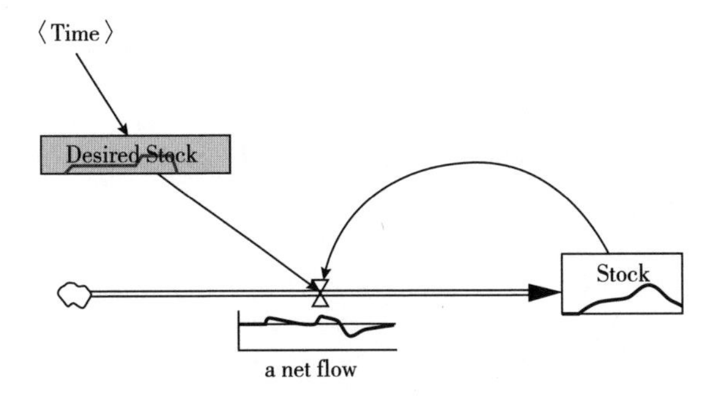

通过公式可以看到，当 desire stock 为 0 时，the net inflow 也为 0。第一阶段，当 Desire Stock = 10 时，Stock 瞬间有一个初始值，但其值随着时间推移而降低，且降低幅度越来越小；第二阶段，当 Desire Stock = 20 时，Stock 有明显

的迅速增长，其后与第一阶段趋势一致；当 Desire Stock = 0 时，Stock 速度迅速降为 0。

（f）黑色实线为 X，即 the net inflow。DT = 0. 01。

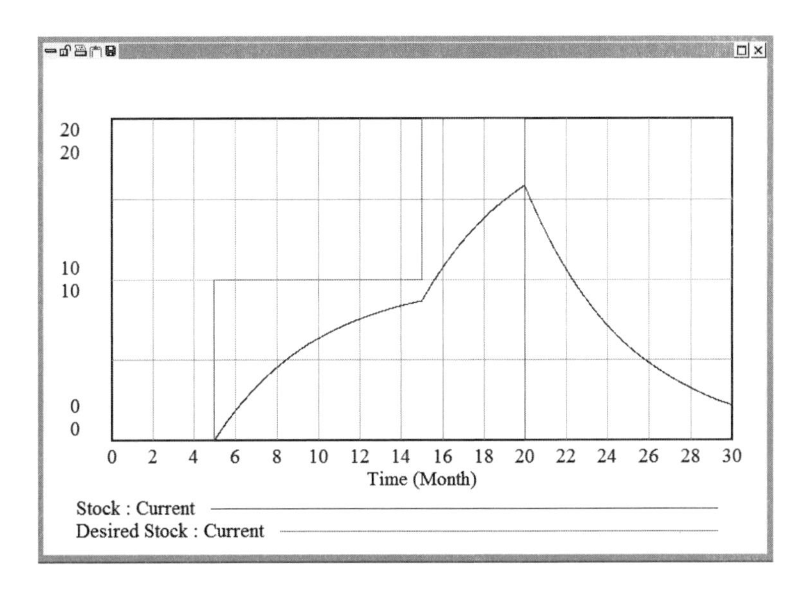

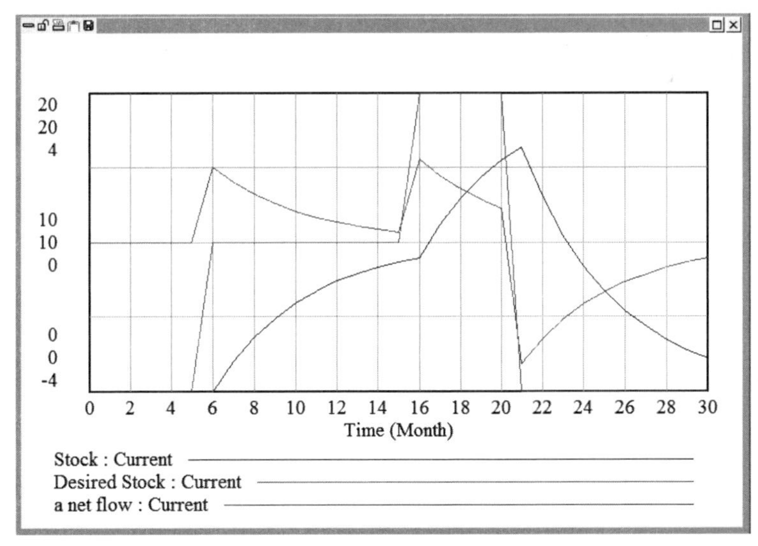

(g)

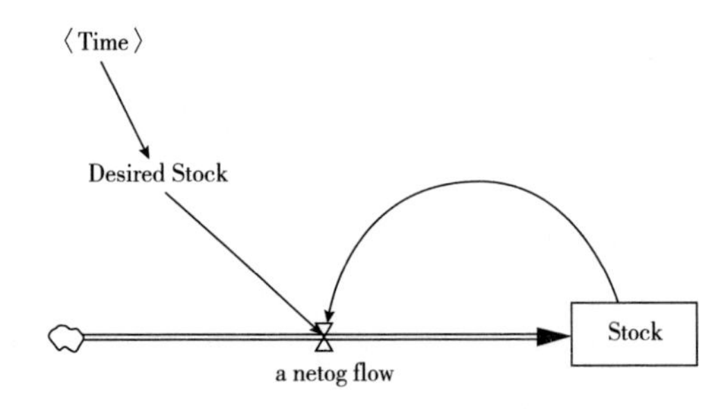

Desired stock = Time

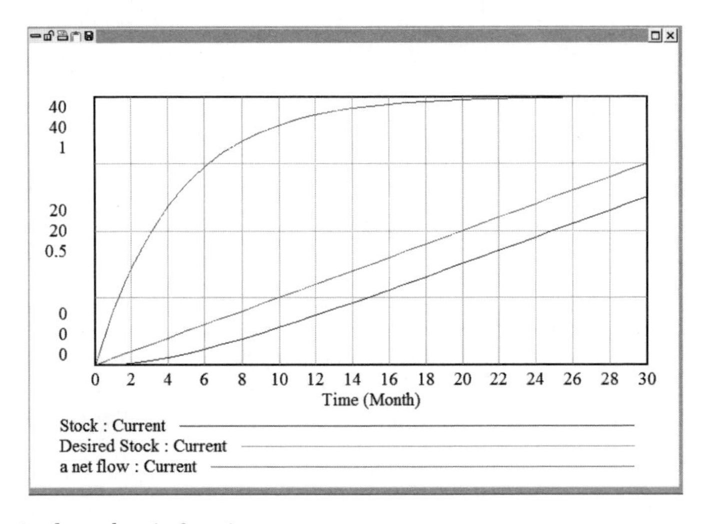

The desired stock = 1. 0 * time

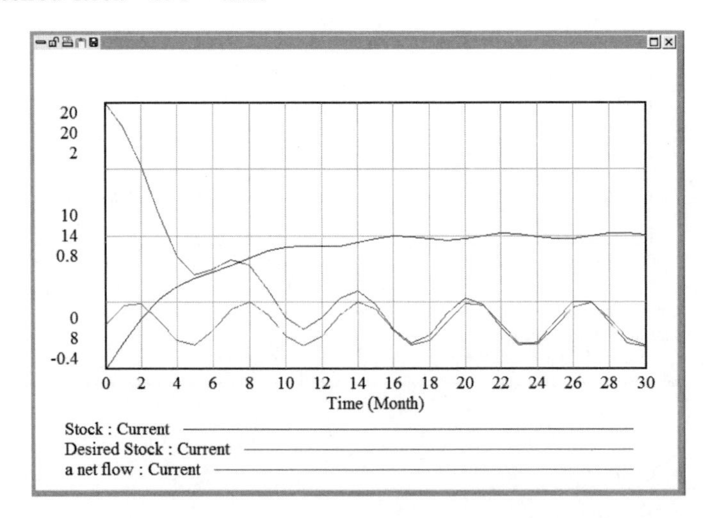

Desired stock = 10+SINWAVE（Amplitude，Period）

3.4

（a）CLD

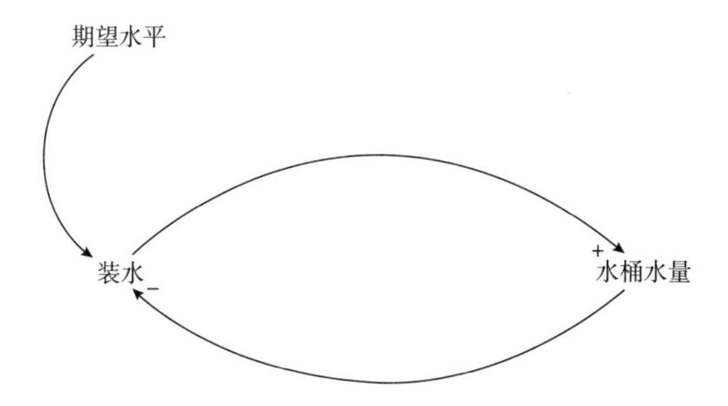

SFD

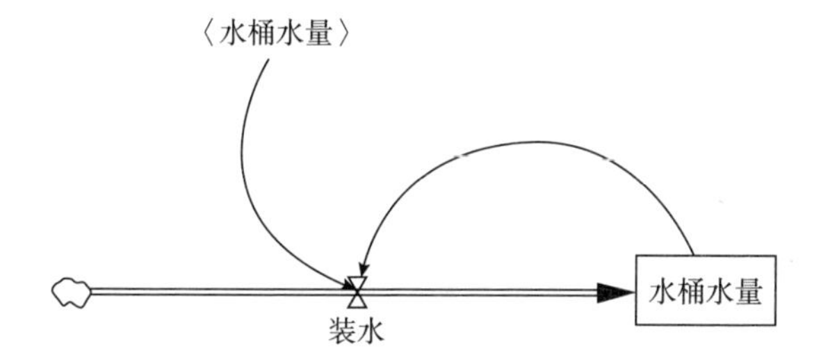

（b）CLD

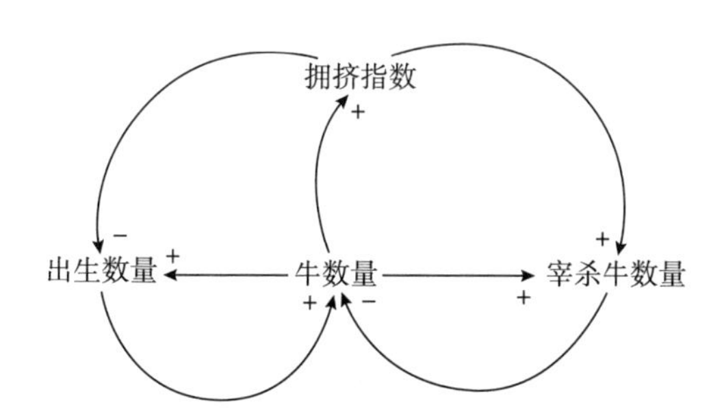

SFD

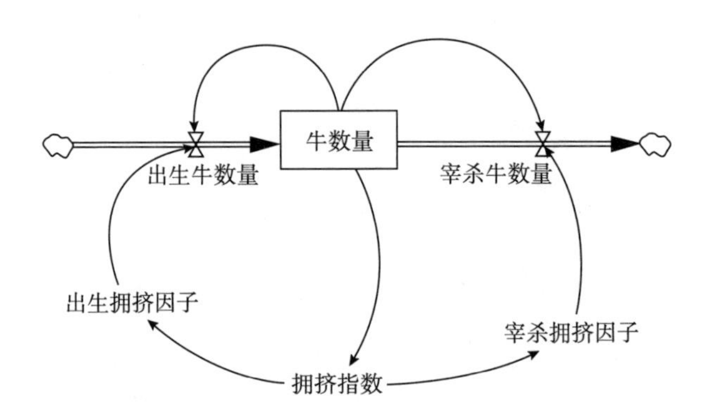

上机实验解答 13

3.5

(a)

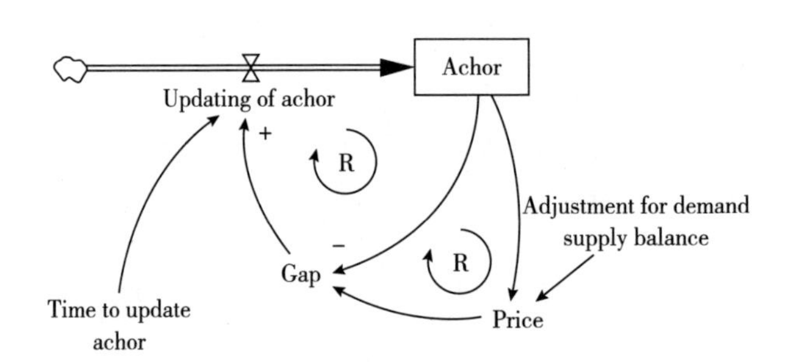

(b)

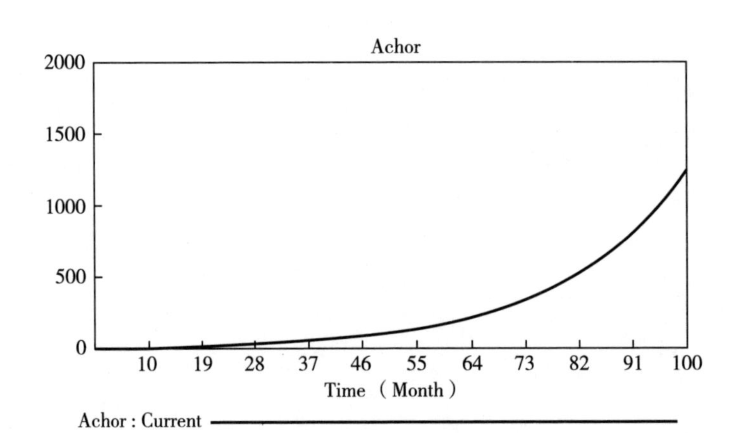

nonlinear. balancing loop and reinforcing loop

One reinforcing loop: Achor->Price->Gap->Updating of anchor->Achor

One balancing loop: Achor->Gap->Updating of anchor->Achor

The equation calibrating the growth is

Updating of anchor = 0. 05 Achor

(c) One reinforcing loop will dominate with this parameter value.

Achor = $e^{0.05t}$

Anchor->Price->Gap->Updating of anchor->Achor

Nonlinear exponential growth.

(d)

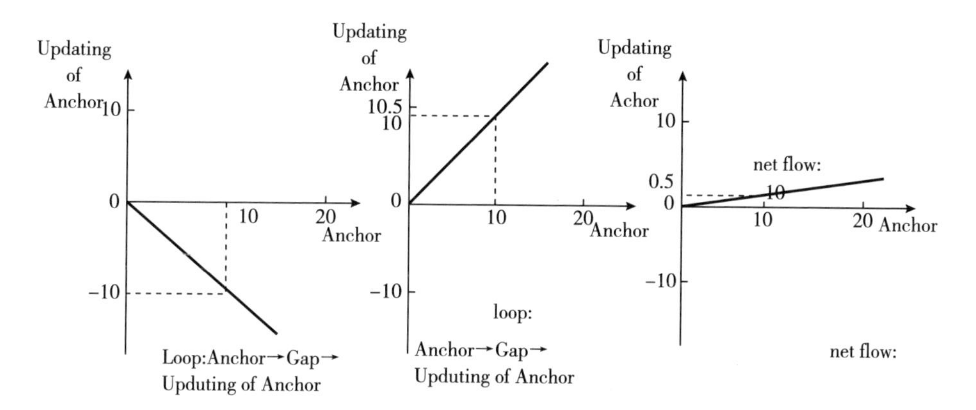

(e) When Adjustment for demand supply balance = 0. 95.

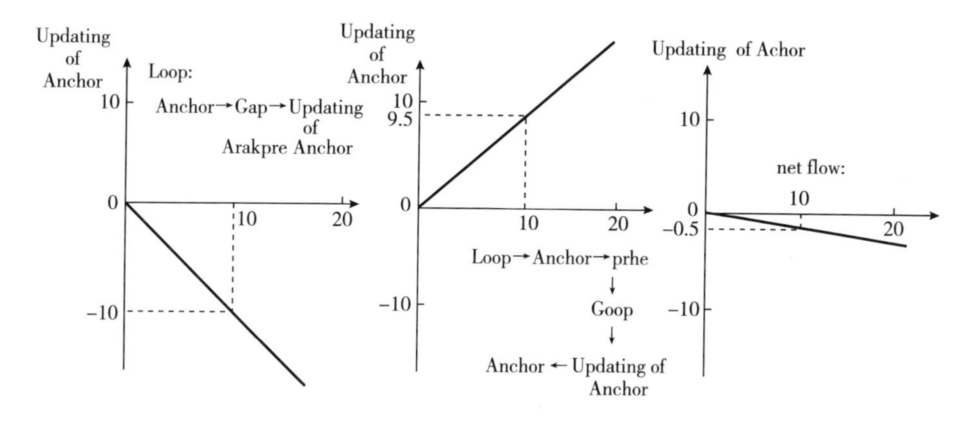

$Achor = e^{-0.05t}$

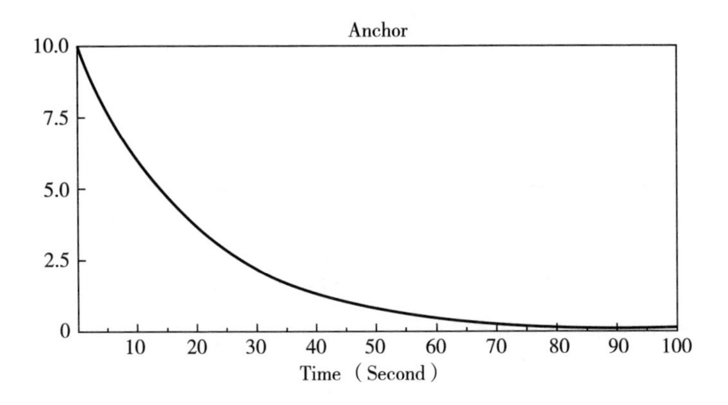

（f）

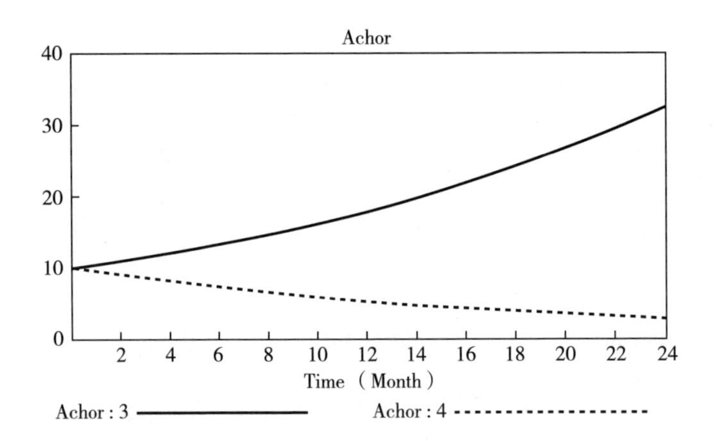

（g）Learnings.

a）Loops jointly create behavior.

b）Reinforcing Loop can creare exponential decay, if the flow id subtracted from the stock and is linearly related to the stock.

c）Loop which influence the system more greatly will dominate the behavior In this case, 0.95 makes the balancing loop dominate and 1.05 makes the reinforcing loop dominate.

4. 1

(a) SFD.

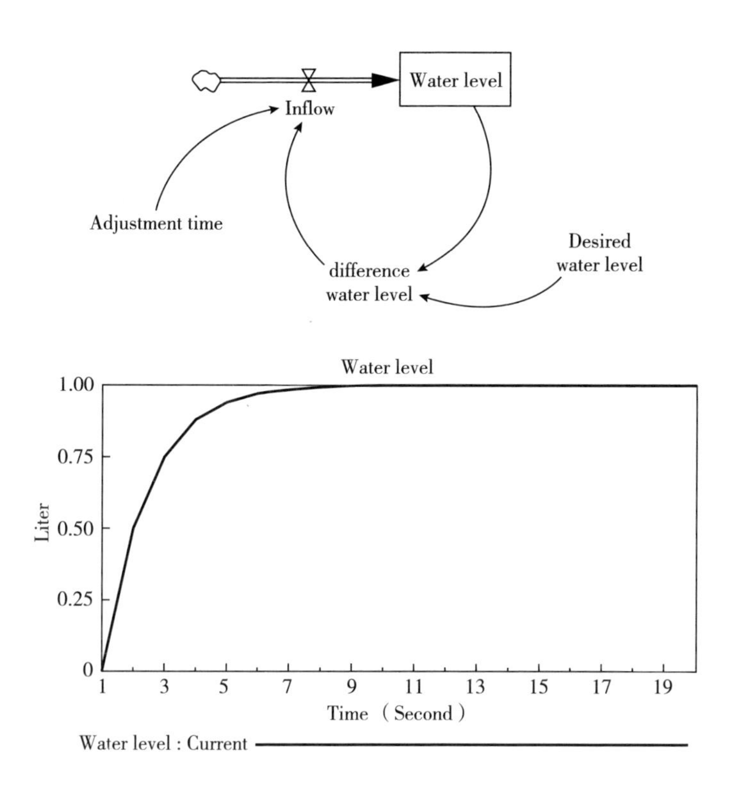

The system is one first order linear system with a closed balancing loop.

The system produce exponential growth, approaching a certain level, which is goal-seeking.

(b) Realism of decision rule.

The equation may be a useful approximation, however it assumeds that the faucet can be opened in no time, and the formulation is unrealistic in the following situation.

With a desired water level of 100 liters and an initial level of 0 liters, the initial inflow rate will be 100 liters/2s. That is 50 liters per second, which is unrealisticly much for a kitchen faucet.

(c) Limited inflow in structure graph.

Structure graph when maximum flow is 0. 1 liters/second.

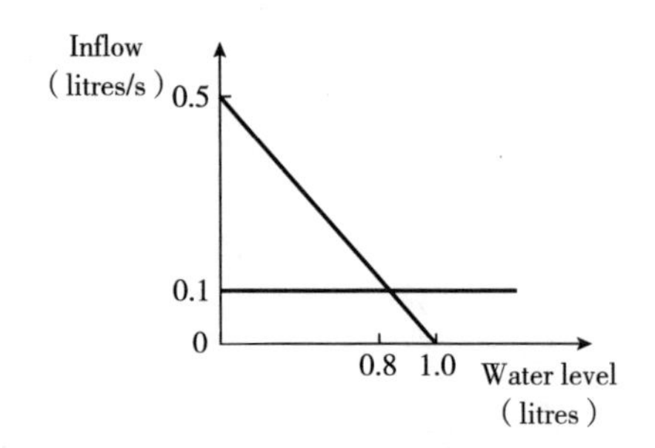

–The red line denotes the inflow with no limitation (no MIN–function)

–The blue line illustrates the limitation (the maximum flow)

–At low water levels (<0.8), a constant inflow produces a linear increase in the stock.

–From 0.8 to 1.0, a counteracting loop operates and gives exponential decay of the remaining gap between desired and actual stock.

(d) Simulation with and without the MIN–function.

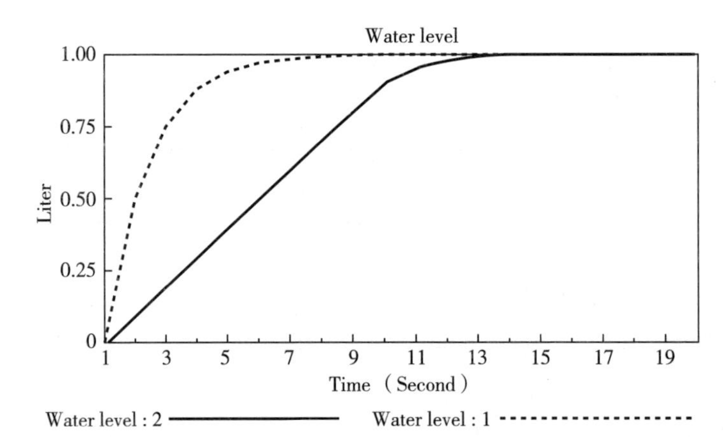

Run1 without MIN–function and Run2 with MIN–function.

(e)

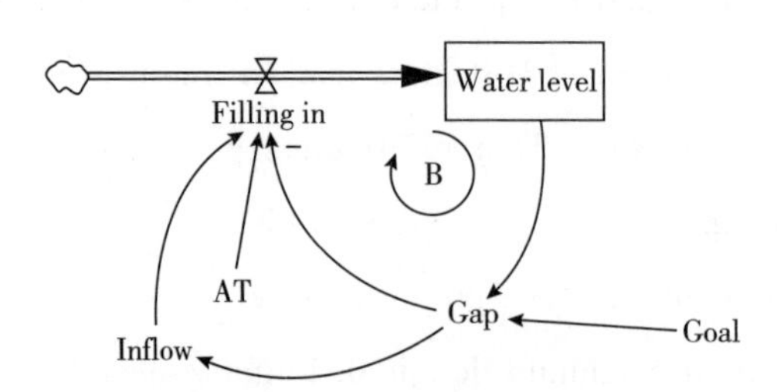

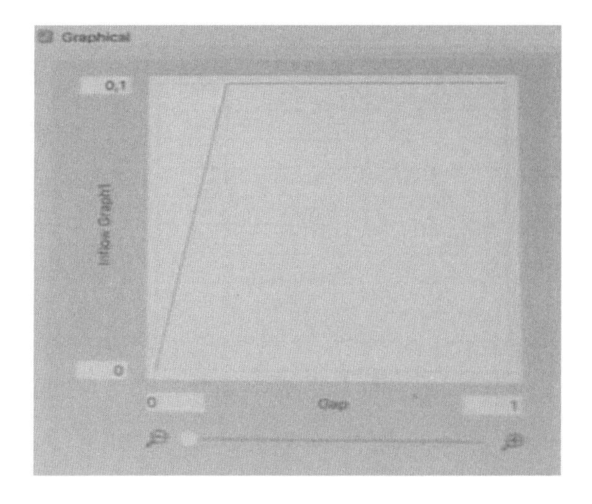

4.2

(a) SFD of the system.

Reindeer = 1250 Lichen　density = linchen/Area　Grazing = Eating per reindeer per year * Reindeer

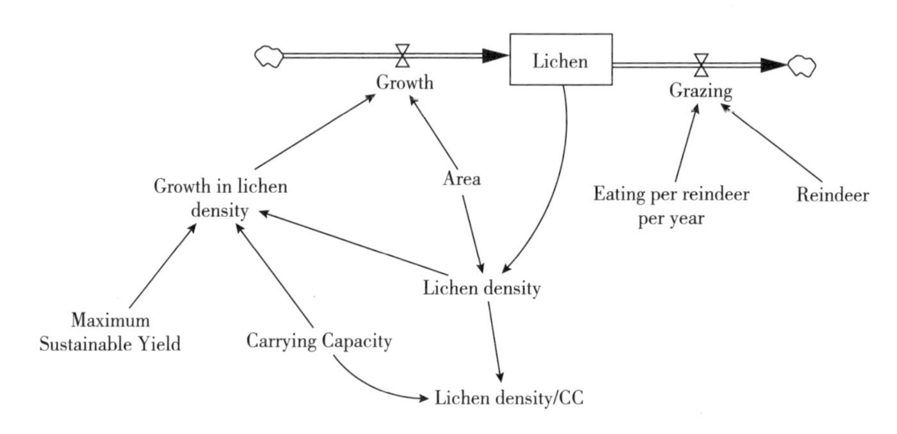

(b) Structure graph for growth of lichen (tonnes/year) over density (tonnes/km^2).

Growth = Area * Growth in lichen density

= 5 * MSY * 4 * (LD/CC) * (1-LD/CC))

a) When CC = 1200 LD = 1200, Growth = 0

b) When CC = 1200 LD = 600, Growth = 500

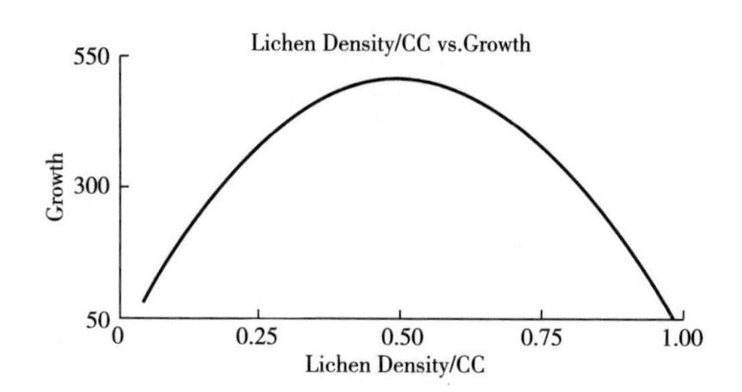

(c) Yes, it makes sense biologically. If there is no lichen, there is no growth, so the curve crosses x->axis at x=0. If there is only a little lichen, there will be more birth of lichen than death, since the resources are abundant for its reproduce. As this trend reaches and exceeds the maximum growth, resources per lichen keeps decreasing, and death grows due to a larger population, which makes death rate meet birth rate at some point. This point is the carrying capacity, a point where total number keeps constant and therefore net growth is 0.

(d) MSY = 100 tonnes/km^2/year

Total growth in 1 year = 500 tonnes

Grazing of 1 reindeer = 0. 4 tonnes

Number of reindeer = 500/0. 4 = 1250

(e) lichen behavior when density is 0 to 300 tonnes/km^2: = linear reinforcing loop and exp.

Around 600 tonnes/km^2: constant flow and = linear growht

Above 900 tonnes/km^2: linear balancing loop and exp. decay.

(f)

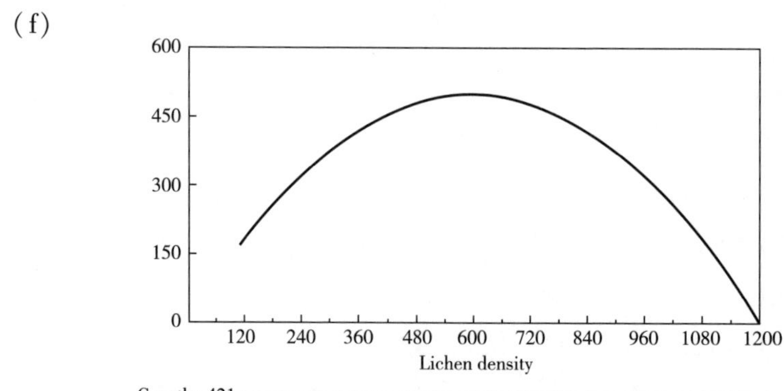

(g) Analysing grazing with structure graph.

Recall that a herd of 1250 reindeer requires lichen growth of 500 tonnes/year.

Grazing is a constant outflow, not dependent of lichen density.

Net growth = Growth − Grazing

Starting below 600 tonnes/km^2 there is a reinforcing loop with a repeller at 600 tonnes/km^2.

Starting above 600 tonnes/km^2, there is a counteracting loop with a seemingly stable equilibrium at 600 tonnes/km^2.

lichen density = 900

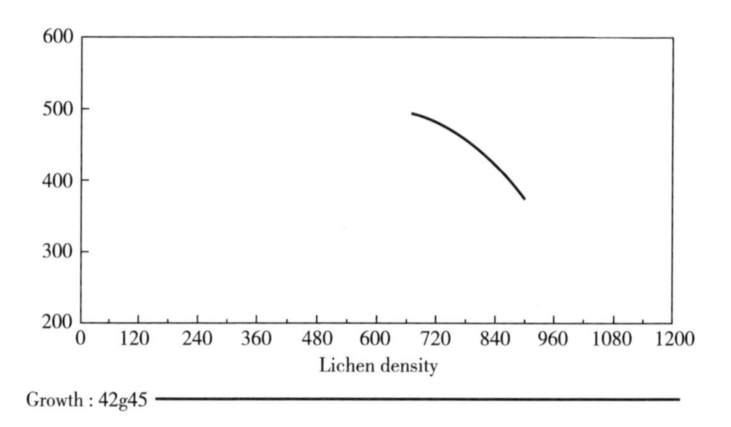

(h) What happens with 1250 reindeer and different initial lichen densities?

When starting below 600 tonnes/km^2 (pink), the constant outflow is all the time target than then inflow, which decreases as lichen density decreases.

Starting above 600 tonnes/km^2 (green), the constant outflow is all the time target than then inflow, which increases as lichen density decrease.

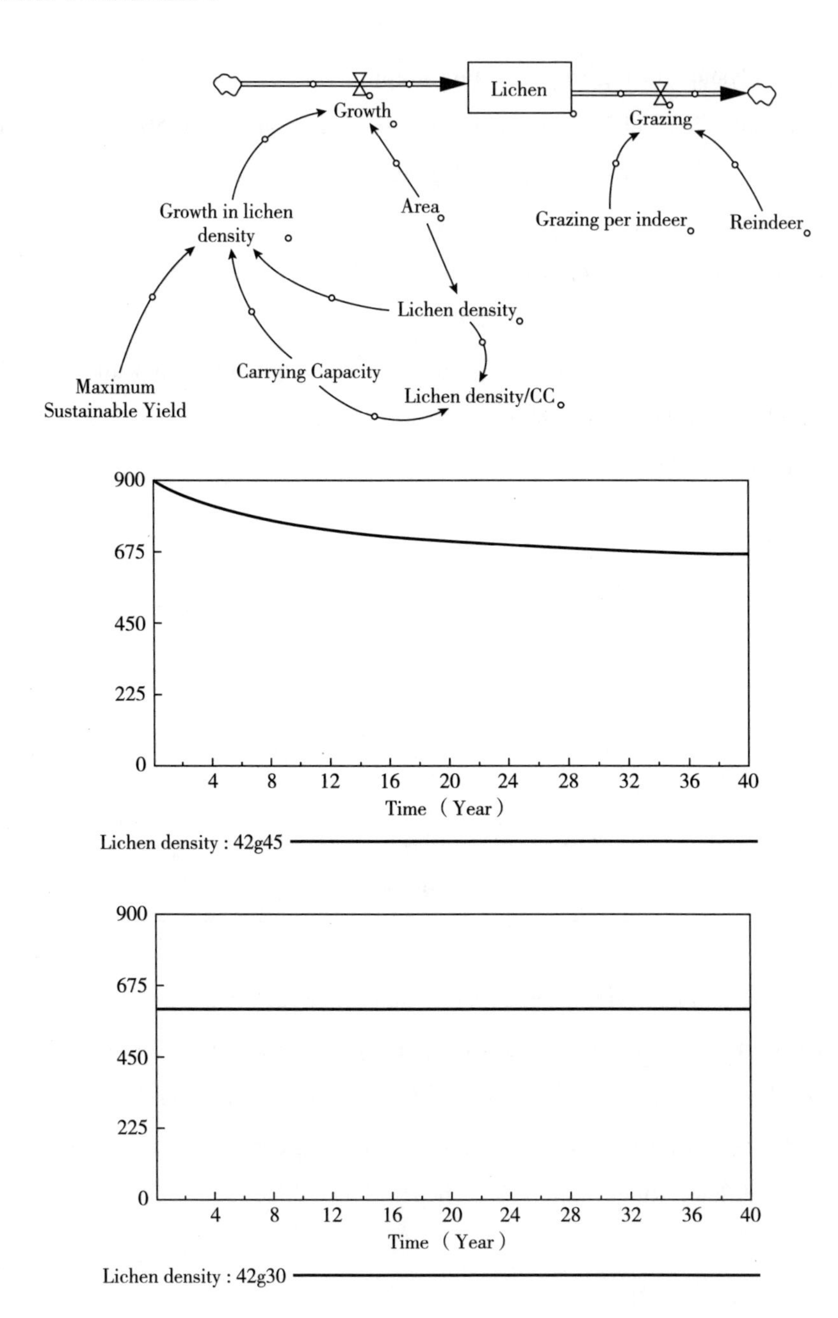

Lichen
Growth
Grazing
Growth in lichen density
Area
Grazing per indeer
Reindeer
Lichen density
Maximum Sustainable Yield
Carrying Capacity
Lichen density/CC
900
675
450
225
0
4
8
12
16
20
24
28
32
36
40
Time（Year）
Lichen density : 42g45
900
675
450
225
0
4
8
12
16
20
24
28
32
36
40
Time（Year）
Lichen density : 42g30

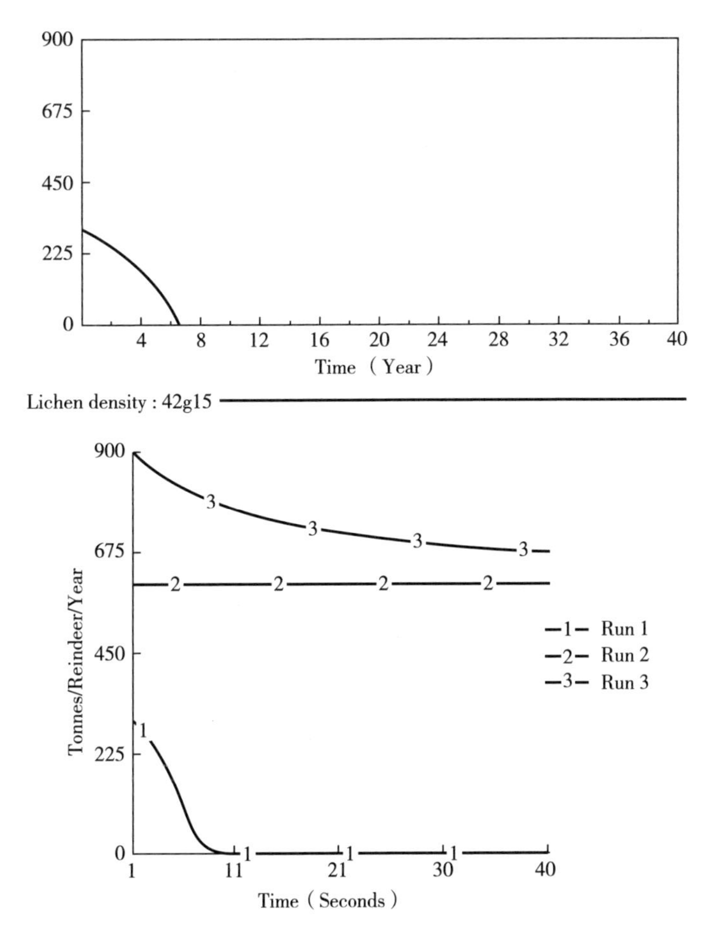

Run 1900；Run 2600；Run 3300.

（i）Using graphical functon to define desired grazing.

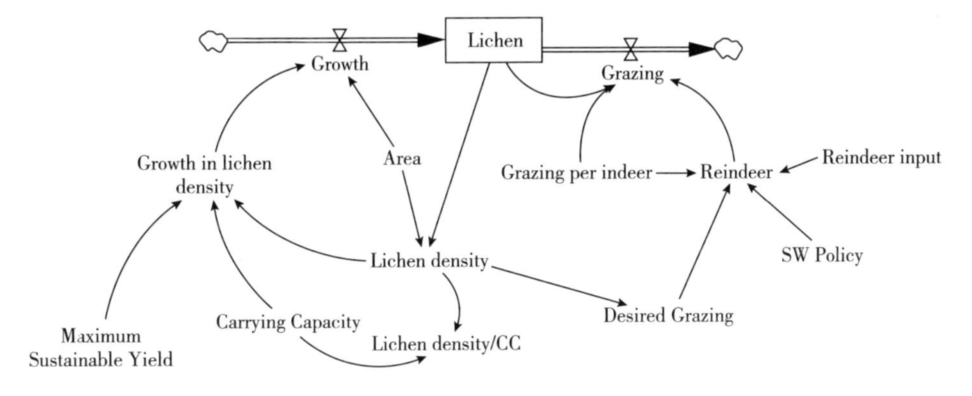

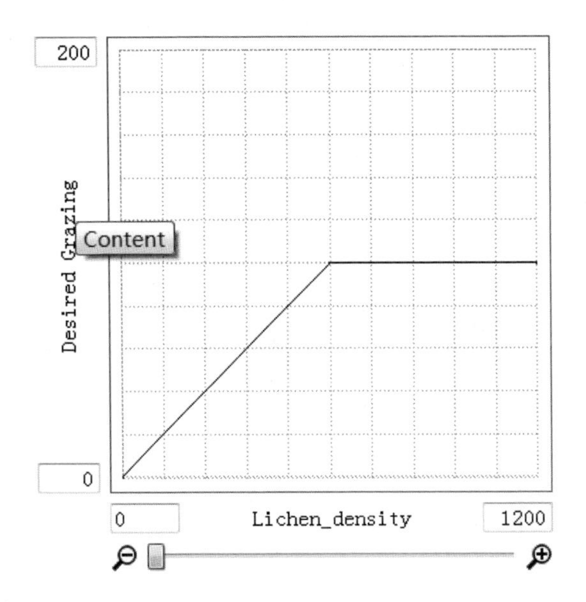

This can also be done using IF function in equation for Desired grazing.

Desired grazing = IF (Lichen density < 600) THEN (Lichen density/6) ELSE (100)

Simulation of the hypothesized policy:

SW policy = 0. 08 lichen = 4500 and 550 lichen density.

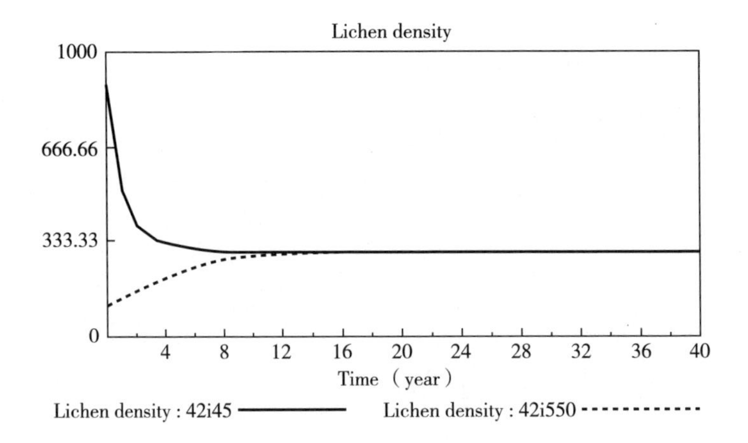

(j)

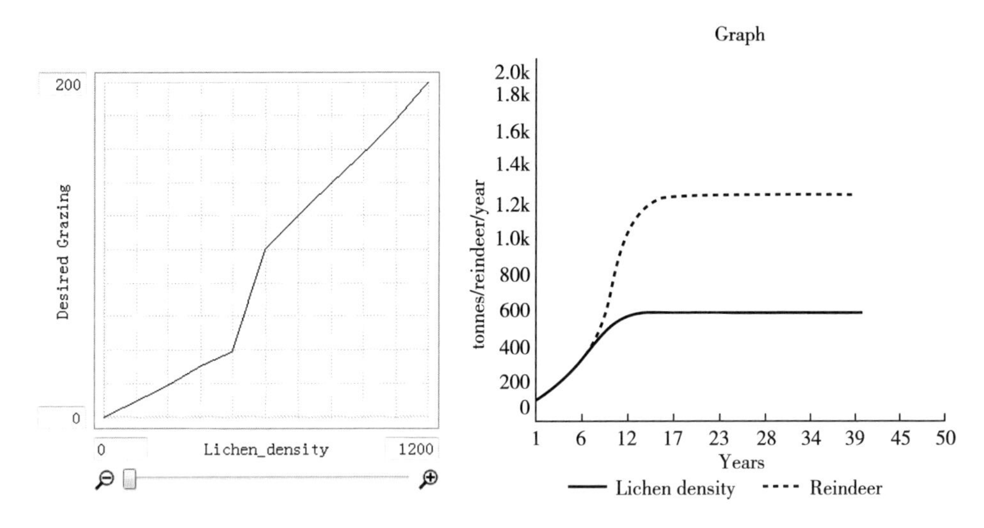

(k)

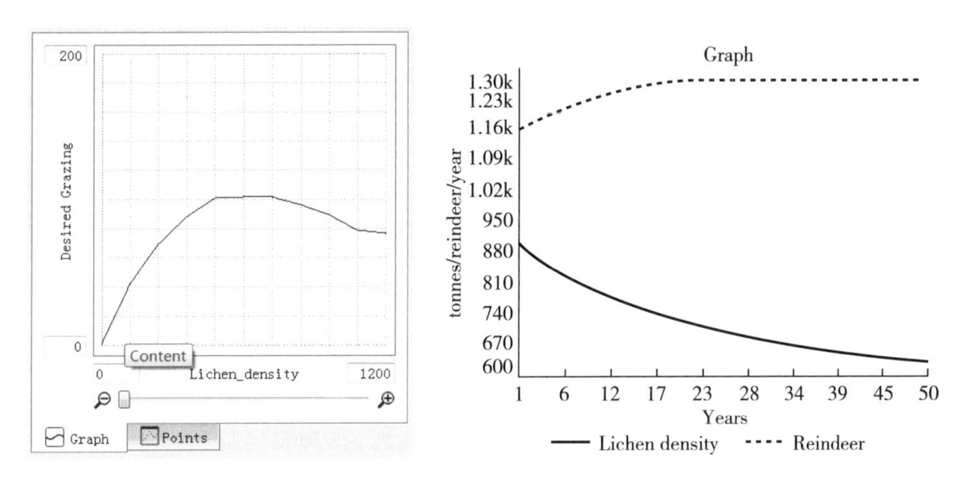

4.3

My version:

P′: Managing a lichen pasture that has been overgrazed by reindeer. The lichen pasture should be recovered to a sustainable condition and stabilize there.

H: Grazing is larger than lichen's Net growth, which leads to a decline in lichen.

A：(Structure) Amount of lichen can accumulate (a stock). Net growth (Birth-Death) can add to lichen's amount (inflow), while Grazing reduces lichen's amount (outflow).

Net growth reaches maximum at a certain lichen stock, and is 0 at carrying capacity. (Behavior) Grazing>Net growth explains the decrease of lichen over time.

P：(Structure) Reduce reindeer and allow the lichen to recover to its maximum growth point. Then in order to stabilize there, the number of reindeer should be adjusted to generate same amount as grazing. (Behavior) Reducing reindeer will cut down the outflow below the inflow and let the lichen stock to grow to the Biomass Maximum Sustainable Yield. When reaching BMSY, a grazing (outflow) that equals to Net growth will make the stock neither to grow nor to decay.

I：Decision makes tend to regard every relationship as instaneous cause and effect, thus they may hope to stop the lichen decay by simply keeping the number of reindeer stable. They should learn that grazing and lichen amount is an accumulation relationship, and only by reducing more aggressively can decay be stopped.

上机实验解答 14

4.4

(a) Causal Loops diagram.

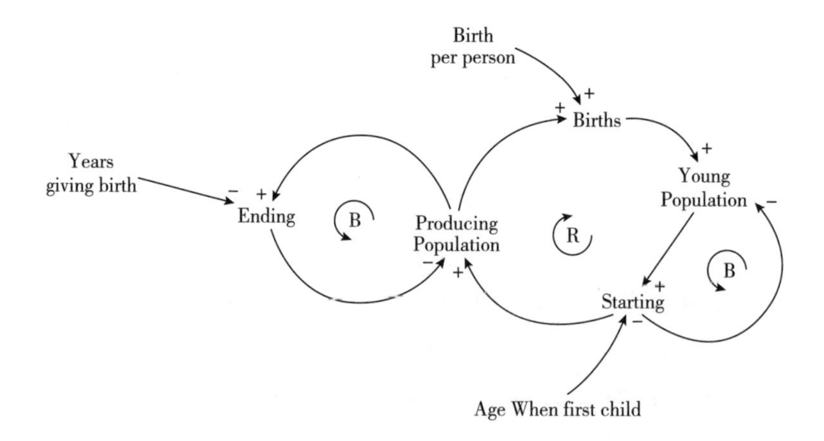

(b) Behavior of Young Population.

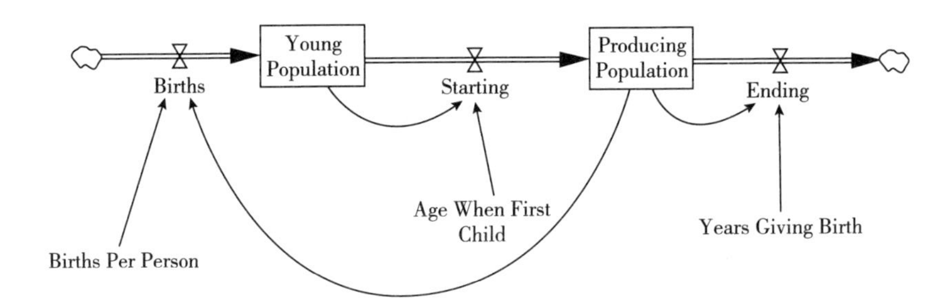

Births = Producing Population * Births Per Person

Starting = Young Population/Age When First Child

Ending = Producing Population/Years Giving Birth

Births Per Person = 0. 5 person per year

Years Giving Birth = 4 years

Young Population = 100 million persons

Producing Population = 22 million persons

Age When First Child = 15 years

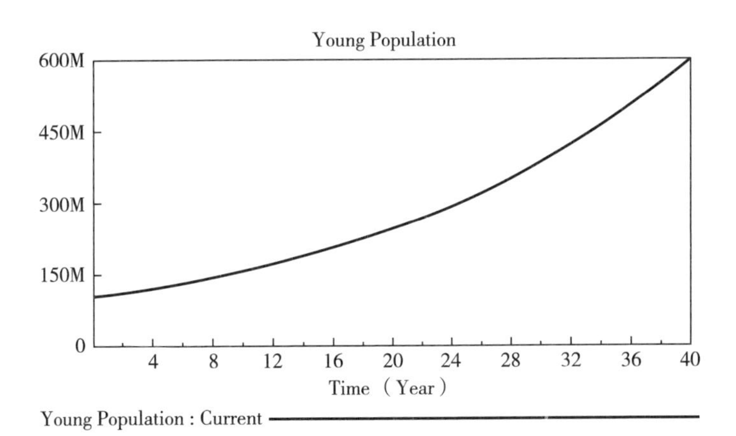

a) It is a nonlinear system. Because every relationship in this system can be defined by a linear equation.

b) Explanation in terms of structure:

-There are 3 feedback loops: one reinforcing and 2 balancing.

-Young population' s behavior looks like exponential growth.

-Loop dominance is the reason that behavior looks like exponential growth. Under this set of parameters, the dominant loop is the reinforcing loop, which leads to exponential growth.

c) Explanation in terms of parameters:

-There are 3 parameters in all: *Age when first child*, *Years giving birth*, and *Birth per person*.

-Hypothesis 1: Births and Ending are inflow and outflow for the entire system. They share a common factor (cause), *Producing population*, and have *Birth per person* and *Years giving birth* as second factor respectively. Only when *Births per person*> (1/*Years giving birth*) can the whole system have an inflow>outflow, which guarantees a growth of Producing population in the later stage. In other word, it guarantees the reinforcing loop to be the dominant loop.

-Sensitivity test 1: altering Birth per person while keep the other 2 stable. When BPP is lower than 0.25 (1/Years giving birth), Young Population will show a decreasing trend in later development stage (and higher for increasing), while BPP = 0.25 makes Young Population stable at last. Keeping Birth per person stable while altering Years giving birth will generate the same thing. Hypothesis 1 is therefore accepted.

-Hypothesis 2: The two stocks, Young population and Producing population, will always coordinate to each other in long run. Age when first child affects at what ration will the two stocks coordinate to each other. The larger this parameter is, the larger Young Population/Producing population will be.

-Sensitive test 2: Reset Births per person = 0.5 and Years giving birth = 4, tune Age when first child from value near 0 to as high as 40 and monitor Young population/Producing population. Behavior of Young population and Producing population both show a growing trend, but their ratio become stable in long run. As Age when first child turns up, this ratio also goes up. Hypothesis 2 is there fore accepted.

(c)

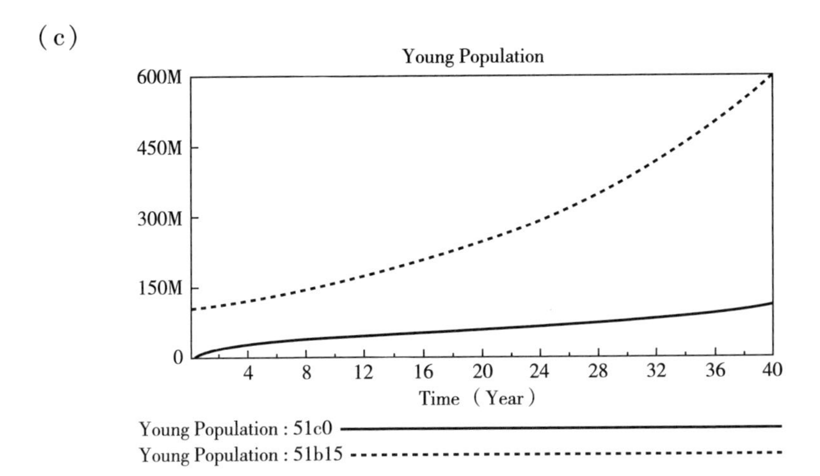

Run 1 Young Population = 100; Run 2 Young Population = 0

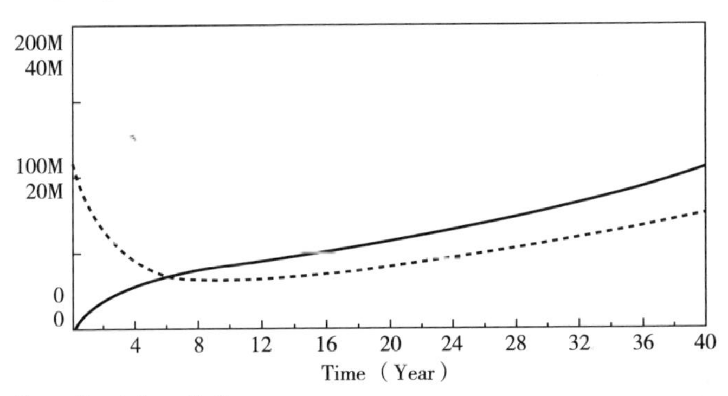

(d)

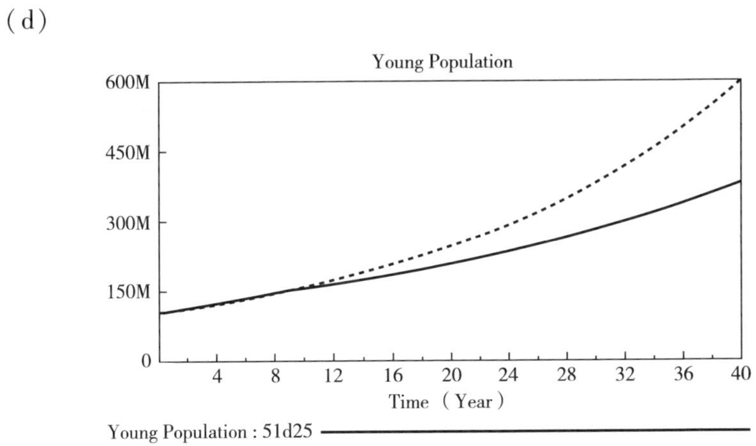

Run 1 Age When First Child = 15; Run 2 Age When First Child = 25

Growth is slower both for young population and producing population, when Age when first child is 25. This means it will take 10 years for each woman to have her first child. For the whole system. Age when first child works like a faucet. A larger number will slow down the flow of the system.

(e) 1 couple = 2 persons.

When Years giving birth = 2.2, 2 persons * 0.5birth/person/year * 2.2years = 2.2 births.

When Years giving birth = 4.4, 2 persons * 0.5birth/person/year * 4.4 years = 4.4 births.

The difference is because there will be a shorter period in which a couple can give birth to babies.

4.5

(a)

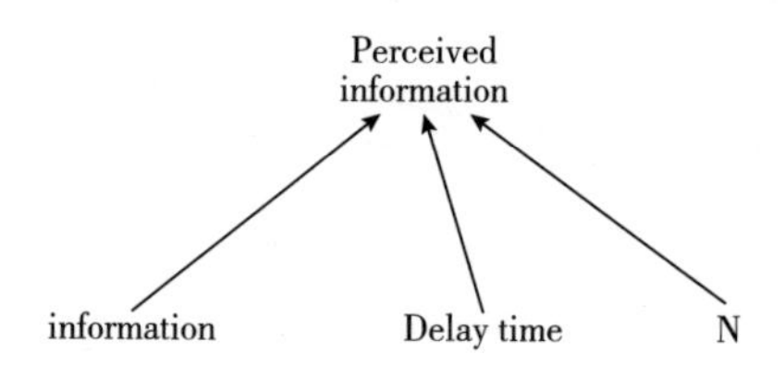

Order N	Delay time
1	Average time
3	Average time/3
6	Average time/6

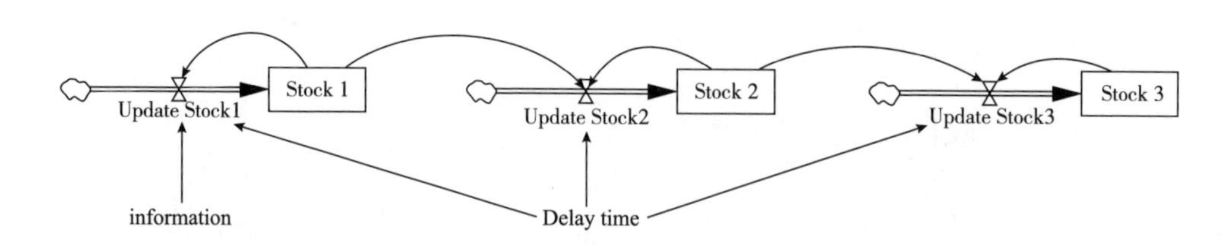

6th order delay is a linear system, for every relationship in it can be defined by a linear equation.

(b) Yes, the choice of Order for a delay have implications for the choice of DT. DT must be smaller than Delay time per stock, otherwise Stella will not be able to calculate every individual 1st order delay.

Reason: In the case N=3, if there is an input of information into the first 1st order delay at t=0, the output from this delay with start from t=DT, which is also the beginning time of input into the second 1st order delay. For the same reason, it takes at least 3DT for the whole system to have output (Perceived information). Therefore, given a Delay time and an Order, DT must be smaller than (Delay time/Order).

(c)

Update Stock 1= (Information-Stock_1) /Delay_time_per_stock

Update Stock 2= (Stock_1-Stock_2) /Delay_time_per_stock

Update Stock 3= (Stock_2-Stock_3) /Delay_time_per_stock

N=1 Delay_time=4 weeks

Information=step (10, 2)

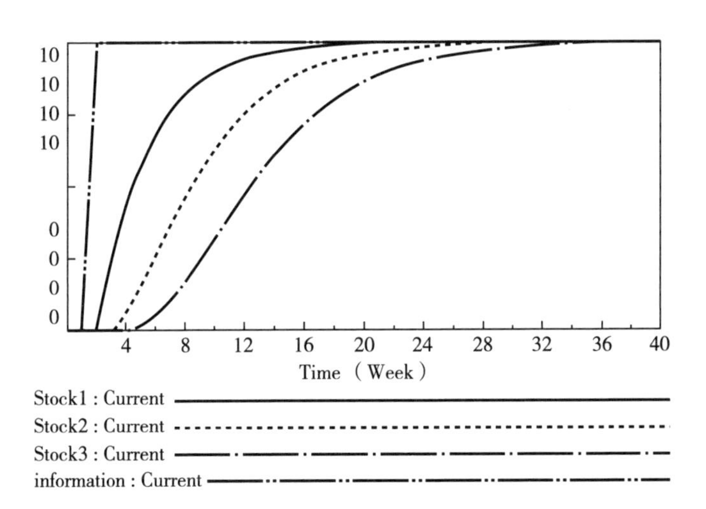

Explanation:

Stock 1: A goal-seeking behavior, for it is a closed balancing feedback loop with a constant goal (information).

Stock 2: Early development is slow for its input (Stock 1) is at a low state. As Stock 1 grows fast, Stock 2 also shows a fast growth in late development. As Stock 1 reaches a plateau, Stock 2 shows a goal-seeking-like behavior. The whole behavior starts to be like S-growth.

Stock 3: Behavior of Stock 3 follows the same pattern as Stock 2. It looks more like S-growth than Stock 2 does.

(d)

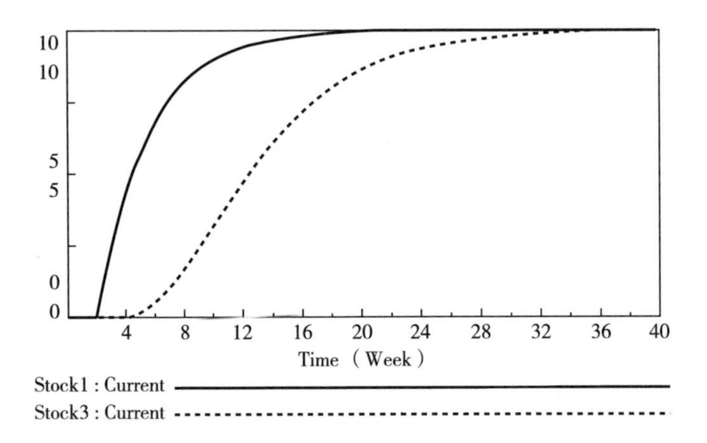

Run1: N=1 Run2: N=3

Not logical. The two Perceived Information should be the same to Information in the long run. The behavior of the 1 st order delay is goal-seeking, of which the goal is information (aka. 10). In the long run, Perceived Information will exponentially approach 10.

The behavior of the 3rd order delay is composed of 3 consecutive 1st order delays. Each later delay uses the former outcome (stock) as a goal to perform goal-seking behavior, so they cannot outnumber the goal, but only approach it.

Delay time will only determine how long a transient behavior will be. However, in the long run, the behavior will be in steady state, whatever the delay time is. Since both systems share the same goal, they will give same Perceived Information in the long run.

(e) Ordering = step (10, 2) −step (10, 3)

Delivering = DELAY1I (Ordering, Material Delay time/N, initial flow)

Material Delay time = 12 week

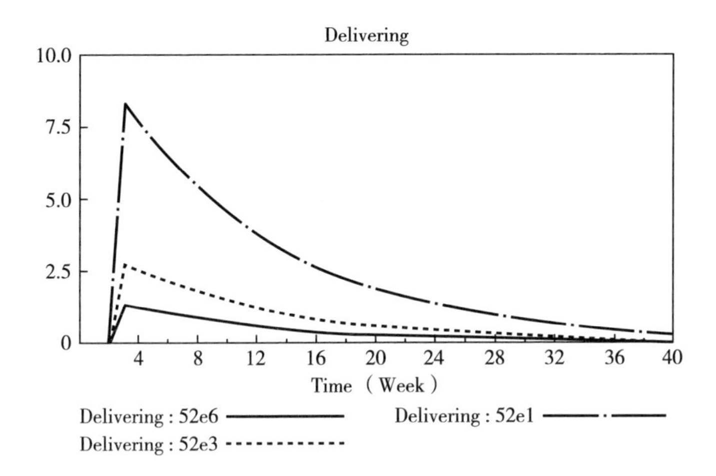

Run1: N = 1 Run2: N = 3 Run3: N = 6

Perceived information takes more time to show a wave crest at N = 6 than N = 1. In decision making, 1st order delay means information is directly and immediately perceived by the decision maker, while higher order delay means the ultimate decision maker can only perceive the information after all his preceding link have perceived it. This also means flat enterprises are more conducive to efficient decision-making.

(f)

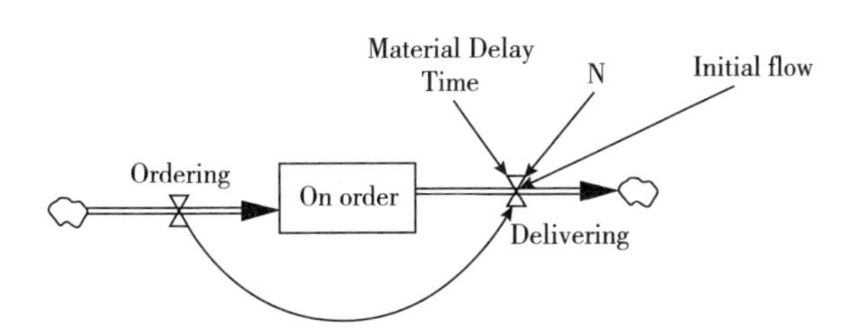

Delivering = DELAYN (Ordering, Material_Delay_Time, N, Initial_flow)

Material Delay Time = 12 N = 6

Ordering = PULSE（100，2，1）

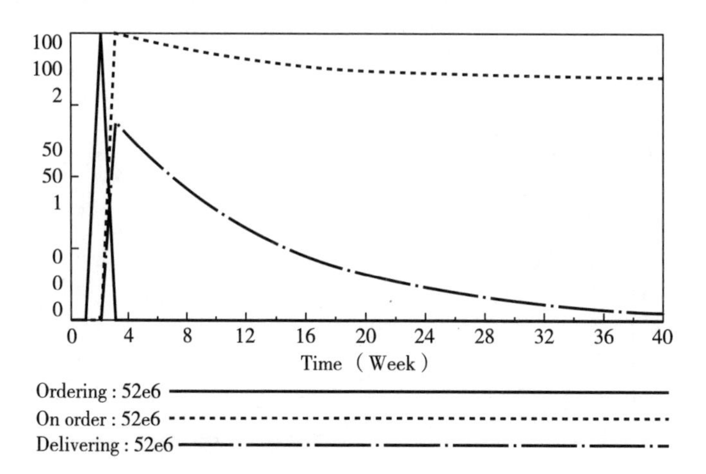

（g）

Ordering = 10 units per week

N = 6

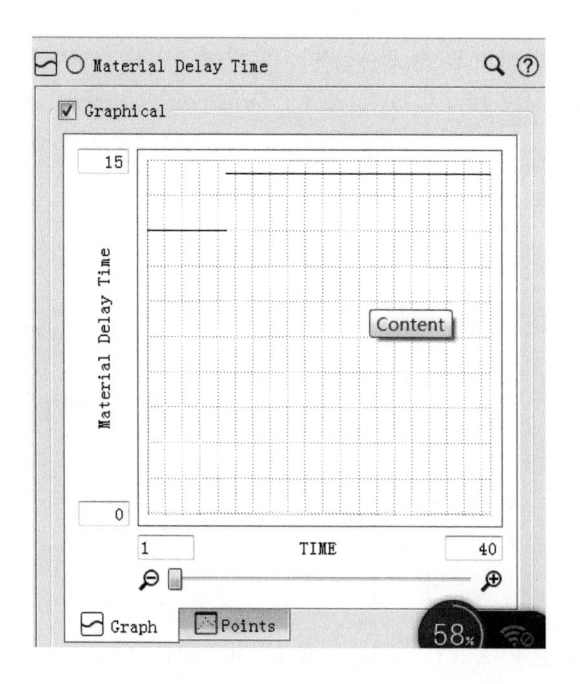

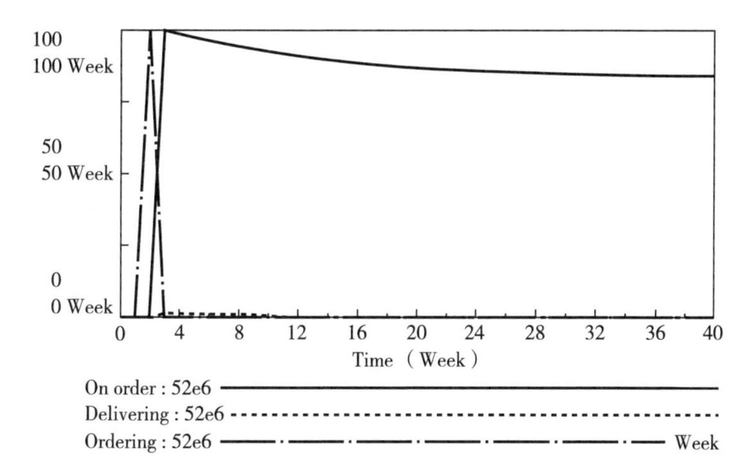

On order contains the sum of all 6 individual stocks in the material delay. Its inflow is Ordering, and outflow is delivering. Therefore, On order is the accumulation of the gap between inflow and outflow, over time. Delivering takes time to reach the same level as ordering, due to material delay. During this time, On order accumulates to a certain amount (transient), and keeps stable (steady-state) after Delivering meets Ordering.

4.6

(a)

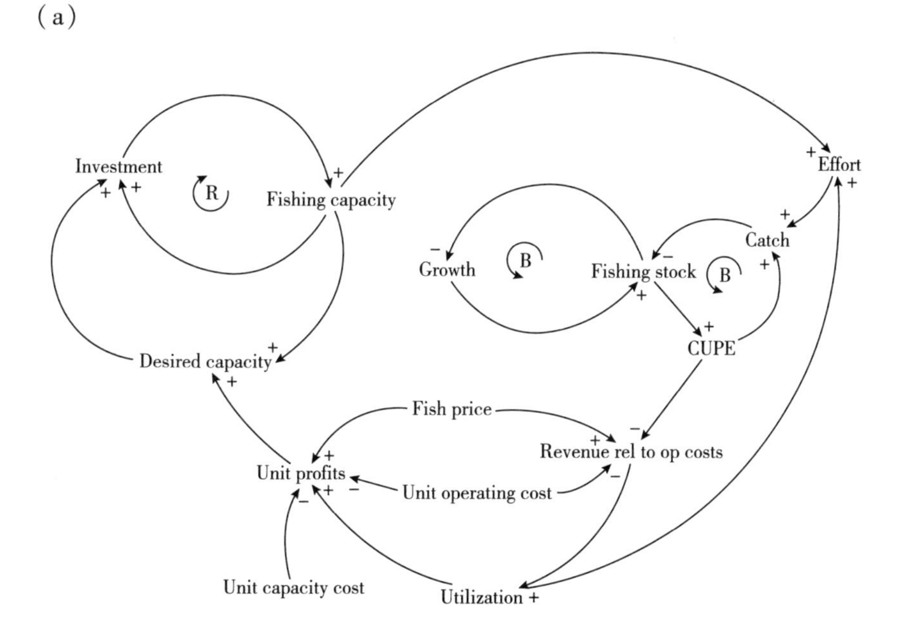

(b) Initially there is only a little fishing capacity and there is a lot of fish (close to carrying capacity). The fishing capacity and the fish stock will develop like the curves in following picture.

At first, there is a lot of fish and only a little fishing capacity. So CUPE is large, and assume that the fish price is higher than the initial unit costs. Thus, the profits is large, as result, most fishers are fishing and fishing capacity is increased in order to catch more fish.

4.7

Alice's Version:

P′: Young people tend to be overly intoxicated than intended when they drink alcohol. The death rate of juveniles caused by overdrinking has been raised. This became a new social challenge to be solved.

H: According to the physiology, when we drink alcohol, it is accumulated in our stomach delaying in transferring into the blood stream. Moreover, young people who do not have much experience drinking are apt to drink alcohol without any considering the delivering time of the uptake alcohol from their stomach into the Blood Alcohol Concentration (BAC). They try to make a decision only relying on the BAC instead of the amount of alcohol in their stomach. As a result, they drink alcohol over their desired drunkenness.

A: This model has two stocks. One is "Alcohol in Stomach" and the other is "BAC". Between two stocks, there is a flow as "Diffusion", which makes delay accumulating alcohol into the stock, BAC. There is feedback loop between BAC and Drinking but there is no feedback between Alcohol in Stomach and Drinking.

This goal seeking model illustrates the overshoot behaviors for drinking alcohol very well. This behavior pattern is caused by no feedback loop between Alcohol in Stomach and Drinking and the delay between two stocks.

The data from experiment also support this model. Participants in the experiment misunderstood the delay time and finally overshot their goal.

P: To avoid overdrinking, new feedback loop between the Alcohol in Stomach and Drinking will be needed.

However, since the Alcohol in Stomach is invisible stock, careful rules to make decision should be considered such as "first time slow and careful drink" rule. In addition, education to fix misunderstanding about the delay is needed to be introduced.

I: Applying new decision rule is not easy. People tend to resist accepting new information. Nurturing and creating new social culture are hard work.

My Comment:

In "Problem", the importance of the overdrunk problem was detailed illustrated and persuasive.

In "Hypothesis", delay was properly mentioned as an explanation to the gap between drinking and getting drunk. The way young people make decisions, namely, relying on BAC rather than stomach alcohol level, was discussed to explain overshooting.

In "Analysis", a goal-seeking behavior with oscillation was elaborated detailed, based on a well-defined two-stock reference model.

Policies discussed before "Implementation" is highly realistic. However, resistance and barriers in implementation can be more diverse, and multi-perspective hypothesis will be welcomed in consulting.

My Version:

P′: Inexperienced drinkers tend to drink more than they want. According to a study in U. S. A., 1600 alcohol related deaths and half million injuries happens per year. Moreover, most campaigns to reduce binge drinking have little effect. "Becoming drunker than intended" get people into risk of consequent physical damages.

H: (Physiology) How much drunk we feel is associated with the level of alcohol in our blood. Physiology mechanism of human body determines that before entering blood, alcohol will stay in stomach and that it always takes time for alcohol in stomach to be absorbed into blood. (Decision rules) Inexperience drinkers tend to use how much drunk they feel as an indicator for stopping drinking. However, after stop drinking, they will end up with overshooting since alcohol in stomach keeps being absorbed into blood.

A: (Structure) Alcohol in blood (a stock) increases by absorption (inflow) and decreases by metabolism (outflow) over time. Alcohol in stomach (a stock) increases by drinking (inflow) and decreases by being taken into blood (outflow). Human's feeling drunk depends on alcohol level in blood. There is a delay between drinking and feeling drunk, since there is another stock (stomach) between them. Therefore, goal-seeking behavior toward an expected drunk degree will occur with oscillation. (Behavior) It follows logically that people do not feel drunk soon after they start to drink and that people get even drunker after they stop drinking. Simulation of a balancing feedback loop with 2 stocks can explain the observed overshooting. Experiments with simulated drinkers also show a high likelihood of becoming overdrunk.

P: To keep away from becoming drunker than intended, man should stop before he feel so drunk as intended. For it is difficult to tell the accurate delay of a specific individual, one should not drink too fast, and need to try several times to have anticipation of the delay. Instructions from doctors who know such delay well will help with preventing people from overshooting. Legislation restricting maximum alcohol content may ensure a delay time long enough for whoever drinks stop timely.

I: Drinking slowly may not conform to social traditions in some cultures. Additional resources will be needed to promote people's perception. Alcohol beverage producers may not be happy with drinker's early stopping.